SELF-ORGANIZATION AND AUTONOMIC INFORMATICS (I)

Frontiers in Artificial Intelligence and Applications

FAIA covers all aspects of theoretical and applied artificial intelligence research in the form of monographs, doctoral dissertations, textbooks, handbooks and proceedings volumes. The FAIA series contains several sub-series, including "Information Modelling and Knowledge Bases" and "Knowledge-Based Intelligent Engineering Systems". It also includes the biannual ECAI, the European Conference on Artificial Intelligence, proceedings volumes, and other ECCAI – the European Coordinating Committee on Artificial Intelligence – sponsored publications. An editorial panel of internationally well-known scholars is appointed to provide a high quality selection.

Volume 135

Recently published in this series

Vol. 134. M.-F. Moens and P. Spyns (Eds.), Legal Knowledge and Information Systems – JURIX 2005: The Eighteenth Annual Conference
Vol. 133. C.-K. Looi et al. (Eds.), Towards Sustainable and Scalable Educational Innovations Informed by the Learning Sciences – Sharing Good Practices of Research, Experimentation and Innovation
Vol. 132. K. Nakamatsu and J.M. Abe (Eds.), Advances in Logic Based Intelligent Systems – Selected Papers of LAPTEC 2005
Vol. 131. B. López et al. (Eds.), Artificial Intelligence Research and Development
Vol. 130. K. Zieliński and T. Szmuc (Eds.), Software Engineering: Evolution and Emerging Technologies
Vol. 129. H. Fujita and M. Mejri (Eds.), New Trends in Software Methodologies, Tools and Techniques – Proceedings of the fourth SoMeT_W05
Vol. 128. J. Zhou et al. (Eds.), Applied Public Key Infrastructure – 4th International Workshop: IWAP 2005
Vol. 127. P. Ritrovato et al. (Eds.), Towards the Learning Grid – Advances in Human Learning Services
Vol. 126. J. Cruz, Constraint Reasoning for Differential Models
Vol. 125. C.-K. Looi et al. (Eds.), Artificial Intelligence in Education – Supporting Learning through Intelligent and Socially Informed Technology
Vol. 124. T. Washio et al. (Eds.), Advances in Mining Graphs, Trees and Sequences
Vol. 123. P. Buitelaar et al. (Eds.), Ontology Learning from Text: Methods, Evaluation and Applications
Vol. 122. C. Mancini, Cinematic Hypertext – Investigating a New Paradigm

ISSN 0922-6389

Self-Organization and Autonomic Informatics (I)

Edited by

Hans Czap
University of Trier, Germany

Rainer Unland
University of Duisburg-Essen, Germany

Cherif Branki
University of Paisley, UK

and

Huaglory Tianfield
Glasgow Caledonian University, UK

Amsterdam • Berlin • Oxford • Tokyo • Washington, DC

ISBN 1-58603-577-0
Library of Congress Control Number: 2005909732

Publisher
IOS Press
Nieuwe Hemweg 6B
1013 BG Amsterdam
Netherlands
fax: +31 20 687 0019
e-mail: order@iospress.nl

Distributor in the UK and Ireland
IOS Press/Lavis Marketing
73 Lime Walk
Headington
Oxford OX3 7AD
England
fax: +44 1865 750079

Distributor in the USA and Canada
IOS Press, Inc.
4502 Rachael Manor Drive
Fairfax, VA 22032
USA
fax: +1 703 323 3668
e-mail: iosbooks@iospress.com

PRINTED IN THE NETHERLANDS

Self-Organization and Autonomic Informatics (I)
H. Czap et al. (Eds.)
IOS Press, 2005

Preface

Today's IT systems with its ever-growing communication infrastructures and computing applications are becoming more and more large in scale, which results in exponential complexity in their engineering, operation, and maintenance. Conventional paradigms for run-time deployment, management, maintenance, and evolution are particularly challenged in tackling these immense complexities. Recently, it has widely been recognized that self-organization and self-management/regulation offer the most promising approach to addressing such challenges. Consequently, a number of autonomic/adaptive computing initiatives have been launched by major IT companies, like IBM, HP, and others.

Self-organization and adaptation are concepts stemming from the nature and have been adopted in systems theory. They are considered to be the essential ingredients of any living organism and, as such, are studied intensively in biology, sociology, and organizational theory. They have also penetrated into control theory, cybernetics and the study of adaptive complex systems. The original idea was to understand complex systems behaviour by understanding the systems' self-organization and adaptation mechanisms, i.e., to understand a system by observing the behaviour of its components and their interactions. However, as stated, the study of self-organization and adaptation has mainly been related to living systems so far.

Computing and communication systems are basically artificial systems. This prevents conventional self-organization and adaptation principles and approaches from being directly applicable to computing and communication systems. The complexity attributes in terms of openness, scalability, uncertainty, discrete-event dynamics, etc. have varied contexts in large-scale complex IT systems, and are too prominent to be solved by procedures pre-defined at off-time. Rather, they have to be tackled by means of run-time perception of the complexity patterns and the run-time enforcement of self-organization and adaptation policies. The current knowledge about large-scale complex IT systems is still very limited, and a framework has yet to be established for their self-organization and adaptation.

The methodology of multi-agent systems and the technology of Grid computing have shed lights for the exploration into the self-organization and adaptation of large-scale complex IT systems. A multi-agent system is one that consists of a collection of autonomous and intelligent agents that have the ability to interact with each other and, thus, may by themselves constitute organizations at run-time. The global behaviour of a multi-agent system stems from the emergent interactions among the agents. Multi-agent systems have been studied widely, not only in computer science, software engineering and artificial intelligence, but even more widely under the heading "systems theory" in economics, management science and sociology. In fact, multi-agent systems permeate social, economic, and technical domains. Essentially, multi-agent systems provide a generic model for large-scale complex IT systems. Exploring and understanding the self-organization and adaptation of multi-agent systems is of profound significance for engineering the self-organization and self-management/regulation of large-scale complex IT systems, comprised of communication infrastructures and computing applications.

Grid computing is the new generation of distributed and networked information and computing systems which have the capacity to enable users and applications, in an emergent manner, to transcend the organizational boundaries and to gain access to the distributed heterogeneous computing resources owned and administrated locally by different organizations. A Grid computing system is by nature a large-scale, complex, and open multi-agent system. Grid computing compounds various areas such as distributed computing resource management, semantic web technology, service-oriented architecture and service management, distributed workflow management, distributed problem solving environment, etc. A Grid commutating system exposes all the complexity attributes typical of large-scale complex IT systems. Investigating the self-organization and autonomic systems for Grid computing has remained a huge challenge.

This book provides in-depth thoughts about the above discussed challenges as well as a range of state-of-the-art methodologies and technologies for the entirely new area, that is, self-organization and autonomic systems in computing and communications. We refer to this newly emerging area as Self-Organization and Autonomic Informatics, which has represented the future generation of IT systems, comprised of communication infrastructures and computing applications, which are inherently large-scale, complex, and open.

The 16 full-length and 17 short papers included in this book are carefully selected from the 58 initial manuscripts based on a rigorous peer review process that was conducted by the 86 technical reviewers worldwide who are experts or active researchers in the related areas. The contents of the book are structured as five parts, i.e.,

Part I: Self-Organization and Adaptation in General;
Part II: Self-Organization/Adaptation of Multi-Agent Systems;
Part III: Self-Organization/Adaptation for Grid Computing;
Part IV: Autonomic Computing in General; and
Part V: Autonomic Communications.

We are sure that you will find the book interesting.

Editors:
Professor Hans Czap, University of Trier, Germany
Professor Rainer Unland, University of Duisburg-Essen, Germany
Professor Cherif Branki, University of Paisley, UK
Professor Huaglory Tianfield, Glasgow Caledonian University, UK

Technical Reviewers

Hamideh Afsarmanesh, Netherland
Klaus-Dieter Althoff, Germany
Cosimo Anglano, Italy
Marc Becker, Germany
Bernhard Beckert, Germany
Ralph Bergmann, Germany
Lukasz Biegus, UK
Ken Birman, USA
Sven Brueckner, USA
Luis M. Camarinha-Matos, Portugal
Mario Cannataro, Italy
Walid Chainbi, Tunisia
Peter Chamoni, Germany
Giovanni Chiola, Italy
Angelo Corsaro, USA
Armin Cremers, Germany
Jonathan Dale, USA
Rohan de Silva, Australia
Philippe De Wilde, UK
Joerg Denzinger, Canada
Giovanna Dimarzo, Switzerland
Prashant J. Doshi, USA
Alexis Drogoul, France
Torsten Eymann, Germany
Marco Fargetta, Italy
Geoffrey Fox, USA
Ulrich Furbach, Germany
Angela Goh E.S., Singapore
Sven Graupner, USA
Steven Guan S.-U., Singapore
Pietro H. Guzzi, Italy
David Hales, Italy
Alexandre Hanft, Germany
Carsten Holtmann, Germany
Tom Holvoet, Belgium
Zhi Jin, China
Matthias Klusch, Germany
Ryszard Kowalczyk, Australia
Helmut Krcmar, Germany
Karl Kurbel, Germany
Winfried Lamersdorf, Germany
Fritz Laux, Germany
Minglu Li, China
Zhen Li, USA
Zushu Li, China
Kecheng Liu, UK
Daniel Livingstone, UK
Peter Lockemann, Germany
Iori Loutcheko, Germany
Evan Magill, UK
Quentin Mair, UK
Tommaso Mazza, Italy
Jörg Müller, Germany
Günter Müller, Germany
Oliver Obst, Germany
Mike Papazoglou, Netherland
Manish Parashar, USA
H. Van D. Parunak, USA
Michal Pechoucek, Czeck Republic
Sara T. Piergiovanni, Italy
Agostino Poggi, Italy
Alexander Pokahr, Germany
Ramana Reddy, USA
Corrado Santoro, Italy
Matthias Schumann, Germany
Guifang Shao, China
Richard Sinnott, UK
Steffen Staab, Germany
Peter Sturm, Germany
Georgios Theodoropoulos, UK
Ingo Timm, Germany
Robert Tolksdorf, Germany
Weiqin Tong, China
Daniel Veit, Germany
Thomas Wagner, Germany
Andrzej Walczak, Germany
Xingyu Wang, China
Zidong Wang, UK
Yugeng Xi, China
Hongji Yang, UK
Franco Zambonelli, Italy
Li Zhang, UK
Shensheng Zhang, China
Xiaochuan Zhang, China
Hong Zhu, UK
Jörg Zimmermann, Germany

Contents

Part I

Self-Organization and Adaptation in General

Self-Organization and Autonomic Informatics (I)
H. Czap et al. (Eds.)
IOS Press, 2005

A Logical Treatment for the Emergence of Control in Complex Self-Organising Systems

Martin RANDLES [a], Azzelarabe TALEB-BENDIAB [b], Philip. MISELDINE [c]
[a, b, c] *School of Computing and Mathematical Science, Liverpool John Moores University, Byrom St. Liverpool, L3 3AF, UK*
[a] *m.j.randles@2004.ljmu.ac.uk,* [b]*a.talebbendiab@ljmu.ac.uk*
[c]*p.l.miseldine@2004.ljmu.ac.uk*

Abstract: In a complex dynamic system the centralised control and local monitoring of system behaviour is not achievable by scaling up simple feedback adaptation and control models. This paper proposes using a variety of concepts from distributed artificial intelligence (DAI) to logically model an abstract system control using adaptable agent federations to induce self-organisation in a swarm type system. The knowledge acquisition and updates are handled through a modal logic of belief for team dynamics and the system as a whole evolves to learn from local failures that have minimal impact on the global system. Self-governance emerges from innate (given) action thresholds that are adapted dynamically to system demands. In this way it is shown that such a system conforms to the prerequisites that have been specified as necessary for a system to exhibit self-organisation and the intrinsic benefits of agent teamwork are established for a robust, reliable and agile system. The approach is illustrated by looking at team formation in a swarm scenario from a proposed NASA project. The Situation Calculus is used to formalise the dynamic nature of such systems with a dynamic logic implementation to reason about the ensuing programs. Subsequently the model is encoded using the Neptune scripting language and compiled to an object-oriented system for its deployment on distributed systems architecture.

1. Introduction: Coping with Complexity

The efficient utilisation of widely spread resources, throughout applications in systems engineering, has led to the requirement for highly distributed autonomous systems to cope with the prevailing heterogeneity; whilst guaranteeing improved dependability and reduced costs. Arguably this cannot be attained without addressing software dependability issues [1], including long standing concerns of; reliability, availability, safety, security, survivability and maintainability [2], for instance. However, the complexity of such systems is itself a major obstacle in achieving such a dependability capability [3]. As reiterated by Rash et al [4], proponents of formal methods argue for the merits of representing such systems as rigorous mathematical models. A formal account for a particular domain gives an abstract specification of the problem. So, for a logical specification, various properties of the system can be proved, completely by logical deduction. Additionally, when the deduction can be performed efficiently, an executable model is available providing a simulator for the system as a code carrying specification.

The optimal methods and paradigms applicable to the governance of computer systems, at the present time, is the subject of much debate, complexity and confusion. Autonomic computing [5] is IBM's proposed solution to the problem, producing systems that function independently with little human intervention. This means that they adapt, repair, optimize and protect themselves in response to environmental triggers, based on a model of the autonomic nervous system that manages our body without conscious effort. However the mechanisms and supporting methods required to achieve such systems are yet to be defined and specified. Current works focus on the 'top-down' approach of the sensor-effector mechanism [6] using feedback and reconfiguration as originally proposed in well-established cybernetic approaches [7], where the key state variables are identified, in advance, with formulae governing their relations and change specified. Whilst, such an approach has shown promising results [8], the complexity of modern systems severely limits the scaling up of these methods. This is because in complex systems it is very difficult to exhaustively identify state variables and their interrelationships. It is for this reason that these approaches have yielded little success in producing a scaleable autonomic solution [9].

In contrast extremely complex natural systems, with intelligent organisation, appear to have emerged and evolved with no external intervention or pre-programming [10]. They regulate and adapt their own behaviour and organise themselves despite providing solutions to problems far more complex than any computer system [11].

Stigmergy [12], swarm intelligence or ant colony, techniques can allow a notion of utility for reactive locally interacting swarms. Here environment sensitive information is dynamically distributed by local agents and may be read and written to by agents present, at that information point, in subsequent situations. Essentially this work advocates an approach to grid computing by addressing the self-organising and adaptive requirements of large, complex, open, multi-agent systems. A generic logical framework is first established to define the base interactions for the formation of the swarm/team, based on knowledge. Then simple rules, for the system, swarm/team or individual agent, are represented in a form of second order logic that does not require quantification over functions. This allows the description of agent team/swarms with minimal rule sets. However second order logic is incomplete, so there is no exhaustive axiomisation or rules of inference to recursively enumerate all the valid sentences for the system. Thus some rules and behaviour arise as logical consequences of the specification whilst others emerge from the incompleteness of the axiomisation.

Accordingly this paper proceeds by a brief investigation of the properties and defining features of complex systems. Section 3 assesses the modelling needs of complex dynamic systems and advocates the merits of a logical sentence based approach leading to a modal representation of knowledge and beliefs in the Situation Calculus. This is used in section 4 to provide a formal account of team formation. A case study is presented in section 5 using the formalism to study team formation whilst assessing the likely program structure through dynamic logic to prove emergent logical features of the implementation. Section 6 discusses the aspects of the system that define its complexity and enforce its self-organisational capabilities. Section 7 details the implementation of the formalism and the paper concludes (section 8) with an evaluation of this approach.

2. Complex Systems

The behavioural rules given to the agents within a system are the basis of the global emergence of system control, abstracting the role of system controller. It is the local interactions caused by the: *if (condition) then do (action)* procedure, for each agent, which distributes control across the whole federation. In turn these rules may be evolvable by a number of methods such as genetic algorithms or adaptation to danger signals. So, for instance, the death of a process may be used to form a receptor that monitors and constrains similar processes to act within safe limits. The properties of complex systems [13] have an impact on the modelling techniques that need to be employed. Aggregation means that an agent federation can be considered as a single meta-agent. Labelling assigns roles to the various system agents. Flow of information and objects is a typical phenomenon of complex systems. Typically energy is used from the environment to increase an object with the resulting entropy being dissipated back into the environment. Nonlinearity is apparent in team functionality being greater than the combined functionality of the participants. Diversity amongst agents promotes evolution so that rules of interaction allow modification by other agents via labelling and by using genetic algorithms, for instance, to form optimal aggregations.

For a complex system to exhibit self-organising behaviour four prerequisites have been rigorously established [14]

(1) There must be two mutually influencing components within the system (mutually causal).

(2) At least one system component must be enhanced by the action of another component (autocatalysis).

(3) The system must take in resources from the environment to enhance itself and dissipate the resulting increase in entropy back to the environment (far from equilibrium).

(4) At least one of the system components must be accessible to random events from the environment (morphogenetic change).

So in order to achieve the required self-organisation, within the system, to permit system control to emerge, base level architecture and interaction frameworks need to be proposed. The overall model ought to include both a top down normative intentional approach to deliberation and a bottom up distributed control mechanism with methods, adapted from pattern recognition, such as Chance Discovery [15], novelty and danger detection, used to indicate emergence via suitable partial observation conditions. The next section advocates the requirement for and proposes a logical approach to emergent control in highly distributed multi-agent systems based on the knowledge an agent acquires in a situation.

3. A Logical Axiomatic Approach

The purpose of this section is to assess and review the modelling and evaluation of behaviour in self-organising systems exhibiting swarm-like properties. This addresses the fundamental dichotomy of approach to the reasoning about control in multi-agent systems where cooperation and coordination, as general interactions, provide the basic functions of the system. In current practice there is little common ground between the model based centralised controller approach and the distributed emergent control of a collective system.

The system controller approach to distributed applications is fairly well established and understood [16]. However an alternative form of control is exhibited in natural systems. An organised pattern of system behaviour emerges that is not appreciated by any single member of that system. This leaves a central controller with a less complex (easier implemented) monitoring observer's role in the system. The knowledge of the local agents is, by its nature, restricted. So a local agent finding itself in a particular state has only partial information on that state. In order to capture this knowledge the Possible Worlds Semantics [17] allows an agent, in a particular state, to believe he may be in one of some set of alternative states with consistent knowledge bases. So in a global setting, where all the variables describing the situation are known, the agent is in a specific situation whereas locally, where only a limited set of situation variables (fluents) are known, the agent may be in one of a number of situations. Thus the utility of actions based on the knowledge base can be modelled and the nonlinear relationship between global and local utility established.

Artificial intelligence techniques seek to model these dynamical collective systems with mathematical models. The theory and practice of such models is captured in the *Knowledge Representation Hypothesis* formulated by Brian Smith [18]: *An embodied intelligent process will consist of structural components that:*

(1) *External observers understand to represent a propositional account of the knowledge that is exhibited in the global process.*

(2) *Play a formal causal role in producing the behaviour that gains that knowledge.*

If this hypothesis is accepted then a number of important consequences follow. The most important, in seeking to model swarm behaviour, is that it is natural to use mathematical logic as a foundation for the *propositional account of the knowledge that is exhibited in the global process.* Also a propositional account is very different from a state based approach because no enumeration of states and transitions is needed, rather what is true in the system and the causal laws that affect it and its environment can be used. Logical entailment views system behaviour as consequences of the propositional account of the domain. That is determining how a system behaves depends on deducing how it must behave given the systems description. This follows from part 2 of the hypothesis stating a causal connection between the propositional sentences and the system behaviour. Another result is that there is a direct method, logical deduction, for establishing correctness properties of the system. This follows because, as stated earlier for a logical specification, the propositional account gives an abstract specification of the problem which can be used to prove, within the logic, properties of the system.

There is an emphasis on belief (knowledge) and how it conditions individual behaviour, to influence group behaviour. The adoption of the Knowledge Representation Hypothesis leads to logical sentences being the fundamental mathematical specification of systems and their underlying agents' beliefs. Then knowledge may be represented in a logical language as sentences that condition behaviour.

The Situation Calculus [19] was specially developed for axiomatising dynamic domains and recent research to incorporate time, concurrency, probability etc. [20, 21 and 22] into the methods for producing efficient implementations, provides techniques for a very useful realistic practical formalism. There is no need for an explicit enumeration of the state space but rather behaviour arises from axiomatic representation. This contrasts significantly with alternative approaches. For instance Finite State Machines require all transitions to be identified and specified at design time. Process algebras have a similar elegance, to logical representations, but, in practice, require suitable modelling and verification methods to fit above the primitive constructs of the formalism [23].

3.1 Knowledge and the Situation Calculus

In recent years the Situation Calculus [24] has gained prominence in modelling dynamical systems [19]. This views situations as action histories. Fluent values are initialised in the starting situation (S_0) and change from situation to situation according to effect axioms for each action. The partial solution to the resultant frame problem [25] gives successor state axioms that largely specify the system together with action precondition axioms and the initial situation [26]. So an initial situation, S_0 is the start of the Situation Calculus representation. An action, a, then changes this situation from S_0 to do(a, S_0) with the next action, a_1 say, changing the situation to do(a_1 ,do(a,S_0)) with a_2 giving do(a_2 ,do(a_1 ,do(a,S_0))) and so on. In this way an action history is formed read from right to left in the situation term. Additionally in S_0 a set of fluents (predicates and functions) has an initial set of values. The set of successor state and action precondition axioms, one for each fluent and action respectively, show the changes in value of the fluents and the possibility of completing an action in each situation accordingly. i.e. A fluent is TRUE in the next situation if and only if an action occurred to make it TRUE or it is TRUE in the current situation and no action occurred to make it FALSE, with precondition axiom poss(a, s) meaning it is possible to perform action a in situation s.

The representation of knowledge and beliefs in the situation calculus is achieved by seeing the world states as action histories or situations with the concept of accessible situations [27]. So if s_1 and s_2 are situations then $(s_1, s_2) \in K_i$ means that in situation s_2 agent i considers s_1 a possible situation with K_i an accessibility relation for agent i. That is all fluents known to hold in situation s_2 also hold in s_1. So an accessibility fluent may be specified: $K_i(s_1,s_2)$ meaning in situation s_2 agent i thinks s_1 could be the actual situation.

So knowledge for agent i (knows_i) can be formulated in a situation as:

$$\text{knows}_i(\phi, s) \equiv \forall s_1(K_i(s_1, s) \rightarrow \phi(s_1)) \text{ [alternatively } \forall s_1 (\neg K_i(s_1, s) \vee \phi(s_1)) \text{]}$$

This gives rise to a fluent to represent knowledge dynamics in the situation calculus that still satisfies the constraints of the solution to the frame problem [28]. However to make any axiom complete it is necessary to establish whether a sensing action has taken place. That is if the action that occurred, to change the situation to its successor, was the perception of the value of a fluent. So the change was a change in the epistemic state of the agent. Thus it is necessary to distinguish sensing actions by writing $SR(sense_\phi, s)$ to denote that the action produced a result for ϕ.

$$SR(sense_\phi, s) = r = \text{value of } \phi \text{ in } s$$

Thus a successor state axiom for K can be stated:

$$K(s_2, do(a, s)) \Leftrightarrow \exists s_1(s_2=do(a, s_1) \wedge K(s_1, s) \wedge poss(a, s_1) \wedge SR(a, s)= SR(a, s_1))$$

This formalism thus allows the representation of knowledge in partially observable domains. This is a vital prerequisite, for the approach presented in this paper, to produce a scaleable methodology to enable deliberation, in a swarm composed of socio-cognitive agents engaged in team working.

4. A Formal Approach for a Collective Swarm Architecture

It is known that as complexity increases and failures tend to occur more frequently teams of agents waste fewer resources and are more robust than solipsistic agents [29]. The formation of an agent federation, through a swarm, is characterised by the concept of a Joint Persistent Goal (JPG). This approach roughly follows the methods of Cohen and Levesque [30, 31]. However there are a number of differences. Firstly more recent results in knowledge dynamics in the Situation Calculus, as outlined above, are used to formalise the concepts of Persistent and Joint Persistent Goals using additional modal operators as defined by Halpern and Moses [32]. Also the swarm is allowed to grow or diminish dynamically by means of member addition or subtraction. Thus swarm members are committed to the team as it is comprised during their membership. The JPG formalises the joint commitment of the swarm. That is the JPG imbues the swarm with the goal-directed behaviour of a single agent. Tasks, to realise completion of the goal, are performed by sub-agents, within the swarm, as part of the agent society. To have a joint intention to perform an action, *a*, the swarm will have a JPG to do *a* within a particular subset of the situation space. The joint intention assumes the initial belief, by all the members, that the swarm is going to complete the intended action next. Following the enactment of a joint intention the agent swarm will mutually come to believe one of three things, from the initial mutual belief that the team members are going to work jointly on the intended action:

(1) That *a* has been done.
That *a* is impossible to do.
That *a* is irrelevant

Firstly considering a single agent a persistent goal can be represented in the situation calculus by:

$$\text{Pgoal}(p, do(a, s)) \Leftrightarrow$$
$$(\text{Pgoal}(p,s)\wedge\neg(\neg B(p,s)\vee \text{knows}(p,s)))\vee(a=\text{setPgoal}(p)\vee \text{knows}(\neg p,s))$$

As previously introduced *knows* is the knowledge operator in the situation calculus with the analogous belief operator B [$B(p) = \neg\text{knows}(\neg p)$]. Obviously with only one agent these are knows_1 and B_1, with the number of agents, n=1. This agrees with an intuitive notion that a goal persists while an agent doesn't know p to be true and doesn't know that p cannot be true. Replacing the belief operator with its corresponding formulation using the knowledge operator in the above formula and working through the expression gives an equivalent formula:

$$\text{Pgoal}(p, do(a, s)) \Leftrightarrow$$
$$(\text{Pgoal}(p,s)\wedge(\neg\text{knows}(\neg p,s)\wedge\neg\text{knows}(p,s)))\vee(a=\text{setPgoal}(p)\vee$$
$$a=SR(\text{sense}_p, s)\rightarrow\neg p)$$

Now to extrapolate this to n agents in a multi agent setting it is necessary to think of the agent swarm as acting as a single entity. However this is not sufficient as it must also be possible for any agent to become aware individually that a goal has been achieved or is impossible to reach. So when a swarm member comes to believe a fact, regarding the goal, which entails the dropping of the commitment by the individual member, the swarm federation must also drop the commitment. Thus each swarm member has as a sort of lesser goal one of three choices. Either it believes p not to be the case but has p as a goal or it believes p does hold and has a goal to make this common knowledge or it believes p to be impossible and has a goal to make this common knowledge. So if we have a lesser individual goal, for each agent i, relative to the other swarm members, $LIgoal_i$:

$$LIgoal_i(p, do(a,s)) \Leftrightarrow LIgoal_i(p,s) \wedge \neg((knows_i(\neg p,s) \wedge goal_i(p,s)) \vee (knows_i(p,s) \wedge goal_i(C(p),s)) \vee (\neg B_i(p,s) \wedge goal_i(C(\neg p),s)))$$

where C is the common knowledge operator [32]. In this way each federation member of the swarm doesn't assume that the other federation members have p as a goal but rather as a lesser individual goal. Thus any swarm/federation member may have discovered p has been achieved or is impossible and be in the process of making this common knowledge in the federation. So a joint persistent goal is established, using the *everyone knows* operator (E) [32] with single agent belief replaced by common knowledge and lesser individual goal as appropriate. i.e.

$$JPgoal(p, do(a,s)) \Leftrightarrow JPgoal(p,s) \wedge (\neg(E(\neg p,s) \vee E(p,s)) \wedge (\exists i(LIgoal_i(p,s)))) \vee (a=setJPgoal(p) \vee E(\neg p,s))$$

This then provides an example mathematical model of the teamwork behaviour, required in the system, to set defining goals for the swarm. In this way a scaleable method of specifying distributed control emerges, from a minimal set of system imperatives for the agents, to determine behaviour by logical entailment and emergent novelty. The next section presents a case study, where agent knowledge plays a key role, including instances of some imperatives and the logical consequences that follow from them.

5. A Case Study

To illustrate the proposed modelling approach a scenario is adapted from the behaviour specification of a worker/ruler robot in the NASA Autonomous Nano-Technology Swarm (ANTS) project [33]. Formulation and deliberation of emergent properties occur at observer level with the necessary checks and guards incorporated as a norm based deliberative model.

5.1 The ANTS Project Description

Briefly stated the ANTS project arises from a class of space exploration missions termed *nanoswarms,* where many cooperating intelligent spacecraft work in teams, based on the efficiency and coordination of hive culture. In particular the project, which is under conceptual development, by NASA, to occur in the 2020s, envisages a thousand picospacecraft working cooperatively to explore the asteroid belt. Once commissioned the teams consist of three classes of spacecraft combined in specific ways to form teams that explore individual asteroids. *Workers* make up 80% of the swarm and carry the instruments to gather data. *Rulers* coordinate data gathering by assembling teams of appropriate *workers*. *Messengers* manage communications between *workers, rulers* and mission control on earth. This work addresses the verification of emergent behaviour through establishing logical consequences of the interactions of the participants and observing the emergent behaviour that occurs. A minimal rule set is formed so that additional behavioural rules follow as a result of the specification. Emergent behaviour arises, from the specification techniques advocated here, as second order logic is incomplete. The framework, described in the previous two sections, is particularly applicable to this NASA project. This is because the mission is knowledge critical. The overall aim is to gain knowledge for use by scientists, back on earth, whilst the epistemic state of the swarm members is crucial in stabilising the behaviour of the swarm to carry out its intended duties.

5.2 Axiomatising the Domain

The goals to be satisfied rely on team-work within the swarm. So without team formation, or the notion of the logical process of team assembly, it is impossible to continue with the process. The workers under each Ruler obviously form a federation, additionally the Rulers can be conceived of as forming a Ruler team that is connected to the workers so that the most appropriate team members can be selected for particular mission tasks. Each mini spacecraft, in a team, is committed to successfully performing its quota of tasks as well as to the success of the team as a whole. So the formalism of the domain can proceed as follows with situation terms added, as the final argument of the fluents, where required:

- worker(w) w is a worker spacecraft
- ruler(r) r is a ruler spacecraft
- W, R are worker and ruler teams with T representing a general team that may consist of heterogeneous units.
- knows(w,p) means p follows from a worker w's knowledge
- G(w,p) means p follows from a sequence of worker W's available actions
- I(w,a) means a is an intention of worker w
- E(T,p) means everyone believes p in team T (eg knows(w_1,p)$\wedge$knows(w_2,p) $\wedge$......knows(w_n,p) where T={w_1, w_2,.....w_n}
- C(T,p) means p is common knowledge in team T [ie E(T, p) $\wedge$E(E(T, p)) $\wedge$.......]. This represents universal mutual belief in that every team member believes that every other team member believes p
- member(r,R) means ruler r is a member of ruler team R
- registered(w,R) means worker w is registered with ruler team R
- connected (w, r) means worker w is connected to the ruler team via ruler r.

Using the formalism of the previous section joint persistent goals can be set in the situation calculus. An augmented variation of dynamic logic [32] will be applied to this Situation Calculus representation to produce a logical implementation, similar to pseudo-code, to provide formal reasoning techniques for the ensuing programs. So:

- p? means give p a valuation (ie true or false)

- $a_1;a_2$ means action a_1 followed by action a_2
- $a_1|a_2$ is non deterministic choice
- Complex action expressions can be used such as IF-THEN and WHILE-DO
- occurs (a*) means a sequence of actions a* ($= a_1,a_2,\ldots.a_n$) is scheduled to happen next. It can be noted that for use in the Situation Calculus $a^*=do(a_n, do(a_{n-1},\ldots\ldots do(a_1, S_0))\ldots)$ where S_0 is the starting situation for the action sequence.
- finished (a*) means the sequence of actions has occurred.

This treatment is mostly achieved using First Order Logic (FOL). There is an appeal to some Second Order Logic (SOL), however this is solely to describe smallest sets with certain properties, there is no requirement to quantify over function variables. This is done in order to more easily semantically state the properties of actions and their consequences. Additionally, as previously stated, emergent behaviour follows from the incompleteness of SOL.

Now it is possible to state some imperatives that the teams must follow. In this way low level interactions are specified that encourage swarming and emergent organisation.

Ruler Team Perspective: Imperative I: When a *worker (w)* registers on the mission, by registering with the *Ruler* team (R), and has not entered any failure state leading to its "death", the *Rulers* possess a team intention to connect with the *worker*, should it ever be disconnected.

So in situation s, after the swarm team is formed, the ruler team swarm have

$$JPgoal([\forall w \exists r \in R(connected(w,r))], s)$$

The *connected* fluent can take the situational value independent of its goal attribution

$$connected(w,r,do(a,s)) \Leftrightarrow (connected(w,r,s) \wedge a \neq [SR(sense_w,s)=nil]) \vee a=connect(w,r)$$

$$poss(connect(w,r),s) \Rightarrow (r \in R) \wedge registered(w,R,s)$$

This converts into a dynamic logic representation:

$$\models \forall w\ [(worker,w) \wedge finished(registered(w, R)?) \Rightarrow (I(R, connect(w, R))]$$

where connect(w, R)= (WHILE registered(w,R) DO
[IF $\neg\exists R$ connected(w, r) THEN
[UNTIL [$\exists$ r$\in$R connected(w, r)] $\forall$ r I(r, connect(w, R))]]

It should be noted that a team intention (ie I(R, connect(w)) in the above formula) is stronger than just a collection of individual team member intentions. It is based on the team joint persistent goal, which requires the team members to have beliefs regarding team membership. This requires team members to hold the lesser individual goals (LIgoal) in certain circumstances. That is if a ruler team member believes that the goal p (say) is not achieved it has an individual goal to see to it that p eventually holds. If it believes the goal has been achieved, cannot be achieved or is of no consequence

then it has an individual goal to make this mutually believed by the ruler team. If it believes there to be a new team member or that a member has left the team then it has a lesser individual goal to make this a belief for all the other team members. So we have LIgoalr([(∀r'∈R)C(R, (r'∈R)],s)∧ LIgoalr(p,s)
This lesser individual goal for r∈R is represented by

$$
\begin{aligned}
&LIgoal(r, R, p) \Leftrightarrow \\
&\quad [[\neg knows(r,p)\wedge G(r,p)]\vee[knows(r,p)\wedge G(r,C(R,p))]\vee \\
&\qquad\qquad [knows(r,\neg p)\wedge G(r,C(R,\neg p))]] \wedge \\
&\quad [[\forall r'\ (knows(r, \exists a(finished(member(r',R))?;a;\neg member(r',R)?)))]\Rightarrow \\
&\qquad\qquad G(R, C(R, \neg member(r',R)))]\wedge \\
&\quad [[\forall r'\ (knows(r, \exists a\ (finished(\neg member(r',R))?;a;\ member(r',R)?)))]\Rightarrow \\
&\qquad\qquad G(r, C(R, member(r',R)))]
\end{aligned}
$$

where a is a single action

A lesser team goal (LTG) means that the ruler team mutually believes that every ruler in the team has the LIgoal. e.g.

$$LTG(R,p)\Leftrightarrow C(R,(\forall r\in R\ LIgoal(r,R, p)))$$

So now the team R has p as a mutual goal when the team mutually believes p is not achieved, there is a mutual goal to achieve p and there is a lesser team goal to achieve p until the team come to believe that p is achieved, unachievable or irrelevant. ie

$$G(R,p)\Leftrightarrow knows(R,\neg p)\wedge G(R,p)\wedge[UNTIL[knows(R,p)\vee knows(R,\neg p)]LTG(R,p)]$$

So finally it can be written:

$$I(R,connect(w)) \Leftrightarrow G(R, finished(\ C(R,\ occurs(R,\ connect(w,R)))?;connect(w,R)) \quad \mathbf{(1)}$$

This gives the power of mutual beliefs (from common knowledge) leading to mutual team goals to give team intentions and shows how the team intention I(R,a), for action a, is syntactically different from the worker intention I(w,a) just because the goal G(R,p) is similarly different from the goal G(w,p).

In general this formalism also holds for any team. So the team under consideration could be the worker team or a heterogeneous team of variable members. This distinction then makes it relatively straight forward to specify team goals, in a similar way to individual goals, but with the proviso that the above team goal formalism holds. So a number of logically provable consequences emerge as a result of **imperative I**. E.g.: When a registered worker becomes disconnected the ruler team have a team commitment to reconnect with the worker throughout the lifetime of the team.
Proof: The assertion can be logically stated as:

$$
\begin{aligned}
&\models\forall w[C(R,\ registered(w,R))\wedge C(R,\ (\neg\exists r\in R\ connected(w,r)))]\Rightarrow \\
&\qquad\qquad G(R,\ (finished(connect(w,\ R))))
\end{aligned}
$$

So assume the left hand side of the implication is true then expanding imperative I using the definition of team intention **(1)** gives a team goal:

G(R, (finished(C(R, occurs(connect(w,R))))?;connect(w, R)))

so let connect(w,R) = pre(w,R);(registered(w,R))?;
[(¬∃Rconnected(w,r))?;connect(w,R)| (∃r∈R connected(w,r))?];post(w,R)

where pre(w,R) represents the previous iterations and post(w,R) denotes the remaining iterations.

So substituting gives the ruler team goal of:

G(R, (finished(C(R, occurs(connect(w,R))))?;pre(w,R));(registered(w,R)?;
[(¬∃r∈Rconnected(w,r))?;connect(w,R)|(∃r∈R connected(w,r))?];post(w,R))

The iteration of interest is the one where the conditional becomes true. So assume the action sequence C(R, (occurs(connect(w,R))))?;pre(w,R);registered(w,R)?; (¬∃r∈R connected(w, r))? has just occurred. Now for actions a and b with a general team T it is known that:

⊨G(T, finished(a;b))∧C(T,finished(a))∧C(T,¬finished(b))⇒G(T,finished(b))

Applying this to connect(w,R) in imperative I gives:

G(R, (finished(C(R, occurs(connect(w,R))))?;pre(w,R));(registered(w,R)?;
[(¬∃r∈Rconnected(w,r))?;connect(w,R)|(∃r∈R connected(w,r))?];post(w,R))⇒
G(R, finished(connect(w,R));post(w,R))

However by assumption C(R,¬connected(w,R)) holds so that members of the ruler team mutually believe that the action [¬∃r∈R connected(w,r)]? has occurred so they then have a team commitment to do connect(w,R). Thus

G(R, finished(connect(w,R));post(w,R)) ⇒G(R, (finished(connect(w,R))))

which was the result to be proven?

In this way logical consequences of the specification emerge as verifiable, provable outcomes for the system. Thus, with a meaningful specification of team work, imperatives can be stated from which emerge additional, not necessarily intended, features of the system. To continue a further imperative for the swarm team might be stated as:

Ruler team: Imperative II: The ruler team must have be made up of at least a specified number of ruler units

JPgoal(numberofRulers(R)>=N, s)

numberofRulers(R,do(a,s))=M⇔(numberofRulers(R,s)=M)∨
[(numberofRulers(R,s)=M-1)∧
∃r a=join(r, R)]∨[(numberofRulers(R,s)=M+1)∧)∧∃r a=leave(r, R)]

poss(join(r,R),s)⇒ruler(r,s)

poss(leave(r,R),s)⇒member(r,R,s)

So the imperative states:

⊨ G(R,numberofRulers(R)>=N)

A number of results follow in a logically provable manner:

E.g.: Rulers as individuals have a commitment to maintain the number of rulers above a specified level so that when one believes that the number of rulers is less than required and believes that it is not mutually believed by the team and believes it is not impossible to establish mutual belief for the team then it has an individual commitment to bring about this mutual belief.

Proof: The assertion states:

⊨G(R,p)⇒∀r∈R [knows(r,¬p∧¬C(R,¬p))∧
¬knows(r, ¬C(R,¬p))⇒G(r,C(R,¬p)∧knows(r,¬p))]
where p=(numberofRulers>=N)
proof: Let r∈R and assume

∀r∈R knows(r,¬p∧¬C(R,¬p))∧¬knows(r, ¬C(R,¬p))

Now from the definition of team goal it can be stated:

knows(r, ¬p)∧¬C(R,¬p)⇒G(r, C(R,¬p))

and G(r, C(R,¬p)) is satisfied because if one member of the team does not believe there is mutual belief then there is no mutual belief. So since the consequent of an implication must remain true until the antecedent or the implication statement becomes false G(r, C(R,¬p)) holds until

r∈R knows(r,¬p∧¬C(R,¬p))∧¬knows(r, ¬C(R,¬p)) doesn't; that is until

¬knows(r,¬p)∨C(R, ¬P)∨knows(r, ¬C(R,¬p)) is true.

So all the conjuncts in the definition of G(r,C(R,¬p)∧knows(r,¬p)) are satisfied proving the result.

6. Discussion

The complex nature of such a system is evident in the aggregation, labelling, data flow, nonlinear responses and diversity, which occurs to provide a meta-monitoring structure from the agents. The entire mission comprises of an aggregated swarm of robots whilst individual workers, for example, consist of aggregated systems for propulsion, monitoring, communication, etc. As such each system can be labelled from worker/ruler down through lower level systems. Data flow is evident in the sensing functions to determine action. The nonlinear response occurs in the teamwork dynamic. Diversity occurs as separate swarms develop and evolve to specialize at separate specific tasks. The prerequisites for self-organising behaviour, as defined above, can be accommodated in the proposed knowledge acquisition framework, namely:

(1) The interaction of robot agents, in the swarm, is mutually causal in that the state of one agent causes an action to be instigated in another agent and vice versa.
(2) Autocatalysis is common as when an agent fails another is influenced to increase itself to take on the extra duties or impel the collective to do so.
(3) The system is open to the environment. Indeed its main concern is to deal with environmental effects taking in resources from the environment, building itself into an ordered structure and feeding back into the environment.
(4) Random variations are very evident in such systems with hardware malfunctions, variable demand, exogenous actions etc.

Hence although a simple system scenario is presented the interaction between components enforce a complexity that engenders self-organising behaviour. Simple norms in the form of imperatives can be laid down which in turn affect the system through the consequences of logical entailment that can be rigorously established. Additionally certain behaviours can emerge that are completely separate from any rules that can be derived logically. For instance certain team formations may emerge as more efficient in carrying out specific tasks. In this paper it is the perspective of the ruler team that has been considered. However the perspective of the observer can be treated in a similar manner to monitor for danger signals [35], proscribe unsafe behaviour and assess emergence for novel benefits.

7. Implementation

This work is primarily concerned with a logical modelling and reasoning system for swarm behaviour. Thus a full description of the implementation is outside the scope of this paper, though details of the implementation framework are available elsewhere [36, 37]. Briefly stated, the implementation is achieved through a specially developed Cloud architecture using the custom designed language Neptune with runtime-adaptable Neptune scripts directly implementing logical constructs.

The use of dynamic logic allows verification and formalization of correctness whilst the situation calculus captures the dynamic nature of swarm behaviour conditions. The Cloud provides the overview of the system incorporating communication and monitoring at runtime. The Neptune objects provide introspective adaptable control modules through which the base level local interactions are prescribed and evolved throughout runtime. The direct implementation of provably correct formal statements is easily achieved due to Neptune's designed logical formulation. Thus automatic code generation can be fairly readily induced. Furthermore runtime introspection and adaptation allow emergence of control to be incorporated into the implementation.

8. Evaluation and Conclusion

This paper has presented an investigation into complex systems and the defining properties of complex systems in an attempt to provide scaleable solutions for the implementation of autonomic systems. The self-organisation exhibited by large teams or swarms of entities is proposed as a means of distributing control across the system thus leaving the traditional system controller object with a simpler observer's role.

The case study showed a scenario of team formation from the NASA ANTS project. The dynamic nature of the domain was easily captured in the Situation Calculus representation. Dynamic logic provided a formal method for reasoning about the ensuing programs. It was shown that a minimal set of imperatives could be stated from which provable logical consequences emerge to regulate and determine behaviour. Suitable domain architectures can be established using general imperatives. Additionally novel emergence is captured through the observer's monitoring role. The resulting implementation is achieved whereas previous methods, to produce such systems, failed because the required complexity of the rule set production system was not scaleable.

It has been established, in formal provably logic, what some base level behaviours might be in such a system. Additionally such an axiomatic approach achieves greater scalability as every event in the systems history does not have to be accounted for with a separate policy. Rather behaviour is determined as a logical consequence of the system's specification. Any circumstances not included in such an approach are detailed to a minimal controller acting as an observer agent. It is suggested that the results are highly applicable as the swarm or teamwork ideas presented can be applied to any linked network. Any network can exhibit dependable behaviour when the system is represented as a formal verifiable model. Additionally the automatic code generation that a logical model produces further reduces the chance of erroneous code. The implementation can be proven to be equivalent to the requirements specification.

References

[1] M.G. Hinchey, J.L. Rash, C.A. Rouff, "Requirements to Design to Code: Towards a Fully Formal Approach to Automatic Code Generation" NASA Technical Report, January 2005

[2] A. Avižiensis, J-C Laprie, B. Randell, C. Landwehr, "Basic Concepts and Taxonomy of Dependable and Secure Computing" IEEE Transactions on Dependable and Secure Computing 1(1) Jan-Mar 2004

[3] R. Sterritt, D. W. Bustard, "Autonomic Computing-A Means of Achieving Dependability?" Proceedings of IEEE International Conference on the Engineering of Computer Based Systems (ECBS'03), pp247-251, Huntsville, April 2003

[4] J.L. Rash, C.A. Rouff, M.G. Hinchey, "Experience Using Formal Methods for Specifying an Agent Based System" Proceedings of Sixth International Conference on Engineering of Complex Computer Systems (ICECCS 2000), Tokyo, Japan, 2000

[5] IBM Research. Autonomic Computing. http://www.research.ibm.com/autonomic/. Accessed August 2005

[6] A.G. Ganek, T.A. Corbi, "The Dawning of the Autonomic Computing Era" IBM Systems Journal, 42(1) pp5-18, 2003

[7] A.G. Laws, A. Taleb-Bendiab, S. J. Wade, D. Reilly (2001) 'From Wetware to Software: A Cybernetic Perspective of Self-adaptive Software'. IWSAS 2001: 257-280

[8] M. Randles, A. Taleb-Bendiab, P. Miseldine, A. Laws "Adjustable Deliberation of Self-Managing Systems." Proceedings of IEEE International Conference on the Engineering of Computer Based Systems (ECBS 2005) pp449-456, Maryland, USA, 2005

[9] M Shackleton;,F Saffre;,R Tateson, E Bonsma, C Roadknight "Autonomic Computing for Pervasive ICT — A Whole-System Perspective", BT Technology Journal, Vol.22 No.3 pp 191-199, July 2004

[10] F. Heylighen "The Science of Self-Organisation and Adaptivity" in Knowledge Management, Organisational Intelligence and Learning and Complexity in The Encyclopaedia of Life Support Systems, Eolss Publishers, Oxford, 2003

[11] K. Kelly "Out of Control: The New Biology of Machines" Addison-Wesley, New York, 1994

[12] Hadeli, P. Valckenaers, M. Kollingbaum, H. van Brussel, "Multi-Agent Coordination and Control using Stigmergy" Computers in Industry 53(1) pp 75-96, 2004

[13] J.H. Holland "Hidden Order: How Adaptation Builds Complexity." Addison-Wesley, Redwood City, CA. 1995
[14] P. Glansdorff, I. Prigogine "Thermodynamic Study of Structure, Stability, and Fluctuations" Wiley New York 1978
[15] Y. Ohsawa (editor), "Proceeding of the 1st International Workshop on Chance Discovery", Japanese Society for Artificial Intelligence, 2001
[16] IBM (2003) "An Architectural Blueprint for Autonomic Computing"
[17] J. Hintikka "Modalities and Quantification" Theoria 27 pp119-128, 1961
[18] B.C. Smith "Reflection and Semantics in a Procedural Language" PhD. Thesis, MIT, Cambridge, Mass., 1982
[19] H. J. Levesque, F. Pirri and R. Reiter "Foundations for the Situation Calculus". Linköping Electronic Articles in Computer and Information Science, Vol. 3(1998): nr 18. http://www.ep.liu.se/ea/cis/1998/018/
[20] R. Reiter "Natural actions, concurrency and continuous time in the situation calculus". Principles of Knowledge Representation and Reasoning: Proc. of the Fifth International Conference, ed: L. C. Aiello, J. Doyle and S. C. Shapiro, pp 2-13. Morgan Kaufmann Publishers, San Francisco, 1996
[21] F. Pirri and R. Reiter 'Some contributions to the metatheory of the situation calculus'. Journal of the ACM 46(3), pp 325-361,1999
[22] R. Reiter, "Stochastic Actions, Probabilities and Markov Decision Processes in Situation Calculus". Proceeding of the 18th National Conference on AI, Edmonton, Canada July 2002
[23] Cerone, G.J. Milne, "Property Verification of Asynchronous Systems", Innovations in Systems and Software Engineering Vol 1 pp.25-40, Springer-Verlag, 2005
[24] J. McCarthy "Situations, Actions and Causal Laws". Technical report, Stanford University, 1963. Reprinted in Semantic Information Processing, ed: M. Minsky, pp 410-417, MIT Press, Cambridge, Massachusetts, 1968.
[25] J. McCarthy and P. Hayes "Some Philosophical Problems from the Standpoint of Artificial Intelligence". Machine Intelligence 4, ed: B. Meltzer and D. Michie, pp 463-502, Edinburgh University Press, Edinburgh, Scotland, 1969
[26] R. Reiter "The Frame Problem in the Situation Calculus: a Simple Solution (Sometimes) and a Complete Result for Goal Regression". Artificial Intelligence and Mathematical Theory of Computation: Papers in Honor of John McCarthy, ed: V. Lifschitz, pp359-380, Academic Press, San Diego, California, 1991
[27] R.C. Moore, "Reasoning about Knowledge and Action", Technical Report, SRI International, 1980
[28] R.B. Scherl, H.J. Levesque, "Knowledge, Action and the Frame Problem", Artificial Intelligence 144, pp:1-39, 2003
[29] N. R. Jenning "Controlling Cooperative Problem Solving in Industrial Multi-Agent Systems using Joint Intentions" Artificial Intelligence. 75(2) pp 195-240, 1995
[30] H.J. Levesque, P.R. Cohen, J.H.T. Nunes "On Acting Together", Proceedings of the Annual Meeting of the American Association for Artificial Intelligence, AAAI-90, 1990.
[31] P.R. Cohen, H.J. Levesque, "Confirmations and Joint Intentions", Proceedings of 12th Intl. Conference on Artificial Intelligence, Sydney, Australia, 1991
[32] J.Y. Halpern, Y. Moses (1990) "Knowledge and Common Knowledge in a Distributed Environment" Journal of the ACM, 37(3) pp: 549-587
[33] C. Rouff, A. Vanderbilt, W. Truszkowski, J. Rash, M. Hinchey. "Verification of NASA Emergent Systems," Proceeding of the 9th ICECCS, pp. 231-238, 2004.
[34] D. Harel, "First-Order Dynamic Logic", Lecture Notes in Computer Science, Vol. 68, Springer-Verlag, New York (133 pp.), 1979
[35] M. Randles, A. Taleb-Bendiab, P. Miseldine "Using Stochastic Situation Calculus to Formalise Danger Signals for Autonomic Computing" Proceedings of the 6th Annual Conference on the Convergence of Telecommunications, Networking and Broadcasting (PGNET2005) pp241-246, Liverpool, UK, 2005
[36] P.Miseldine "Cloud Architecture Schematic" Technical Document http://www.cms.livjm.ac.uk/2nrich/ (Accessed 12th October 2005)
[37] P.Miseldine " The Neptune User Guide" Technical Document http://www.cms.livjm.ac.uk/2nrich/ (Accessed 12th October 2005)

Towards a Methodology for Engineering Self-Organising Emergent Systems

Tom De Wolf, Tom Holvoet
AgentWise@DistriNet Research Group
Department of Computer Science, KULeuven
Celestijnenlaan 200A, 3001 Leuven, Belgium
{Tom.DeWolf, Tom.Holvoet}@cs.kuleuven.be
http://www.cs.kuleuven.be/~tomdw/

Abstract When designing self-organising emergent multi-agent systems (MASs), a fundamental engineering issue is to achieve a macroscopic behaviour that meets the requirements and emerges only from the behaviour of locally interacting agents. Agent-oriented methodologies today are mainly focussed on engineering the microscopic issues, i.e. the agents, their rules, how they interact, etc, without explicit support for engineering the required macroscopic behaviour. As a consequence, the macroscopic behaviour is achieved in an ad-hoc manner. This paper proposes a way to define a full life-cycle methodology based on an industry-ready software engineering process, i.e. the Unified Process, which is customised to explicitly focus on engineering macroscopic behaviour of self-organising emergent MASs. As such, the MAS paradigm is integrated into an existing and widely accepted methodology and we can systematically develop a solution that exhibits the required macroscopic behaviour.

Keywords. self-organisation, emergence, methodology, multi-agent systems, software engineering, unified process

1. Introduction

Modern distributed systems exhibit an increasingly interwoven structure [13] (e.g. ad-hoc networks, transportation systems, etc.). Different subsystems depend on and interact with each other in many, often very complex, dynamic, and unpredictable ways. Also, more and more systems require and exhibit a completely decentralised structure (e.g. ad-hoc networks) and need to achieve their requirements autonomously [16].

A promising approach is to use a group of agents that cooperate to autonomously achieve the required system-wide or macroscopic behaviour using only local interactions, local activities of the agents, and locally obtained information. Such decentralised multi-agent systems (MASs) exhibit *self-organising emergent behaviour* [5]. A system exhibits emergence when there are coherent properties at the macro-level (i.e. of the system as a whole) that dynamically arise from the interactions between the parts at the micro-level. Such *macroscopic properties* are novel w.r.t. the individual parts of the system. Self-organisation is a dynamical and adaptive process where systems acquire and maintain structure themselves, without external control. A combination of emergence and self-organisation is often used and/or perceived in many systems [5].

It is a common problem that the overall system tends to exhibit macroscopic properties for which it is practically infeasible to derive them when analysing the individual entities in isolation (e.g. pheromone paths in ant optimisation algorithms [7], degree of connectivity or service success ratio in mobile ad-hoc networks [22]). A fundamental problem is the lack of a step plan that allows to systematically specify desirable macroscopic properties, map them to the behaviour of individual agents, build the system, and verify it to guarantee the required macroscopic properties, i.e. *a full life-cycle engineering process or methodology.*

Agent-oriented methodologies today are mainly focussed on engineering the microscopic issues, i.e. the agents, their rules, how they interact, etc., without explicitly engineering the required macroscopic behaviour [36]. Engineering the macroscopic behaviour is currently done in an ad-hoc manner because there is no well-defined process that guides engineers to address this issue. This paper proposes a way to define a full life-cycle methodology based on an existing industry-ready software engineering process, i.e. the Unified Process [15]. The UP process is customised to explicitly focus on engineering macroscopic behaviour of self-organising emergent MASs. As such, we are not reinventing the wheel by proposing a completely new agent-oriented methodology. MAS should be allocated a correct role within mainstream software engineering, rather than positioning MASs as a radically new approach [29]. The MAS paradigm is integrated into a widely accepted methodology and the research focus can be on issues for developing a self-organising emergent solution with desired macroscopic properties.

This paper has no intention to describe a full-fledged completed methodology. Instead, the paper is a starting point in which future work can be integrated to reach a usable full life-cycle methodology. The paper is structured as follows. Section 2 discusses why existing agent-oriented methodologies only focus on the microscopic issues of a MAS. Then section 3 describes the Unified Process and where customisation is needed to explicitly address the macroscopic behaviour of self-organising emergent MASs. The following sections discuss in detail the different customisations. Section 7 illustrates the process with a case study and section 8 concludes.

2. A Shortcoming of Today's Agent-Oriented Methodologies

This section describes a shortcoming of today's agent-oriented methodologies with respect to engineering the macroscopic behaviour of self-organising emergent MASs. The authors of [36] make a distinction of methodologies based on different scales of observing a system:

- *Micro-scale:* The focus here is on systems of a limited number of agents and the engineer details the features of each agent (i.e. agent action-selection architectures, the rules of each agent, knowledge representation), the mechanisms underlying each agent's interaction, and each interaction of the agents with their environment (i.e. protocols, indirect/direct coordination, action models), etc.

- *Macro-scale:* At this scale, the collective behaviour of the system is what matters. The focus is on understanding and controlling the macroscopic behaviour of systems with a very large number of interacting agents, possibly distributed over a network in dynamic and uncontrollable environments.
- *Meso-scale:* When micro-scale systems are used in a larger macro-scale system, two new issues are important: (i) the impact on the micro system of being used in an open and complex scenario and (ii) the impact on the macro-scale system of incorporating the micro-scale system.

As argued in [36], most current approaches to agent-oriented software engineering mainly focus on development of small-size MASs (micro), and on the definition of suitable models, methodologies, and tools for these kind of systems. Meso-scale and especially macro-scale issues are mostly disregarded.

For example, in Gaia v.2 [35] models are constructed for the individual entities in the system, i.e. environmental model, role model, and interaction model. Also, the organisational model, intended to cover the structure of the system as a whole, is only concerned with relations between individual roles. The performance of the organisation of agents with respect to certain macroscopic properties is not considered explicitly. Also in MaSE [33] solely micro-scale issues are addressed: capturing goals, capturing use cases and refining roles with tasks, creating agent classes, constructing conversations, specifying how agents are deployed, etc. And as a final example, Tropos [12] uses actors, goals, tasks, resources, organisational theory from analysis through design which implies a strong focus on the micro-scale and less on the macro-scale issues. Although this methodology also emphasises the importance of social design patterns, those patterns are also focussed on how individual agents interact and relate and not on what macroscopic property should result from a certain social pattern.

Micro-scale issues are nonetheless important. However, for self-organising emergent MASs the lack of explicitly dealing with macro-scale issues is a serious shortcoming. Until now the desired macroscopic behaviour has been engineered in an ad-hoc way because no support exists to systematically engineer it. To address this problem, the remainder of this paper focusses on how and where existing techniques and knowledge on macro-scale issues can be integrated into an existing software engineering process.

3. A Customised Unified Process for Self-Organising Emergent MAS

Typically, engineering MASs means having 99% of the effort go to conventional computer science and only 1% involves the actual agent paradigm [10]. Therefore, to engineer self-organising emergent MASs, developers should exploit conventional software technologies and techniques wherever possible [29]. Such exploitation speeds up development, avoids reinventing the wheel, and enables sufficient time to be devoted to the value added by the multi-agent paradigm [34]. Instead of devising new entire methodologies, which was the trend until now [29], it is a

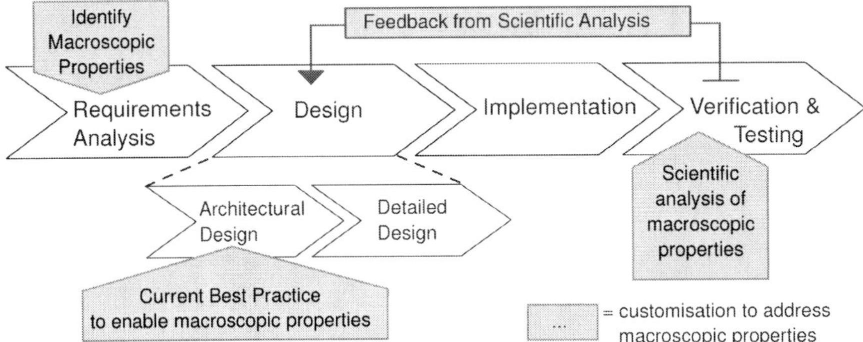

Figure 1. A traditional engineering iteration annotated with customisations for macro-scale issues of self-organising emergent MASs.

good start to follow existing traditional software engineering processes, such as the Unified Process (UP) [15], as much as possible.

A key practice in both the UP and most other modern processes is iterative development [18]. Development is organised into a series of short mini-projects called iterations. The outcome of each iteration is a tested, integrated, and executable partial system. Each iteration typically includes a number of disciplines: requirements analysis, design, implementation, and testing-and-verification. Each discipline consists out of a number of activities that result in artifacts (documents, code, models, etc.). As such, a solution is successively refined, with cyclic feedback from verification to design and adapting the solution accordingly as core drivers to converge upon a suitable solution.

The idea is to follow this iterative process, however, customisation is needed to cope with the specific issues of constructing a coherent macroscopic behaviour that is autonomously maintained by a self-organising emergent MAS. Therefore, in each discipline of the engineering process one should focus on how to *address the desired macroscopic properties* of the system. Of course, also other agent specific issues are important such as the micro-scale issues described in section 2. Most of them are to be incorporated into the design. For this paper, the emphasis is on addressing the macro-scale issues.

The following sections outline the most important disciplines in one iteration of the UP process and indicate which customisations are needed that address the macroscopic behaviour. A schematic overview is given in figure 1. After that, starting from section 4, each customisation is covered in more detail.

3.1. Requirements Analysis

Requirements analysis emphasises an investigation of the problem with functional and non-functional requirements, rather than the solution [18]. This is achieved using standard techniques such as use cases, feature lists, and a domain model that reflects the problem domain. It is important to realise that after the requirements are known it is not always useful to use a self-organising emergent MAS. The decision to use such a solution is made in the design based on the fact that (mostly non-functional) requirements indicate that a self-organising emergent MAS solution is suitable and promising. Section 4 gives an overview of the typical characteristics of the problem that ask for a self-organising emergent MAS. To facilitate that decision, the requirements analysis can be customised to give *special attention to issues that typically lead to self-organising emergent MASs with macroscopic behaviour*:

- Explicitly check if the characteristics that call for self-organising emergent MAS solutions (see section 4) are present or not.
- A self-organising emergent MAS typically 'maintains' certain macroscopic properties. Therefore, explicitly identifying those requirements that are 'ongoing', i.e. that have to be adaptively maintained, is important. Typical examples are the so-called self-X properties: self-optimising, self-healing, self-configuring, etc [16].
- How to measure the system performance? Typically, these measurements are related to macroscopic properties of the system that have to be maintained and explicitly engineered (e.g. the throughput in many systems is a performance measure and a certain throughput level has to be maintained).

3.2. Design

The design emphasises a conceptual solution (in software and hardware) that fulfills the requirements, rather than its implementation [18]. There are two design phases: architectural and detailed design.

3.2.1. Architectural Design.

In early iterations of the engineering process the design focusses more on coarse-grained structural decisions that determine the software architecture. The software architecture of a program or computing system is the structure or structures of the system, which comprise software elements, the externally visible properties of those elements, and the relationships among them [1]. The architectural design is mainly driven by non-functional requirements because they apply on the system as a whole. The overall structure of a system influences significantly how well these requirements are achieved. However, functional macroscopic properties of a self-organising emergent MAS are also system-wide and can significantly influence the system architecture.

The author of [18] states that architectural design is partially a science and partially an art. The *science* of architecture is the collection and organisation of information about the architectural significant requirements (e.g. non-functional and macroscopic functionality). The *art* of architecture is making skillful choices and constructing a solution that meets these requirements, taking into account trade-offs, interdependencies, and priorities. The 'art' of architectural design is the creative step where designers use knowledge from a variety of areas (e.g. architectural styles and patterns, technologies, pitfalls, and trends) as a guide to reach a suitable solution.

As mentioned in section 3.1 the first decision to be made is if self-organising emergent MAS is a suitable solution. If that is the case, then *knowledge and experience from current best practice* in engineering macroscopic behaviour of self-organising emergent MASs is to be integrated into the architectural design of the engineering process in order to creatively address the required macroscopic properties. Section 5 elaborates on this.

3.2.2. Detailed Design.

In later iterations the architecture converges more and more to a fixed structure and the more fine-grained design issues are resolved, i.e. mostly micro-scale issues. For example, with respect to self-organising emergent MASs the choice of the internal action-selection architecture [27] of the agents can be addressed.

3.3. Implementation

The implementation realises the design in code of a specific language and on hardware. No customisations for macro-scale issues are needed because the implementation is completely microscopic. However, existing frameworks, tools, and middleware [2,24,25] specific for building self-organising emergent MASs can support the implementation.

3.4. Testing-and-Verification.

Self-organising emergent MASs will only be acceptable in an industrial application if one can give *guarantees about the macroscopic behaviour*. In other words, when verifying whether the system meets the requirements it is important to explicitly verify if the macroscopic behaviour evolves as required. Such a behaviour is typically system-wide and required to 'dynamically' maintain certain macroscopic properties over a long period in time.

In traditional software engineering the testing-and-verification discipline is typically unit-based [18], i.e. one formally proofs or verifies that the behaviour of a single unit in the system meets the specifications of that unit. It is clear that this is not suitable for verifying system-wide behaviour. In addition, so called integration tests verify if for a certain trigger, a chain of successive unit executions ends in the required post-conditions. This is also not suitable because, like unit-tests, they only test whether a particular result is obtained at a particular moment in time. Because macroscopic behaviour needs to be verified over a period in time, another approach is needed.

Although, the macroscopic behaviour can still be specified formally, it is practically infeasible to formally verify or proof the correctness of the macroscopic behaviour. The existing unit-tests and integration-tests are still useful but need to be complemented by an empirical approach that allows to verify the required evolution of the macroscopic properties of the system. It is important that such an empirical approach allows to systematically analyse the behaviour and obtain results that are scientifically founded, i.e. more than mere observation of simulation measurements. The results are used as feedback for the design. Section 6 discusses why formal approaches are practically infeasible and gives an example of an empirical approach and the kind of feedback that can result from it.

4. When to Use a Self-Organising Emergent MAS?

In the design one decides if a self-organising emergent MAS solution is suitable. Agents and self-organisation have been used in a wide range of applications, but are not a universal solution. For many applications, conventional software development paradigms are enough and more appropriate [34]. The requirements should indicate if a self-organising emergent MAS is a promising solution or not. First of all, there should be a need for a certain **autonomous behaviour** otherwise agents in general and self-organising solutions in particular are not suited. Based on literature [20,23,36,29,30], the following list gives an tentative overview of what we belief to be the most important problem characteristics that promote the use of self-organising emergent MASs:

1. **Decentralised or Distributed Information Availability.** The information that is needed for automated decision making is only available in a decentralised manner, i.e. no central aggregation is possible. Such problems tend to force that only local interactions are possible. Typical application scenarios in which this occurs are: restrictions on information sharing in competitive and no-cooperative domains (e.g. supply chain management), geographical distribution of the information (e.g. logistics, mobile and collective robotics, ...), and partial or temporal communication inaccessibility (e.g. ad-hoc networks).
2. **High Dynamics: failures and frequent changes.**
   - *Failures: Robustness.* A robustness requirement can impose that a central planner or controlling entity should be avoided as much as possible to eliminate a single point of failure (e.g. military mission-critical applications). Or a decentralised coordination mechanism should ensure that a failing planner is immediately replaced. Robustness also implies that the desired macroscopic behaviour has to be maintained while changes occur, i.e. self-organisation.
   - *Frequent Changes: Flexibility and Adaptability.* The flexible way in which agents operate and interact (both with each other and the environment) makes them suitable for dynamic and unpredictable scenarios. Because agents are typically embedded or situated in the environment where all the dynamics take place, agents are able to rapidly respond to local stimuli to handle problems instantly. Such local reconfigurations enable constant replanning in the face of frequent changes.

There are other characteristics mentioned in literature that promote to use multi-agent systems: simulation and modelling, openness, resource-constrained domains, complex problems, discrete domains, etc. Not all of them however imply the need for self-organising emergent solutions. We refer to [20,23,36] for a description of these characteristics and for a discussion on which of them indeed call for multi-agent systems and which of them are not really that essential (e.g. complex problems are equally hard to solve with agents or a self-organising solution).

5. Design: Exploit Current State-of-the-art

After deciding if a self-organising emergent MAS solution will be used, the creative design can explicitly address the development of macroscopic behaviour by exploiting existing best practice and experience. There exists a lot of work that indicates how macroscopic properties, that arise solely from local interactions between agents, can be achieved:

- On the one hand there are **general guidelines or 'design principles'** such as "Think in terms of Flows rather than transitions", "Keep agents small, focus on interactions", and "Provide a local fitness measure for agents to react on that is correlated with the global situation" [23].
- Then there are more concrete guides like a **reference architectures** for autonomic computing [14] or specific MASs [30]. Such architectures embody know-how on a good MAS structure in terms of software elements.
- There is also a body of **decentralised mechanisms that allow coordination** between a group of agents to achieve desirable macroscopic properties. Often these mechanisms are inspired by biology [21]. For example, a commonly used coordination mechanism is a "digital pheromones infrastructure" [3] in which a group of agents coordinate by dropping synthetic pheromones (i.e. data that evaporates over time) in the environment for others to find and react on. The inspiration comes from ant colonies that forage for food based on trails of pheromones. "Co-fields" [19] are another example which put a kind of gradient map into the environment for agents to perceive and follow up or down hill according to the coordination goal. All these mechanisms have there advantages and disadvantages. According to the macroscopic properties that have to result from the coordination between the agents one can use these 'patterns' as a guide for building a suitable coordination architecture. We refer to [4,21] for a literature study of such decentralised mechanisms.

A lot of work has to be done to actually integrate and exploit these guidelines and mechanisms. For example, translating guidelines into concrete engineering support and providing tools and specific guidelines for using each decentralised coordination mechanism are possibilities. An important issue here is to understand for which kind of macroscopic behaviour each coordination mechanisms is useful, which is more appropriate in which situation, etc. For example, digital pheromones are typically very useful in a dynamic context where the macroscopic structures have to adapt to changing circumstances. The reason for this is that the evaporation mechanism ensures that old and obsolete parts of the structure disappear and the structures are only reinforced to reflect the current situation. As such, macroscopic properties that are subject to so called 'truth maintenance' can be engineered. Also for other coordination mechanisms these kind of observations have to be made explicit as a guide for engineers in the design.

6. Test-and-Verification: Guaranteeing the Desired Macroscopic Behaviour

The goal is to verify if the desired macroscopic behaviour, identified during the requirements analysis, is achieved. This section first discusses why formal approaches are not suited to guarantee the macroscopic behaviour and then gives an example of an empirical approach that allows to verify the evolution of the macroscopic behaviour of self-organising emergent MASs.

6.1. *Formal versus Empirical Verification*

The main question here is how to know if a certain self-organising emergent MAS maintains the required macroscopic behaviour. Firm guarantees could be obtained if the system is modelled formally and the required macroscopic behaviour is proofed analytically. However, constructing a formal model and correctness proof of a complex interacting computing system is infeasible. Wegner [28] proves this based on the fact that computing systems using interaction are more powerful problem solving engines than mere algorithms. Wegner shows that one cannot model all possible behaviour of an interaction model and thus formally proving correctness is not merely difficult but impossible.

The alternative to formal proof is to use an empirical method to verify the macroscopic behaviour. Macroscopic properties are typically quantified with measurable variables which we define as *macroscopic variables*. The variables are then measured from simulations and the measurements are used to obtain statistical results. Although such results remain useful, one should make sure that the results have a scientific or mathematical foundation. As advocated in [9], we need an empirical but scientifically founded method to verify the macroscopic behaviour. Then, more valuable and advanced verification results can be obtained that are supported by mathematical or scientific algorithms. In section 6.2 such a method is described. Note that this does not exclude formal models completely. Formal models are less suited for verification, but are still useful to unambiguously specify the macroscopic behaviour and acquire more understanding on how self-organising emergent MASs work (e.g. [11] and [26]).

6.2. *Example Verification Approach: Equation-Free Analysis*

To achieve objective and scientifically founded results, it is often a good idea to use mathematical methods. There exist a lot of numerical algorithms that allow to analyse the system dynamics. For example, a bifurcation algorithm allows to analyse if there are qualitative changes in the behaviour of the system as a result of a changing parameter. Consider a mobile ad-hoc network. The communication range used by each node is an important parameter. For certain values of that parameter the network can maintain itself as one connected network. For other values, the network can be constantly partitioned or completely disconnected. Bifurcation analysis allows to detect which type of behaviour is achieved for which communication ranges. Such numerical algorithms are typically applied on (differential) equation models that model the evolution of the macroscopic variables. This is illustrated in the left of figure 2. The analysis algorithm repeatedly eval-

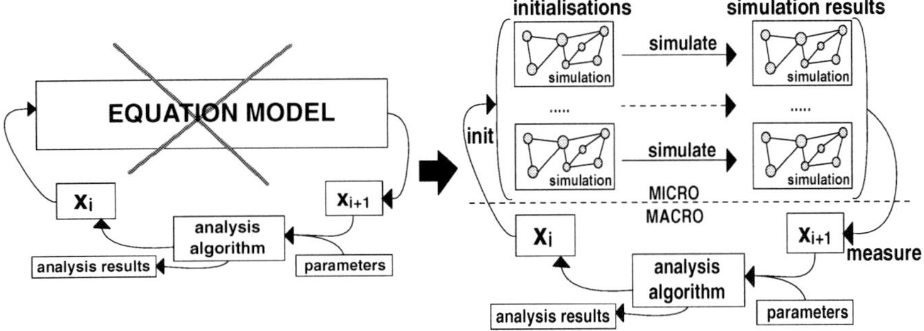

Figure 2. Equation-Free Accelerated Simulation guided by the analysis algorithm

uates the equation for certain initial values (X_i in figure 2) for the variables to obtain the value some time steps later (X_{i+1} in figure 2). As such the behaviour evolution is analysed on the fly by the algorithm.

However, as discussed in section 6.1, in complex interacting systems, deriving an equation, which is a formal model, is often not possible. Even if one derives an equation, one risks results that differ significantly from the real system behaviour (e.g. [31]). An approach that solves this problem is called *"equation-free" macroscopic analysis* [17]. The observation is that most numerical algorithms have no need for the equation itself; they only need a routine that *evaluates* the equation for a given value in order to return a value some time steps later. The idea is to replace the equation evaluations with realistic agent-based *simulations* (see figure 2). Simulations are initialised to reflect the given values (X_i) of macroscopic variables (*init* in figure 2). After simulating for the requested time (*simulate* in figure 2), the new values (X_{i+1}) are measured from the simulation state (*measure* in figure 2). The results are scientifically and mathematically supported because mathematical algorithms decide on the accuracy of the result and on how the behaviour is analysed. As such all numerical analysis algorithms can be applied (e.g. stability analysis, newton steady state detection, etc.) and useful feedback is obtained for the next design iteration of the engineering process.

A big challenge for this approach is to find a suitable set of macroscopic variables that reflects the evolution of the macroscopic properties under study, and an initialisation operator to initialise simulations based on these variables. An advantage of the approach is that the cost for running simulations is limited to the simulation runs that are essential for the numerical algorithm. For example, the newton algorithm to detect to which value a variable converges typically only needs less than 10 rather short simulation runs, even when in a begin-to-end simulation only converges after a long period in time. A disadvantage is that the application of this approach by an engineer is very difficult and requires a rather significant background knowledge about numerical mathematics. We refer to [6] for a more detailed discussion about the challenges, the advantages, and the disadvantages of the approach, and some results on an Automated Guided Vehicle transportation system.

6.3. Iterative Development with Feedback

The required macroscopic properties, identified during requirements analysis, are engineered in an iterative process. This means that the design discipline uses the test-and-verification results to get feedback on the performance with respect to those macroscopic properties and adapt the design to steer the solution towards the required behaviour. The design typically has a different focus in early iterations (architecture design) compared to later iterations (detailed design). As a consequence, to have useful feedback for the design, there should be a varying focus in the test-and-verification discipline also (see figure 3), some possibilities:

- **Non-functional and macroscopic performance:** How good is the performance of the solution for non-functional requirements and macroscopic properties?
- **Comparison of coordination mechanisms:** Which of the mechanisms performs best with respect to the functional/non-functional requirements?
- **Parameter-tuning:** Which range of values for a certain parameter allows to achieve the requirements?
- **Characterisation of macroscopic behaviour:** When a solution is (nearly) completed, it is important to characterise that solution for different possible deployment scenario's: performance for a heavy utilisation load, what utilisation load can it cope with, etc.

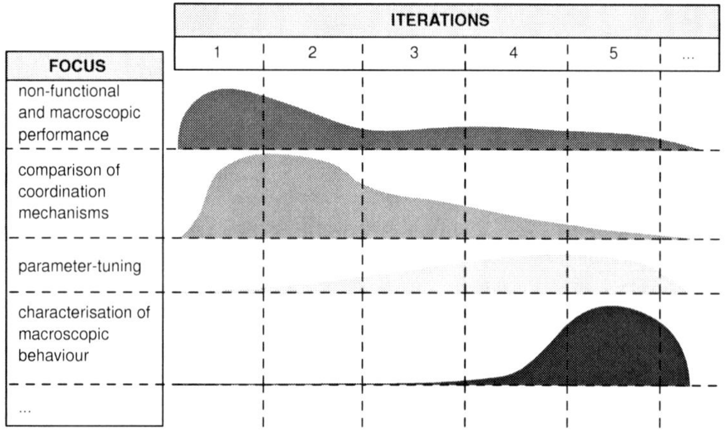

Figure 3. Scientific analysis: different focus in early iterations versus later iterations.

In *early iterations* the focus is more on the 'coarse grained' decisions with respect to the architecture of the solution. The influence of certain decisions on the performance of non-functional properties is analysed. Also, deciding which coordination mechanism will be used in each coordination scenario can influence the performance of macroscopic properties significantly. A comparison of different alternatives for coordination and other architectural decisions is necessary in the early iterations [18].

In *later iterations* one analyses the solution in more detail. For example, a self-organising emergent multi-agent solution typically consists of a lot of parameters that can be set or changed online to adaptively react to changing situations (e.g. adapting the communication range in ad-hoc networks to react on an increase in node density). Analysing which parameter value set is most suitable in which situation serves as important feedback to know how to adapt. Also, when a solution is completed and has to be deployed in different operational conditions it is important to know how it will behave. For example, knowing what the minimum number of network nodes should be in a mobile ad-hoc network in order to still reach the required degree of connectivity determines where the boundaries are of that solution.

As such, an integration of a scientific and empiric verification approach into an iterative engineering process enables to systematically develop a self-organising emergent solution based on feedback about the macroscopic behaviour.

7. Example: Mobile ad-hoc networks

This section discusses an example case study, showing how the proposed engineering process can be applied in practice. First, section 7.1 introduces mobile ad-hoc networks. Then sections 7.2, 7.3, and 7.4 address the most important disciplines of the engineering process, i.e. requirements, design, and verification. The focus is on the customisations of the process for self-organising emergent systems.

7.1. Problem Description

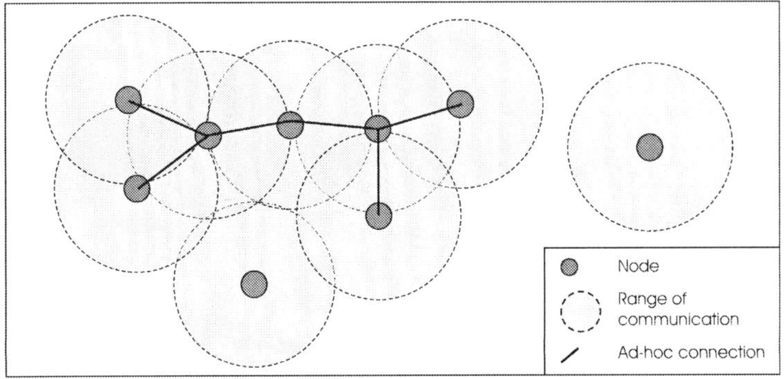

Figure 4. A mobile ad-hoc network where two nodes are not connected to the network.

A mobile ad-hoc network is a local area network or other small network of portable and mobile devices, in which some of the network devices are connected to the network only while in some close proximity to the rest of the network (see Figure 4). The devices or nodes are free to move randomly and organise themselves arbitrarily; thus, the network's wireless topology my change rapidly and unpredictable. The goal of managing such a network is to maintain the connectivity between the nodes and to handle the routing of communication between nodes in he face of changes (e.g. nodes can leave, join, and fail at every moment).

7.2. Requirements Analysis

First of all the problem has to be described in detail by enumerating the requirements for the system to build. On the one hand traditional techniques such as domain models are used. For example, Figure 5 shows a simple domain model in UML notation of an ad-hoc network that allows to capture the problem domain.

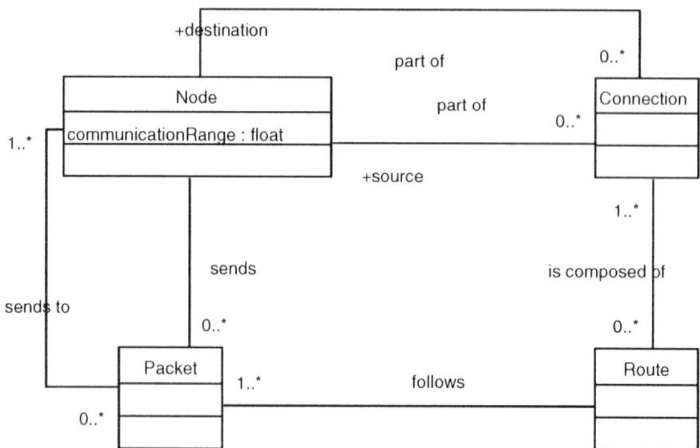

Figure 5. Simple Domain Model for Mobile Ad-hoc Network

During the requirements analysis special attention is given to issues that typically lead to self-organising emergent solutions (see section 3.1). We should think explicitly about properties that have to be 'maintained' by the system as a whole or about how the performance of the system is measured. For example, this can lead to the following 'macroscopic' requirements:

- Adaptively optimise the routing of packets from sources to destinations.
- Adaptively maintain a connected network or a minimum degree of connectivity, i.e. average number of connections per node.
- The system has a good performance if the ratio of communication successes compared to failures is high. Thus, the optimisation of the communication success is another macroscopic requirement.

Another issue to look at for self-organising emergent systems is if the characteristics that ask for such a solution are present or not. First of all, ad-hoc network management is too complex to handle manually. Therefore, *autonomous management* or 'self'-organisation of the network is needed. Secondly, central control is not desirable because the information needed for making management decisions is completely *decentralised and distributed.* Also, partial or temporal communication inaccessibility can occur frequently. Finally, the problem domain is *highly dynamic*, e.g. nodes can fail, enter, or leave at any moment; connections can fail at any moment; there is a very dynamic and unpredictable topology and communication load. In such a dynamic context it is impossible to aggregate enough information about a network centrally. By the time the information would have been aggregated something has already changed which makes the information obsolete. Also, if a central controller fails, the whole network fails. Hence a decentralised solution is needed that allows local reconfigurations to enable constant replanning in the face of frequent changes.

7.3. Design

The first issue to handle in the design is if a self-organising emergent solution is suitable. From the requirements we can see that the problem domain of ad-hoc networks has many of the characteristics that ask for a self-organising emergent solution (see section 4). Hence, such a solution seems promising and as such a first design decision can be to build a self-organising emergent solution.

Secondly, to achieve a self-organising emergent solution the macroscopic requirements have to be achieved by a coordinated group of agents. The task is now to creatively construct a decentralised solution exploiting existing best practice and experience (see section 5). For example, one requirement is that the packets on the network are adaptively routed. A coordination mechanism is needed that allows to adaptively create routes for packets according to the changing topology of the network. A suitable mechanism can be constructed using "digital pheromones" and a swarm of simple agents that construct pheromone paths or routes throughout the network (e.g. [8]). Pheromones evaporate if they are no longer reinforced, i.e. old routes that are no longer valid gradually disappear. Also, pheromone paths tend to optimise to the shortest path. As such, the system adapts to the changing topology. An alternative coordination mechanism is a so called co-field or gradient map approach. Each node maintains a distributed computational field that guides network packets. The field attracts packets to move uphill on the destination's field towards that destination node.

Also for the other requirements a coordinated solution has to be constructed, using experience and existing best practice such as coordination mechanisms from the self-organising emergent systems community.

7.4. Verification

Finally, one needs to verify if the required 'macroscopic' properties are indeed achieved by the system. In earlier iterations more coarse-grained decisions are made. For example, choosing with which coordination mechanism a certain requirement is best achieved is an important design decision. To make this decision both alternatives have to be verified by simulating the system and observing its performance. For example, for routing in ad-hoc networks, one can choose between packets guided by pheromone paths or by gradient maps.

Many of the requirements involve the maintenance of a certain property over a longer period in time. Thus, a verification method has to be applied that allows to verify the evolution of the system behaviour over time. In section 6.2 an example method is described that allows to apply an existing set of numerical algorithms using simulations efficiently. As such, advanced verification results can be obtained.

For example, there exists and algorithm called 'bifurcation analysis' which allows to analyse if there are qualitative changes in the behaviour of the system as a result of a changing parameter. One parameter for ad-hoc networks is the communication range used by nodes. For certain values the network will be able to maintain itself as one large connected network. For other values, the network can be constantly partitioned or completely disconnected. Bifurcation analysis allows

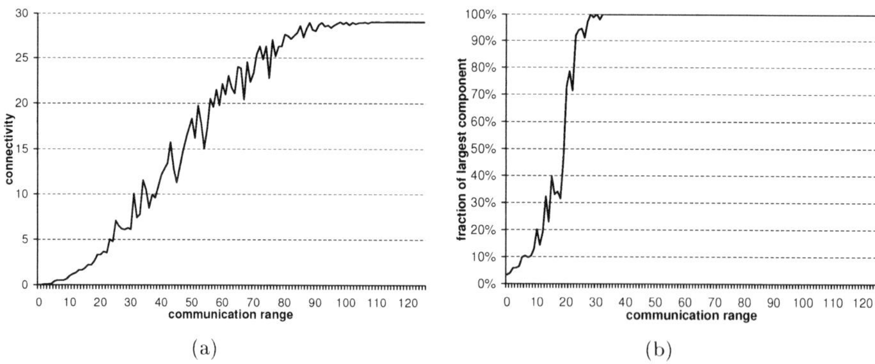

Figure 6. Bifurcation diagrams with a varying communication range. For the average over time of (a) connectivity: average connections of all nodes; and of (b) the fraction of largest connected network component.

to detect which type of behaviour is achieved for which communication ranges. To illustrate this, results from a simple simulation of an ad-hoc network [32] of 30 nodes, randomly moving in a space of size 100x100, are shown in Figure 6. On the horizontal axis the parameter-values (i.e. varying communication range) are depicted and the vertical axis shows the performance measure, i.e. connectivity or the average number of connections per node (figure 6(a)), and fraction of all nodes in the largest connected component of the network (figure 6(b)). This shows that a communication range of 32 is enough to maintain a completely connected network and an average of 15 connections per node can be achieved with a communication range of 45. The result of such a bifurcation analysis is then used in the next iteration of the design to tune the parameters to the values which give the best performance.

8. Conclusion and Future Work

Today's agent-oriented methodologies have a serious shortcoming. They do not deal explicitly with a fundamental issue when building self-organising emergent MASs, i.e. engineering the desired macroscopic behaviour. To come to a full life-cycle methodology that allows to systematically engineer the macroscopic behaviour, one should not re-invent the wheel. One should follow a conventional software engineering process, i.e. UP, as much as possible. This paper proposes to customise the UP process by *explicitly addressing the macroscopic behaviour* throughout the process.

This paper is only a starting point. A lot of future work is needed to achieve a full-fledged and usable methodology. Future work should study each guideline and coordination mechanism in detail in order to provide concrete support to exploit them (e.g. design guidelines, tools, design models, etc.). Also refinement of the empirical verification approach (e.g. a guide for choosing macroscopic variables) and applying it extensively is important.

Acknowledgements.

This work is supported by the K.U.Leuven research council as part of the Concerted Research Action on Agents for Coordination and Control - AgCo2.

References

[1] L. Bass, P. Clements, and R. Kazman. *Software Architecture in Practice.* 2003.

[2] F. Bellifemine, A. Poggi, and G. Rimassa. Jade, A FIPA-compliant agent framework. In *Proc. of PAAM'99*, London, 1999.

[3] Sven Brueckner. *Return From The Ant - Synthetic Ecosystems For Manufacturing Control.* PhD thesis, Humboldt-Universität, Berlin, 2000.

[4] Tom De Wolf and Tom Holvoet. Towards Autonomic Computing: agent-based modelling, dynamical systems analysis, and decentralised control. In *Proc. of the 1st Int. Workshop on Autonomic Computing Principles and Architectures*, 2003.

[5] Tom De Wolf and Tom Holvoet. Emergence and Self-Organisation: a statement of similarities and differences. In *Proc. of the 2nd Int. Workshop on Engineering Self-Organising App.*, 2004.

[6] Tom De Wolf, Giovanni Samaey, Tom Holvoet, and Dirk Roose. Decentralised Autonomic Computing: Analysing Self-Organising Emergent Behaviour using Advanced Numerical Methods. In *Proceedings of IEEE Int. Conf. on Autonomic Computing (ICAC'05)*, Seattle, USA, June 2005.

[7] M. Dorigo, V. Maniezzo, and A. Colorni. Ant System: Optimization by a colony of cooperating agents. *IEEE Transactions on Systems, Man, and Cybernetics-Part B*, 26(1):29–41, 1996.

[8] Frederick Ducatelle, Gianni Di Caro, and Luca Maria Gambardella. Ant agents for hybrid multipath routing in mobile ad hoc networks. In *Proceedings of Second Annual Conference on Wireless On-demand Network Systems and Services (WONS'05)*, pages 44–53, 2005.

[9] Bruce Edmonds and Joanna J. Bryson. The Insufficiency of Formal Design Methods - the necessity of an experimental approach for understanding and control of complex MAS. In *Proc. of the AAMAS'04 Conference*, pages 938–945, New York, July 2004. ACM Press.

[10] O. Etzioni. Moving Up the Information Food Chain: Deploying Softbots on the World Wide Web. In *Proc. of the 13th Int. Conf. on Artificial Intelligence*, 1996.

[11] L. Gardelli, M. Viroli, and A. Omicini. On the Role of Simulations in Engineering Self-Organizing MAS: the Case of an Intrusion Detection System in TuCSoN. In *3rd Int. Workshop "Engineering Self-Organising Applications" (ESOA)*, 2005.

[12] P. Giorgini, M. Kolp, J. Mylopoulos, and M. Pistore. The Tropos Methodology: An Overview. In *Methodologies And Software Engineering For Agent Systems.* 2004.

[13] K. Herrmann, G. Mühl, and K. Geihs. Self-Management: The Solution to Complexity or Just Another Problem? *IEEE Distributed Systems Online*, 6(1), 2005.

[14] IBM. An architectural blue-print for autonomic computing. Tech. Report, 2004.

[15] I. Jacobson, G. Booch, and J. Rumbaugh. *The unified software development process.* Addison Wesley, 1999.

[16] Jeffrey O. Kephart and David M. Chess. The vision of autonomic computing. *IEEE Computer Magazine*, 36(1):41–50, Jan 2003.

[17] Ioannis G. Kevrekidis, C. William Gear, James M. Hyman, Panagiotis G. Kevrekidis, Olof Runborg, and Constantinos Theodoropoulos. Equation-free, coarse-grained multiscale computation: enabling microscopic simulators to perform system-level analysis. *Comm. in Mathematical Sciences*, 1(4):715 – 762, 2003.

[18] Craig Larman. *Applying UML and Patterns: An Introduction to Object-Oriented Analysis and Design and Iterative Development.* Prentice Hall, 3rd edition, 2005.

[19] Marco Mamei, Letizia Leonardi, and Franco Zambonelli. Co-fields: Towards a unifying approach to the engineering of swarm intelligent systems. In *Proc. of the 3rd Int. Workshop on Engineering Societies in the Agents World (ESAW) 2002*, number 2577 in LNAI. Springer, 2003.

[20] Vladimir Marik Michal Pechoucek, Martin Rehak. Expectations and deployment of agent technology in manufacturing and defence: Case studies. In *Proc. AAMAS'05 – Industry Track*, volume 3, Utrecht, Netherlands, July 2005. ACM.

[21] Radhika Nagpal. A Catalog of Biologically-inspired Primitives for Engineering Self-Organization. In *Engineering Self-Organising Systems, Nature-Inspired Approaches to Software Engineering*, volume 2977 of *LNCS*, pages 53–62. Springer, 2004.

[22] H. Van Dyke Parunak and Sven A. Brueckner. Analyzing Stigmergic Learning for Self-Organizing Mobile Ad-Hoc Networks (manet's). In *Proc. of the 2nd Int. Workshop on Engineering Self-Organising Applications*, 2004.

[23] H. Van Dyke Parunak and Sven A. Brueckner. Engineering Swarming Systems. In *Methodologies and Software Engineering for Agent Systems*, volume 11 of *Multiagent Systems, Artificial Societies, and Simulated Organizations.* Springer, 2004.

[24] Repast. http://repast.sourceforge.net, 2003.

[25] K. Schelfthout and T. Holvoet. An environment for coordination of situated multi-agent systems. In *1st Int. Workshop on Environments for MAS (E4MAS)*, 2004.

[26] Jan Sudeikat and Wolfgang Renz. Mesoscopic Modeling of Emergent Behavior - A Self-Organizing Deliberative Minority Game. In *The Third International Workshop on Engineering Self-Organising Applications (ESOA'05)*, 2005.

[27] T. Tyrrell. *Computational Mechanisms for Action Selection.* PhD thesis, University of Edinburgh, 1993.

[28] Peter Wegner. Why Interaction is More Powerful than Algorithms. *Commun. of the ACM*, 40(5):80–91, May 1997.

[29] D. Weyns, A. Helleboogh, E. Steegmans, T. De Wolf, K. Mertens, N. Boucké, and T. Holvoet. Agents are not part of the problem, agents can solve the problem. In *Proc. of the OOPSLA Workshop on Agent-Oriented Methodologies*, 2004.

[30] D. Weyns, K. Schelfthout, T. Holvoet, and T. Lefever. Decentralized Control of E'GV Transportation Systems. In *Proc. AAMAS'05 – Industry Track*, 2005.

[31] W.G. Wilson. Resolving Discrepancies between Deterministic Population Models and Individual-Based Simulations. *American Naturalist*, 151(2):116–134, 1998.

[32] T. De Wolf. Netlogo model: Mobile ad-hoc network. online at http://www.cs.kuleuven.be/~tomdw/software/netlogo/MANET, DistriNet, KULeuven, Belgium., 2005.

[33] M. F. Wood and S. A. DeLoach. An Overview of the Multiagent Systems Engineering Methodology. In *Agent-Oriented Software Engineering*, volume 1957 of *LNCS*. Springer, 2001.

[34] Michael Wooldridge and Nicholas R. Jennings. Software engineering with agents: Pitfalls and pratfalls. *IEEE Internet Computing*, 3(3):20–27, 1999.

[35] F. Zambonelli, N. Jennings, and M. Wooldridge. Developing multiagent systems: The Gaia methodology. *ACM Trans. Software Eng. Meth.*, 12(3):417–470, 2003.

[36] F. Zambonelli and A. Omicini. Challenges and Research Directions in Agent-Oriented Software Engineering. *Autonomous Agents and Multi-Agent Systems*, 9:253–283, 2004.

Self-Organization and Autonomic Informatics (I)
H. Czap et al. (Eds.)
IOS Press, 2005

Moving Nature-Inspired Algorithms to Parallel, Asynchronous and Decentralised Environments

Enda Ridge [a,1], Daniel Kudenko [a], Dimitar Kazakov [a] and Edward Curry [b]
[a] *The Department of Computer Science, The University of York, U.K.*
[b] *The Department of Information Technology, The National University of Ireland, Galway*

Abstract. This paper motivates research into implementing nature-inspired algorithms in decentralised, asynchronous and parallel environments. These characteristics typify environments such as Peer-To-Peer systems, the Grid and autonomic computing which demand robustness, decentralisation, parallelism, asynchronicity and self-organisation. Nature-inspired systems promise these properties. However, current implementations of nature-inspired systems are only loosely based on their natural counterparts. They are generally implemented as synchronous, sequential, centralised algorithms that loop through passive data structures. For their successes to be relevant to the aforementioned new computing environments, variants of these algorithms must work in truely decentralised, parallel and asynchronous Multi-Agent System (MAS) environments. A general methodology is presented for engineering the transfer of nature-inspired algorithms to such a MAS framework. The concept of pheromone infrastructures is reviewed in light of emerging standards for agent platform architecture and interoperability. These ideas are illustrated using a particularly successful nature-inspired algorithm, Ant Colony System for the Travelling Salesman Problem.

Keywords. Ant Colony Algorithms, Ant Colony System, decentralised, parallel, asynchronous, Multi-Agent System, pheromone infrastructures

1. Introduction and Motivation

Nature-inspired algorithms such as genetic algorithms [1], particle swarm optimisation [2] and ant colony algorithms [3] have achieved remarkable successes. Indeed, they are the state-of-the-art solution technique for some problems. The algorithms share the characteristic of being loosely based on a natural metaphor such as evolution's search through the vast space of potential organisms, the movement of flocks of birds through a 3D space or the self-reinforcing chemical trails laid by ants while searching for a route through a 2D space.

The natural systems on which these algorithms are based possess many desirable properties that we would like to transfer to our computer systems.

[1]Correspondence to: Enda Ridge, The Department of Computer Science, The University of York, Heslington, York YO10 5DD, UK. Tel.: +44 1904 43 2722; Fax: +44 1904 43 2767; E-mail: ERidge@cs.york.ac.uk.

- They typically contain large numbers of relatively simple participants.
- They are completely decentralised.
- They operate in parallel and asynchronously.
- They use relatively simple signals
- Their desired functionality emerges from the interactions of their participants. This is achieved despite (and because of) their simple participants that have no global information.

These characteristics make natural systems robust to loss of members, parallel in 'execution', and adaptable to a changing problem domain.

However, the efficient pursuit of increasingly accurate nature-inspired solutions that can compete with more traditional algorithms has lead researchers to resort to quite 'unnatural' designs for their nature-inspired algorithms. The result is that the most successful versions of these algorithms are often centralised and sequential implementations that are highly tuned to a particular problem. They bear only a tenuous resemblance to their successful counterparts in nature. This renders them brittle in the face of the dynamism of changing problem specifications and operating conditions and limits their usefulness to industry's direction of increasing distribution, decentralisation and adaptability.

The pursuit of improved performance at optimisation and search is a very worthwhile endeavour. Yet the pursuit of these goals in a nature-inspired algorithm, to the exclusion of all else, sacrifices the real strengths of natural systems—their robustness, adaptability, decentralisation and parallelism.

It is these very properties that are coming to the fore in emerging computer environments such as autonomic computing [4], ubiquitous and pervasive computing, Peer-to-Peer systems, the Grid [5] and the Semantic Web [6]. These environments demand systems that are robust to failures, adaptable to changing requirements and deployment scenarios, composed of relatively simple components for ease of development and maintenance and are preferably decentralised and parallel.

Multi-agent Systems (MAS) [7] provide a platform on which many independent, parallel and asynchronous software entities can interact using standard protocols and ontologies. It is clear that if nature-inspired algorithms are to be useful in the aforementioned environments, then they must embrace the characteristics of a MAS-type platform. It is not sufficient for researchers to claim their algorithms are robust or parallel while implementing them as sequential, synchronous and centralised loops through passive data structures. For example, cellular automata, which can be considered a sequential abstraction of MASs, yield different results when different approximations to parallel update are used [8]. It is reasonable to suspect that approximations of other characteristics such as asynchronicity do not accurately predict the true system behaviour but merely hint that asynchronicity may be a *factor* in performance. Approximations are not sufficient for drawing conclusions. They are only the first stage of a complete research process [9]. Furthermore, decentralised agents can exhibit other phenomena such as hyperactivity [10] and diversity [11] and their designers must address the real-world costs of messaging [12]. Experiments with MAS implementations of a genetic algorithm have yielded unexpected results that were partly due to the cost of messaging [13]. Equally unexpected results may be uncovered with asynchronous and parallel implementations of other nature-inspired algorithms.

The similarities between natural environments and emerging computing environments motivate disciplined scientific and engineering investigations into the successful transfer of these *algorithms*, *techniques* and *infrastructures* into such environments.

This paper studies Ant Colony System (ACS) [14], a very successful nature-inspired algorithm from the ant colony family of algorithms. It explores how ACS might be transferred to such emerging computing environments as can be represented by a truly decentralised, asynchronous and parallel Multi-Agent System, in a way that complies with emerging standards for agent interaction and architecture [15].

The next section gives an overview of the ACS algorithm for the Travelling Salesman Problem. Section 3 reviews the idea of pheromone infrastructures in light of maturing agent standards. Section 4 presents a general strategy for transferring nature-inspired algorithms to MAS-type platforms and illustrates this with reference to the ACS algorithm. The paper concludes with a review of related work, our conclusions and our direction for future work.

2. Ant Colony System for the Travelling Salesman Problem (ACS-TSP)

We briefly review Ant Colony System for the Travelling Salesman Problem (ACS-TSP) for completeness and to draw attention to the issues with centralised, sequential and synchronous nature-inspired algorithms. The reader is directed to other works [3,16] for a comprehensive background, description and discussion of this algorithm.

The Travelling Salesman Problem (TSP) [17] involves finding the shortest route through a given set of cities such that no city is visited twice. The graph representation of the problem is constructed as a set of nodes and edges where nodes represent the cities and edges join every node (city) to every other node. Activities of the ants then select a subset of these edges that represents a valid travelling salesman tour. The TSP is heavily researched and has a well-established set of benchmark problems with solutions [18]. Furthermore, it is an abstraction of a large class of discrete combinatorial optimisation problems that is easily visualised and understood by non-experts. These types of problems are particularly relevant to Grid and Autonomic computing as efficient task and resource allocation are recurring themes in these environments.

The Ant Colony System algorithm [14] for solving the TSP is loosely based on natural ant colony behaviour when foraging for food. Ants wander away from their nest in search of food. After finding food, they return to their nest, laying a chemical marker called a pheromone. Subsequent ants leaving the next are more inclined to follow strong pheromone trails. This positive reinforcement builds efficient trails to food sources in a decentralised fashion. The Ant Colony System algorithm adapts this process to a graph structure. Figure 1 gives the pseudocode for the algorithm.

Initially, every edge in the problem is given the same pheromone level τ_0. A number of ants m are randomly assigned to their starting cities such that no more than one ant occupies any city. The following main loop then proceeds for a maximum of $t_{\max}$ iterations where t is the count of the current iteration. Each ant builds its own tour by repeatedly applying a decision rule to determine the next city it will visit. The ant's list of cities that remain to be visited is stored in a list J. The ant first creates a random number q. This q is compared to an exploration/exploitation threshold q_0. When q exceeds the threshold, the ant's choice of next city favours choosing less frequented trails

```
/******* Initialisation ********/

Set the pheromone level on each edge to τ_0

Place ant k on a randomly chosen city such that there is no more than one ant per city.

Let T+ be the shortest tour found so far and L+ its length.

/*************** Main Loop ***************/

For t = 1 to t_max do

    For k = 1 to m do

    /**** Build ant's tour by applying the following rule n-1 times ****/

        If there exists at least one unvisited city on the candidate list then:

            Choose a random value for the exploration variable q

            Choose next city j ∈ J_i from candidate list using
```

$$j = \begin{cases} P & \text{if } q > q_0 \\ \arg\max_{u \in J_i} \left\{ \tau_{iu}^{\alpha} \cdot \eta_{iu}^{\beta} \right\} & \text{if } q \le q_0 \end{cases}$$

```
            where city P is chosen using the following probability function:
```

$$p_{ij} = \frac{[\tau_{ij}]^{\alpha} \cdot [\eta_{ij}]^{\beta}}{\sum_{l \in J_i} [\tau_{il}]^{\alpha} \cdot [\eta_{il}]^{\beta}}$$

```
            and i is the ant's current city.

        Else

            Choose the next closest city

        End If

    /******** Local Pheromone Update ********/

        With the edge (i, j) just chosen by the ant, update pheromone as follows:
```

$$\tau_{ij} = (1 - \rho_{\text{local}})\tau_{ij} + \rho_{\text{local}}\tau_0$$

```
    End For

    If an improved tour is found then

        Update the values of T+ and L+

    End If

    /******** Global Pheromone Update ********/

    For every edge (i, j) on the best tour T+ Do

        Update pheromone trail by applying the rule
```

$$\tau_{ij} = (1 - \rho_{\text{global}})\tau_{ij} + \rho_{\text{global}}\frac{Q}{L^{+}}$$

```
    End For

End For

/******* End of Program ********/
```

Figure 1. Pseudocode for ACS-TSP algorithm (adapted from [3]) (exploration). When q does not exceed the threshold, the ant favours well-established trails (exploitation). When at a current city i, the decision rule for a possible next city j is a function of the pheromone intensity τ_{ij} on the edge (i, j) and η_{ij}, the inverse of the length of edge (i, j). Each city has a candidate list of preferred cities to which the decision rule is applied. If all cities in the candidate list have already been visited, an ant does not use its decision rule and simply chooses the next nearest city. The decision rule to choose the next unvisited city $j \in J_i$ from the candidate list is given by

$$j = \begin{cases} P & \text{if } q > q_0 \\ \arg\max_{u \in J_i} \left\{ [\tau_{iu}]^{\alpha} \cdot [\eta_{iu}]^{\beta} \right\} & \text{if } q \leq q_0 \end{cases}$$

where P is chosen from the candidate list using probabilities p_{ij} generated with the following function:

$$p_{ij} = \frac{[\tau_{ij}]^{\alpha} \cdot [\eta_{ij}]^{\beta}}{\sum\limits_{l \in J_i} [\tau_{il}]^{\alpha} \cdot [\eta_{il}]^{\beta}}$$

When an ant moves between two cities, it immediately performs a *local pheromone update* on the edge joining those two cities. Once all ants have built their tours, any improvement on the best tour found is recorded. A *global pheromone update* is then performed on the current best tour. This is one of the most obvious centralised components of the ACS algorithm. The main loop then repeats.

ACS-TSP was competitive at finding shorter tours than other techniques such as a genetic algorithm, evolutionary programming and simulated annealing [14]. Comparisons were made on three benchmark problems [18] of size 50, 75 and 100 cities. However, for larger problems, a local search technique had to be introduced to the algorithm (ACS-3-opt) [14]. Again, we see the tendency for nature-inspired algorithms to drift away from their original natural counterparts. This research uses the original ACS-TSP.

3. Pheromone Infrastructures

In this section, we suggest how to implement a pheromone infrastructure suitable for the TSP and relate it to emerging agent standards for interoperability and architecture [15]. Pheromone infrastructures provide a truly parallel, asynchronous and decentralised environment that can support an ant colony nature-inspired system. The adoption of agent standards is an identified short-term goal of the agent community [19].

3.1. Overview of Pheromone Infrastructures

The idea of a 'pheromone infrastructure' has been proposed in the literature in the context of manufacturing control [20]. A pheromone infrastructure is an agent environment that supports some, if not all, of the 4 basic pheromone operations of *aggregation*, *evaporation*, *propagation* and *sensing* [21]. The environment is represented by a collection of *environment agents* restricted to local information and interaction. This facilitates building a distributed and decentralised environment that can run its own environment processes independently. These environment agents are termed *Place Agents* in keeping with the original literature [20]. Place agents manage the topology of the problem and the pheromone functions.

We term the other agents in the system *Solution Agents*. Solution agents move about on the pheromone infrastructure, interacting with the Place agents to perform pheromone operations such as sensing, deposition and perhaps pheromone removal.

3.2. FIPA-compliant Pheromone Infrastructures

Maturing agent standards facilitate applying standard protocols and ontologies to the original pheromone infrastructure idea. The Foundation for Intelligent Physical Agents (FIPA) is an 'IEEE Computer Society standards organization that promotes agent-based technology and the interoperability of its standards with other technologies.' This standardisation is necessary to fulfil one of autonomic computing's 8 requirements[1] [4]. Calculations show that there is a significant messaging cost in enforcing such standards.

3.2.1. Directory Facilitators as Place Agents

The FIPA Agent Management Specification [22] provides for so-called Directory Facilitator (DF) agents. These offer a 'yellow pages' service by allowing other agents to advertise their particular service(s) with the DF. Other agents can then query a DF about its advertised services and subscribe for notification about the appearance of new services.

We can use a collection of DF agents to build a decentralised pheromone infrastructure of Place agents. Each Place (DF) agent offers its own service description as a Place agent. This description includes a description of its location in the environment. Neighbouring Place agents register their service descriptions with one another. These registrations between place agents form the topology of the environment.

Solution agents occupy a location in the environment by registering themselves with the Place agent representing that location. The solution agents can query their current Place agent for its advertised neighbours. This provides Solution agents with a 'visibility' of the environment. The Solution agents 'move' by deregistering from their current Place agents and registering with one of its advertised Place agent neighbours.

Each Place agent maintains a record of the deposited pheromone types and their associated concentrations. Solution agents can query this pheromone record and modify it by depositing and removing pheromones. The Place agents perform the pheromone functions of evaporation and propagation independently of the Solution agents.

3.2.2. Flexibility of Pheromone Infrastructures

This pheromone infrastructure approach is flexible enough to represent both graph-type and discrete space environments, the typical problem representations to which ant colony algorithms are applied. These environments are illustrated in Figure 2. In a discrete space partitioned into tiles, each Place agent represents a tile in the space. Adjacent tiles are listed in the registered service descriptions of the Place agents. Tiles typically describe themselves with an index. The approach for a graph-type environment is very similar. Each Place agent represents a node in the graph. 'Adjacent' nodes that share a common edge with the given node register their Place agent service.

---

[1] "An autonomic computing system cannot exist in a hermetic environment. While independent in its ability to manage itself, an autonomic computing system must function in a heterogeneous world and implement open standards—in other words, an autonomic computing system cannot, by definition, be a proprietary solution." ..

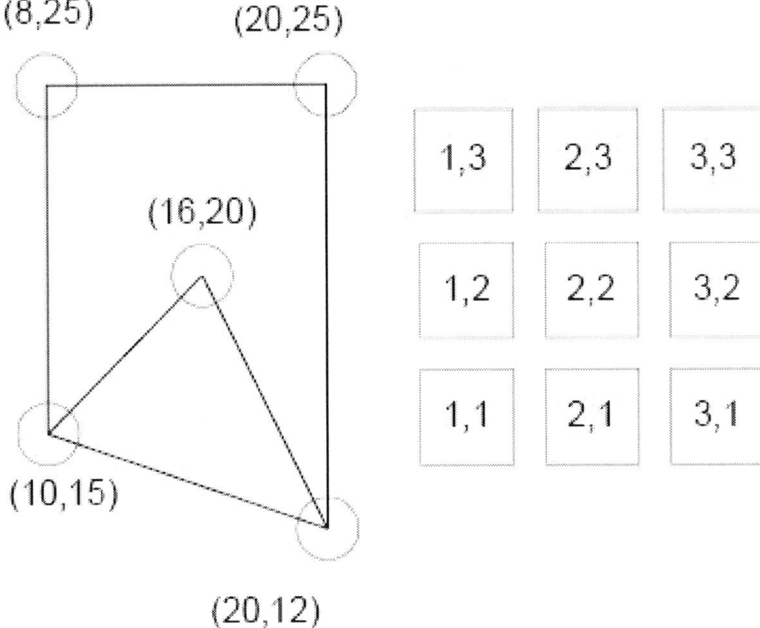

Figure 2. Graph-type (left) and discrete tile (right) environments

3.2.3. General Protocols

From the previous discussion, we can identify the necessary protocols and ontological concepts that would be common across pheromone infrastructures for all ant colony systems. These are summarised in Table 1.

4. A Methodology for Transferring Nature-Inspired Algorithms to Parallel, Asynchronous and Decentralised Environments

In this section, we describe a general methodology for transferring a nature-inspired algorithm to a truly parallel, asynchronous and decentralised MAS-type environment. The strategy is illustrated with reference to the ACS-TSP algorithm described in Section 2. The steps in the methodology are as follows:

- Identifying solution agents
- Identifying global components
- Identifying parameters
- Identifying standard protocols
- Estimating messaging cost
- Deciding on 'termination' criteria

Table 1. FIPA standard protocols for a general pheromone infrastructure

Initiator	Participant	Scenario	Standard Protocol	Comments
Place agent	Place agent	Place agents register their Place service description with one another. This is a standard DF registration	FIPA Request Interaction Protocol [23]	
Place agent	Place agent	Place agents deregister their Place service description from one another. This is a standard DF deregistration.	FIPA Request Interaction Protocol [23]	
Solution agent	Place agent	A Solution agent asks its current Place agent for a list of neighbouring Place agents. This is a standard DF search.	FIPA Request Interaction Protocol [23]	
Solution agent	Place agent	A Solution agent asks its current Place agent for information on the local pheromone infrastructure.	FIPA Query Interaction Protocol [24]	query-ref and inform-result version is used.
Solution agent	Place agent	A Solution agent deposits pheromone with its current Place agent.	FIPA Request Interaction Protocol [23]	agree is omitted. inform-result is used to conclude.

4.1. Identifying Solution Agents

The first step is to identify the Solution Agents in the system. This identification should try to map agents to 'physical entities' rather than system functions [25]. This is because physical entities are relatively simple and have a locality of interaction and information whereas functions can be complicated and are usually defined globally. Other authors use the example of factory scheduling and draw the distinction between the physical machines on a factory floor and the global function of machine scheduling for the entire factory [25]. Solution agents must support the fipa-agent-management ontology and the associated protocols (Table 1) so that they may interact with and move around the pheromone infrastructure described in Section 3.2.

In ACS-TSP, the obvious assignment is a Solution Agent for each ant.

4.2. Identifying Global Components

As mentioned in the motivation, many nature-inspired algorithms have a centralised global component. The exact nature of such a component varies hugely between various algorithm implementations. However, it very presence prevents a decentralised implementation. This component must be removed or replaced with some decentralised equivalent.

In ACS-TSP, there is a global pheromone update stage after all the ants have completed their tours. This is global in two ways. Firstly, the ants must share some knowledge of the best tour found between them. Secondly, the ants must know that they have all finished their tours.

This can be avoided in a decentralised MAS by having ant Solution agents write their trails to some common blackboard. If the Solution agent sees that its tour is the best so far, the agent backtracks over that trail performing the 'global' update procedure.

4.3. Identifying Parameters

The categorisation of parameters is important. It allows us to clearly determine how decentralised our implementation can be while remaining true to the original algorithm.

Agent model parameters divide into three categories: problem parameters, solution parameters and environmental parameters [26]. We propose further dividing the Solution parameters category into *Global* and *Agent* Solution parameters.

1. **Problem parameters:** these vary the problem for which the algorithm is developed. For example, in the Travelling Salesman Problem, the number of cities is a problem parameter.
2. **Solution parameters:** these vary the characteristics of the algorithm itself. They are tailored during deployment.
   (a) **Global Solution parameters:** Global solution parameters can only be seen and modified from an omniscient viewpoint. For example, in a swarm, the number of agents is a global solution parameter.
   (b) **Agent Solution parameters:** Agent parameters could conceivably be applied to each individual agent and have different values for each agent. These parameters determine the diversity of swarm members.
3. **Environmental Parameters:** these vary the deployment scenario. Examples might include communication delays and computation node failures.

Preferably, as many solution parameters as possible should fall into the Agent category rather than the Global category. The ultimate goal of a fully decentralised system demands no global solution parameters. Table 2 and Table 3 summarise the global and agent solution parameters respectively for ACS.

Table 2. Global Solution Parameters in Ant Colony System

Symbol	Description	Comments
m	Number of ants	
N/A	Maximum number of ants per city initially	This was not an explicit parameter in the original account of ACS [14]

Note that although the ACS algorithm mentioned a single parameter ρ for use in both local and global pheromone updates, we distinguish between two respective values of ρ in keeping with our research motivation. Note also that all but one of the agent solution parameters is assigned to the Solution agents rather than the Place agents. This is desirable because it keeps the Place agents as general as possible.

Table 3. Agent Solution Parameters for Ant Colony System

Symbol	**Description**	**Comments**
NCmax	Max number of cycles of the algorithm	A Solution agent builds this number of solutions before reporting its best solution found.
α	Trail importance. The relative importance of trail intensity in the transition probability.	This was not explicitly used in Ant Colony System. We include it in this research with a value of 1.0 for consistency with related algorithms like Ant System [27].
Q	Global pheromone deposition constant.	Again, this was not explicitly used in ACS so we set it to a value of 1.0.
β	Visibility importance. The relative importance of visibility in the decision probability.	
q_0	Exploitation probability threshold. A parameter between 0 and 1 that determines the relative importance between exploration and exploitation in the decision rule.	
ρ_{local}	Decay constant used in local pheromone updates.	
ρ_{global}	Decay constant used in global pheromone updates.	
τ_0	The initial intensity of trail set on all edges at the start of a trial.	This is a parameter for the Place agents in the pheromone infrastructure.

4.4. *Protocols*

The use of agents and their associated messaging introduces the need for protocols. Preferably, these protocols should conform to published standards such as those of FIPA.

Since ants in the ACS algorithm interact only with the environment and not with one another, there is no need to add to the protocols provided with the Place agents and their pheromone infrastructure (Section 3.2). This simplicity of protocols is a clear advantage of nature-inspired MASs over MASs that rely on complicated market mechanisms and negotiations to self-organise.

4.5. *Messaging Cost*

The cost of messaging and its effect on MAS dynamics is a real engineering concern. This has been highlighted in related work with a MAS implementation of a genetic algorithm [13].

Table Table 4 and 5 present an estimation of the messaging needed in a MAS implementation of ACS for a problem of size $n = 50$ cities with $m = 20$ Solution agents. This is typical of the type of problem size on which ACS was first tested [14]. Table Table 4 shows the messaging required by the setting up of the pheromone infrastructure. Table 5 is one iteration of the main loop. Furthermore, there will occasionally be a further step where a Solution agent must backtrack to perform the 'global' update.

In the concrete example, setting up the pheromone infrastructure requires 5180 messages. The equivalent of one iteration of the main loop requires 8040 messages. ACS was

Table 4. Estimated Messaging for setting up the pheromone infrastructure

Stage	Item	Protocol	Msgs per protocol	Total msgs	Value
1	Each Place agent registers with the global DF so that other Place agents can find it.	Request Interaction	2	$2 \times n$	100
2	Each Place agent searches the global DF for its neighbouring Place agents	Request Interaction	2	$2 \times n$	100
3	Each Place agent registers its Place service with its neighbouring Place agents in the topology.	Request Interaction	2	$2 \times n \times (n-1)$	4900
4	Each Solution agent searches the global DF for its starting Place agent.	Request Interaction	2	$2 \times m$	40
5	Each Solution agent registers with its starting Place agent	Request Interaction	2	$2 \times m$	40

Table 5. Estimated Messaging for the equivalent of a single run of the 'main loop'

1	Each Solution agent queries its current Place agent for representation of the local pheromone infrastructure.	Query Interaction	2	$2 \times m \times n$	2000
2	Each Solution agent begins its move by deregistering from its current Place agent.	Request Interaction	2	$2 \times m \times n$	2000
3	Each Solution agent finishes its move by registering with its destination Place agent	Request Interaction	2	$2 \times m \times n$	2000
4	Each Solution agent performs a local pheromone update.	Request Interaction	2	$2 \times m \times n$	2000
5	Each Solution agent reports its tour and receives the length of the best tour.	Request Interaction	2	$2 \times m$	40

run for iterations of the order of 1200. For the given problem example, this would require approximately 9.6 million messages. This is clearly a heavy messaging load. Some of our exploratory studies of messaging in JADE [28], a popular MAS platform, on a desktop machine[2] yielded an average message rate of 550 per second. This would mean an equivalent MAS runtime of about 5 hours. . This is the discouraging reality of a direct MAS implementation of a nature-inspired algorithm when restricted to a single machine. We emphasise that we are investigate nature-inspired systems that will run on large numbers of machines at a scale that can harness the power of real ant colonies. There is no way to predict whether such an implementation will require either a smaller or greater number of 'runs' to achieve a similar solution quality to its algorithm equivalent. This justifies our motivation for studying implementations in truly parallel, asynchronous and decentralised environments.

It is also worth noting the effect of not using the standard Request Interaction protocol. If Solution agents did not require an agree return message then the number of messages per protocol would drop to 1 in steps 2-4 of Table 5. This reduces the corre-

[2] Intel Pentium 4, 2.80 GHz, 512 Mb RAM

sponding total message load to 6 million and the run time to 3 hours. There is a cost to conforming to standards.

4.6. Termination

A fundamental difference between an algorithm and a MAS is the idea of open-endedness. An algorithm typically steps through a number of instructions and either terminates or loops. Termination criteria include executing a certain number of iterations or reaching an acceptable level of solution quality. MASs, by contrast, are intended to be continually running systems. This poses the question of how we can know when our MAS has done an equivalent amount of work to the algorithm on which it was based.

This is easily solved in ACS by having each Solution Agent track the number of tours it has built and by giving the agent a maximum number of tours to build as an Agent solution parameter. Agents then know when to report their result. Agents deactivate after performing the required number of runs. This is, of course, a contrived situation so that a valid comparison can be made with the original ant algorithm. The ultimate goal is to realise a continuously running and adapting system. In such circumstances, it would be more realistic for agents to report their best solutions at some regular interval.

5. Related Work

There has been some related work on the transfer of nature-inspired algorithms to MAS-type platforms, infrastructures for nature-inspired MASs, and the issues when approximating parallelism in sequential and synchronous systems.

Several authors have investigated parallel implementations of Ant Colony algorithms [29,30]. However, these were in the vein of a master/slave type approach. A similar tactic is well established in the Evolutionary Computation community's use of the 'Island' genetic algorithm. Clearly, the master node in these implementations is a centralised component with global knowledge. This approach does not make sense in the context of the Peer-to-Peer type systems for which we are developing.

Smith *et al* [13] have investigated an implementation of a genetic algorithm using IBM's Aglets platform[3]. This approach is different from the vast majority of evolutionary computation (EC) work. Generally, EC algorithms are a centralised program storing individuals as data structures and imposing genetic operations to generate successive populations. In an agent system, there is no concept of a generation since agents interact asynchronously and in parallel. In Smith *et al*'s framework, agents exchange genetic information with messaging and can perform their own internal genetic operations on that information. The authors implemented equivalents of tournament selection and an elitist strategy. The authors tested their framework on the OneMax problem—a simple standard problem in the EC field. Counter to EC intuition, the elitist scheme converged much more slowly than the tournament scheme in real time due to the overhead of the greater number of message exchanges required. We have seen such a message overhead in our calculations on a MAS implementation of ACS (Section 4.5). This highlighted the general point that established conclusions of traditional EC research might not necessarily transfer to a MAS implementation.

[3] www.trl.ibm.com/aglets/

Brueckner's PhD dissertation proposed the original idea of *pheromone infrastructures* [20]. A decentralised MAS for manufacturing control was built on top of this infrastructure. Although there was a clear mention of a yellow pages service and certain ontological concepts, these were not related to any agent standards.

Cornforth has experimented with different update schemes in cellular automata (CA) [8,31]. Cellular automata can be considered as highly abstract MASs. A cellular automaton is a lattice of discrete sites such that each site has a finite set of possible values. Sites change in discrete time steps according to the same deterministic rules and these rules incorporate information from a limited neighbourhood of sites. Cornforth investigated 6 update schemes for site changes in a 1 dimensional cellular automaton. These schemes were: synchronous, random independent, random order, cyclic, clocked and self-sync. Each scheme produced dramatically different results in the cellular automaton. This emphasised that the update scheme in a sequential system is a factor in the system's performance and raises the question of which, if any, of the schemes is the best at approximating a truly parallel and asynchronous system.

6. Conclusions

We have argued the need for disciplined scientific and engineering research into how the successes of nature-inspired algorithms might be transferred to emerging computing environments such as the Grid, Peer-to-Peer systems, Autonomic Computing and the Semantic Web. Such research is of benefit to two communities. On the one hand, researchers of nature-inspired algorithms might see some performance benefits in the parallelism, asynchronicity, decentralisation and distribution of such environments. Furthermore, it gives these researchers a new and very relevant application domain for their algorithms. On the other hand, researchers looking for robust, decentralised, self-organising systems to operate in emerging computing environments should be interested in the promise of nature-inspired methods. These methods have been dramatically successful in their original application domains.

We have framed the idea of pheromone infrastructures within maturing agent standards, making specific reference to the protocols proposed by FIPA [15].

We then outlined a general methodology for transferring nature-inspired algorithms to emerging computing environments using pheromone infrastructures. This involved 6 steps that were illustrated by reference to a suggested MAS implementation of Ant Colony System. The Solution agents that operate on the pheromone infrastructure are identified. The global components of the original algorithm are removed or replaced. The parameters of the algorithm are categorised as solution parameters, problem parameters or environment parameters. We proposed a refinement of the solution parameter category to distinguish between agent and global solution parameters. The necessary protocols are chosen, preferable from suitable existing standard protocols. The cost of messaging is estimated. Some decision is made on when the MAS would have done an 'equivalent' amount of work to its algorithm counterpart. The estimate of the cost of messaging highlighted the price to be paid for the use of standard protocols for interoperability.

7. Current and Future Work

We are currently verifying our program code for our exploratory studies of the ACS algorithm. These experiments begin shortly. We expect these studies to reveal parallelism and asynchronicity as factors in the algorithm's performance and to provide quantitative data that will justify testing a fully-fledged MAS implementation.

Our longer-term goal is to develop a FIPA compliant set of Place agents that can support a pheromone infrastructure using standard protocols and ontologies. We would like to develop ant colony systems to run on this infrastructure and make well defined and statistically grounded assessments of the benefits and costs of such implementations. This will determine bounds on the usefulness of established results from the ant colony algorithm community when transferred to emerging computing environments.

References

[1] John H. Holland. *Adaptation in Natural and Artificial Systems: An Introductory Analysis with Applications to Biology, Control, and Artificial Intelligence.* The MIT Press, Cambridge, Massachusetts, USA, 1995.

[2] J. Kennedy, R. C. Eberhart, and Y. Shi. *Swarm Intelligence.* Morgan Kaufmann, 2001.

[3] Eric Bonabeau, Marco Dorigo, and Guy Theraulaz. *Swarm Intelligence.* Oxford University Press, 1999.

[4] IBM. Autonomic Computing: IBMs Perspective on the State of Information Technology. Technical report, IBM Research, October 2001.

[5] Ian Foster and Carl Kesselman, editors. *The Grid: Blueprint for a New Computing Infrastructure.* Morgan Kaufmann Publishers, 1999.

[6] Dieter Fensel, Wolfgang Wahlster, Henry Lieberman, and James Hendler, editors. *Spinning the Semantic Web: Bringing the World Wide Web to Its Full Potential.* The MIT Press, 2002.

[7] Gerhard Weiss, editor. *Multiagent Systems: A Modern Approach to Distributed Artificial Intelligence.* The M.I.T. Press, Cambridge, Massachusetts, U.S.A.

[8] David Cornforth, David G. Green, and David Newth. Ordered asynchronous processes in multi-agent systems. *Physica D*, 2005.

[9] Paul R. Cohen. *Empirical Methods for Artificial Intelligence.* The MIT Press, Cambridge, Massachusetts, 1995.

[10] H. Van Dyke Parunak, Sven A. Brueckner, Robert Matthews, and John Sauter. Pheromone Learning for Self-Organizing Agents. *IEEE Transactions on Systems, Man and Cybernetics, Part A*, 35(3):316– 326, 2005.

[11] Tucker Balch. Hierarchic Social Entropy: An Information Theoretic Measure of Robot Group Diversity. *Autonomous Robots*, 8:209–237, 2000.

[12] Krzysztof Chmiel, Dominik Tomiak, Maciej Gawinecki, Pawel Karczmarek, Michal Szymczak, and Marcin Paprzycki. Testing the Efficiency of JADE Agent Platform. In *Proceedings of the Third International Symposium on Parallel and Distributed Computing*, pages 49–56. IEEE Computer Society, 2004.

[13] Robert E. Smith, Claudio Bonacina, Paul Kearney, and Walter Merlat. Embodiment of Evolutionary Computation in General Agents. *Evolutionary Computation*, 8(4), 2000.

[14] Marco Dorigo and Luca Maria Gambardella. Ant Colony System: A Cooperative Learning Approach to the Traveling Salesman Problem. *IEEE Transactions on Evolutionary Computation*, 1(1):53–66, 1997.

[15] The Foundation for Intelligent Physical Agents, 2005. Available from: `www.fipa.org/`.

[16] Marco Dorigo and Thomas Stutzle. *Ant Colony Optimization.* The MIT Press, Massachusetts, USA, 2004.

[17] E. L. Lawler, J. K. Lenstra, A. H. G. Rinooy Kan, and D. B. Shmoys, editors. *The Traveling Salesman Problem - A Guided Tour of Combinatorial Optimization.* Wiley Series in Discrete Mathematics and Optimization. John Wiley and Sons, New York, USA.

[18] Gerhard Reinelt. TSPLIB - A traveling salesman problem library. *ORSA Journal of Computing*, 3:376–384, 1991.

[19] Michael Luck, Peter McBurney, Onn Shehory, and Steve Willmott. Agent Technology Roadmap: Overview and Consultation Report. Technical report, AgentLink, December 2004.

[20] Sven Brueckner. *Return from the Ant: Synthetic Ecosystems for Manufacturing Control.* Phd, Humboldt University, 2000.

[21] H. Van Dyke Parunak, Sven A. Brueckner, Robert Matthews, and John Sauter. How to Calm Hyperactive Agents. In *Proceedings of the Second International Joint Conference on Autonomous Agents and Multiagent Systems*, pages 1092–1093. ACM Press, 2003.

[22] FIPA Agent Management Specification. FIPA Specification SC00023K, Foundation for Intelligent Physical Agents, 18 March 2004.

[23] FIPA. FIPA Request Interaction Protocol Specification. Technical report, Foundation for Intelligent Physical Agents, 2002.

[24] FIPA. FIPA Query Interaction Protocol Specification. Technical report, Foundation for Intelligent Physical Agents, 2002.

[25] H. Van Dyke Parunak and Sven A. Brueckner. Engineering Swarming Systems. In Federico Bergenti, Marie-Pierre Gleizes, and Franco Zambonelli, editors, *Methodologies and Software Engineering for Agent Systems*, volume 11 of *Multiagent Systems, Artificial Societies, and Simulated Organizations*. Kluwer, 2004.

[26] Sven A. Brueckner and H. Van Dyke Parunak. Information-driven Phase Changes in Multi-Agent Coordination. In *Poster in the Proceedings of the second international joint conference on Autonomous Agents and Multi-Agent Systems*, pages 950–951. ACM Press, New York, NY, USA, 2003.

[27] Marco Dorigo and Alberto Colorni. The Ant System: Optimization by a colony of cooperating agents. *IEEE Transactions on Systems, Man, and Cybernetics Part B*, 26(1):1–13, 1996.

[28] Fabio Bellifemine, Agostino Poggi, and Giovanni Rimassa. JADE: a FIPA2000 compliant agent development environment. In *Proceedings of the Fifth International Conference on Autonomous Agents*, pages 216–217. ACM Press, New York, NY, USA, 2001.

[29] Marcus Randall and Andrew Lewis. A Parallel Implementation of Ant Colony Optimization. *Journal of Parallel and Distributed Computing*, 62(9):1421–1432, 2002.

[30] Thomas Stutzle. Parallelization Strategies for Ant Colony Optimization. In A. E. Eiben, Thomas Back, Marc Schoenauer, and Hans-Paul Schwefel, editors, *Proceedings of the Fifth International Conference on Parallel Problem Solving from Nature*, volume 1498 of *Lecture Notes in Computer Science*, page 722. 1498 edition, 1998.

[31] David Cornforth, David G. Green, David Newth, and Michael Kirley. Do Artificial Ants March In Step? Ordered Asynchronous Processes and Modularity in Biological Systems. In Standish, Bedau, and Abbass, editors, *Proceedings of the Eighth International Conference on Artificial Life*, pages 28–32. 2002.

Self-Organization and Autonomic Informatics (I)
H. Czap et al. (Eds.)
IOS Press, 2005

Opponent Modeling in Adversarial Environments through Learning Ingenuity

Arash AFKANPOUR and Saeed BAGHERI SHOURAKI
Computer Engineering Department, Sharif University of Technology
Tehran, Iran
{afkanpour, sbagheri}@ce.sharif.edu

Abstract. In Multiagent systems there are several agents with cooperative or competitive goals. Here, we are especially interested in zero-sum games which contain exactly two players with fully opposite goals. We describe a method based on Maximum-Expected-Utility [7] principle that learns the ingenuity of the opponent based on the moves of the opponent through a game and exploits this knowledge to play better against that opponent. Then we demonstrate an application of proposed method in the popular board game of Connect-4. The results show that the proposed method is superior compared to previous methods for adversarial environments especially when there is not adequate training for appropriate adaptation against an opponent.

Keywords. Multiagent Systems, Opponent Modeling, Possibility Distribution

1. Introduction

In reality there is no single agent environment. Agents must interact with each other in order to achieve their goals. When agents in an environment have opposite goals, some kind of competition is inevitable. Acquiring more resources or winning in a game are examples of opposite goals which also happen in humans everyday life.

In competitive situations agents that learn about other agents will have an advantage. By using what is learned about others, the agent can exploit weaknesses of other agents and make its own way easier to the goal.

Some efforts have been done to design agents that learn to play better from experience. It means that the agent's level of playing will be improved through playing more and more. But here we are not looking for strategies to improve skill of playing; rather we are interested in identifying opponent weaknesses and exploiting this knowledge to gain advantage against this particular opponent. Although this method does not improve the level of playing of an agent, it will help the agent to play better against opponents with general weaknesses (disabilities).

2. Definitions

Games are standardization of environments in which more that one agent is involved. If just two agents are involved in a game, then it is called a *two-player* game.

Turn-taking games are games in which agents' actions must alternate. On the other hand, concurrent games are games in which agents must play simultaneously.

In a multiagent system, every state has at least one Nash equilibrium. Nash equilibrium is a set of policies $\pi_1,\ldots,\pi_n$ for n agents in the environment such that no agent can improve its expected payoff by unilaterally changing its own policy [4]:

$$R_i(\pi_1,\ldots,\pi_n) \geq R_i(\pi_1,\ldots,\pi_{i-1},\pi'_i,\pi_{i+1},\ldots,\pi_n) \quad \forall i \tag{1}$$

Informally a *policy*, which is also called *strategy* in the domain of Game theory, determines how a player acts in different states. Suppose that S is the set of states and A is the set of actions for a player. A policy π is a mapping that defines the probability of selecting actions in each state [1]. Formally, $\pi : S \times A \rightarrow [0,1]$ where,

$$\forall s \in S \quad \sum_{a \in A} \pi(s,a) = 1$$

In game theory, games in which agents have totally conflicting goals are called *zero-sum games*. Mathematically it means that the utility values of making actions for players are equal and opposite. In other words, the payoffs in each cell of the payoff matrix of players sum to zero. For example the payoff matrix for the game of two-finger Morra [7] is shown in Table 1. This game is a zero-sum game.

Table 1. Payoff matrices for two players of two-finger Morra

	O: *one*	O: *two*
E: *one*	+2	-3
E: *two*	-3	+4

	O: *one*	O: *two*
E: *one*	-2	+3
E: *two*	+3	-4

For example, if 'Even' player chooses action *one* and 'Odd' player chooses action *two*, then Even will receive a payoff of -3 while the Odd will receive +3 which is equal and opposite of what Even receives. According to these definitions, in two-player zero-sum turn-taking games, two players with completely opposite goals play with alternate moves against each other.

3. Related Work

The optimal policy in zero-sum games can be found using Maximin technique. In this technique it is supposed that the opponent always chooses its best action to maximize its own payoff. As opponent's payoff is equal and opposite of agent's payoff, it means that the opponent always chooses an action which minimizes player's payoff. The Maximin strategy recommends that the player plays the move which maximizes its payoff considering that the opponent plays its optimal move.

The Maximin strategy in turn-taking zero-sum games can be found using minimax tree. In this tree the agent tries to maximize its payoff, so it is called the MAX player.

On the other hand the opponent tries to minimize the agent's payoff in order to maximize its own payoff, so it is called the MIN player. The minimax-value of each node in the tree is computed as below [7]:

$$MINIMAX-VALUE(n)=\begin{cases} UTILITY(n) & \text{if } n \text{ is a terminal state} \\ \max_{s\in Successors(n)} MINIMAX-VALUE(s) & \text{if } n \text{ is a MAX node} \\ \min_{s\in Successors(n)} MINIMAX-VALUE(s) & \text{if } n \text{ is a MIN node} \end{cases}$$

In zero-sum games that agents must act simultaneously, the optimal policy can be found as a mixed strategy. Here the agent must find a probability distribution π on its actions in a way that it maximizes the value V as defined in [3]:

$$V=\max_{\pi\in PD(A)}\min_{o\in O}\sum_{a\in A} R_{o,a}\pi_a \tag{2}$$

In which $R_{o,a}$ is the payoff to the agent when it chooses action a and its opponent chooses action o. Linear programming (see e.g. [5, 6, 9]) is a technique for solving these kinds of problems. Note that the optimal policy in concurrent zero-sum games cannot be a pure strategy, because any pure strategy in such games can be defeated. The optimal policy in these games is called maximin equilibrium and it is also Nash equilibrium [7].

Games with multiple states can be considered as an extension of game theory to MDP's. These games are called Markov games or stochastic games [3]. In each state a payoff matrix is defined for each player. The optimal behavior of agent in each state of a Markov game can be found using a combination of Maximin technique and Bellman equation. Suppose that $V(s)$ is the expected reward of the optimal policy starting from state s, and $Q(s, a, o)$ is the expected reward for taking action a when the opponent chooses o from state s and continuing optimally thereafter. Then the value of $V(s)$ and $Q(s, a, o)$ can be computed as Eqs. (3) and (4) indicates:

$$V(s)=\max_{\pi\in PD(A)}\min_{o\in O}\sum_{a\in A} Q(s,a,o)\pi_a \tag{3}$$

$$Q(s,a,o)=R(s,a,o)+\gamma\sum_{s'} T(s,a,o,s')V(s') \tag{4}$$

In which $T(s, a, o, s')$ is the transition function that indicates the probability of transiting from state s to state s' by action pair (a, o). The optimal policy is π that maximizes $V(s)$ for every state of the game.

In cases where the payoff matrices for states of the game are not known a priori, an algorithm called Minimax-Q [3] can be used to learn optimal policy in zero-sum games. In this technique, after receiving reward r for moving from state s to state s' via agent's action a and opponent's action o, the Q value and V value are updates as:

$$Q(s,a,o) = (1-\alpha)Q(s,a,o) + \alpha[r + \gamma V(s')] \tag{5}$$

$$V(s) = \max_{\pi \in PD(A)} \min_{o \in O} \sum_{a \in A} Q(s,a,o)\pi_a \tag{6}$$

Nash equilibrium points in zero-sum games are a kind of equilibrium called adversarial equilibrium [4]. An adversarial equilibrium is an equilibrium point in which no player is hurt by any change in other players' policies:

$$R_i(\pi_1,...,\pi_n) \le R_i(\pi'_1,...,\pi'_{i-1},\pi_i,\pi'_{i+1},...,\pi'_n) \qquad \forall i \tag{7}$$

A nice property of adversarial equilibria is that the agent will not lose anything if the policies of other agents change. It means that a minimum value of reward for converging to an adversarial equilibrium in a zero-sum game is guaranteed for agent. At first it seems promising, because it somehow makes the payoff of agent independent of action of opponent. But this payoff is obtained assuming that the opponent always acts optimally. In fact methods based on maximin concept are too conservative and pessimistic. The main problem with these methods is that they play the optimal policy no matter who they are playing against.

Humans usually are able to recognize and exploit their opponent's weaknesses during competitions. And in fact it becomes the success key in many situations. What helps us to recognize opponent's weaknesses is our prior knowledge of the environment, e.g. a game. Using this knowledge we can take advantage of our opponent's weaknesses and get better results. Some researchers propose methods to make a model of opponent and then exploiting acquired knowledge without improving the level of play (see e.g. [8]). On the other hand in some methods both knowledge about the game (skill level) and knowledge about the opponent are improving simultaneously (see [11]).

Using second attitude, it seems that methods for learning optimal policy in single-agent MDP's (i.e. reinforcement learning methods [2, 10]) also work in multi-agent environments. For example it may be possible to use traditional single agent Q-learning [12] to learn optimal policy in two-player zero-sum games. In this way, the opponent cannot be considered directly by the agent; rather it will be considered as part of the environment. The existence of another agent will have an impact on both transition and reward functions. If the policy of other agents does not change over time, it will be possible to use a single agent learning algorithm like traditional Q-learning to learn optimal stationary policy even in multiagent environments. But in cases where other agents can learn and adapt their behavior, converging to an optimal stationary policy is not possible due to changes in transition and reward functions of environment which are results of changes in other agents' policies.

But again it seems that online learning would be useful, i.e. using a single-agent algorithm that never stops learning. In this way the agent can always adapt itself even if changes occurred in other agents' policies. The fact is that if other agents' rate of adaptation is more that the agent's rate of learning, then using a learning algorithm like Q-learning would be useless because the agent can never adapt itself to other agents.

As mentioned before, some methods have been proposed to model opponent's behavior explicitly. In [11] it is assumed that reward and transition functions are not known to the agent a priori. So the proposed method is based on reinforcement learning in which the behavior of the opponent is also modeled. In this method, it is assumed that the opponent is Markov, i.e. the way that the opponent arrived in its current state is unimportant. This method called Opponent modeling Q-learning [11], records the number of times that the opponent chooses each action in each state in order to make a probabilistic model of opponent's behavior in that state. This gives the probability distribution of opponent's action in each state, $P(a_o|s)$. This information then is used to calculate the best action for our agent in each state. The Q table for our agent's actions in state s is updated as:

$$Q(s,a_m)=\sum_{a_o} P(a_o \mid s)Q(s,a_m,a_o) \tag{8}$$

The agent in each state chooses an action with the highest Q value:

$$a=\arg\max_{a_m}\{Q(s,a_m)\} \tag{9}$$

In this method, the learning process should not stop. The reason is the same as mentioned for Q-learning, i.e. after the end of learning process, if the opponent can adapt itself (learning opponent) then learned policy would be useless.

In [8] the agent just learns a model of the opponent. This method is proposed for zero-sum turn-taking games and it is assumed that attributes of the environment (the game) are known a priori (through evaluation function). So the goal is just to learn the opponent's model and exploit acquired information to take advantage against weaker opponents. What are learned by agent are conditional probabilities of opponent's actions. It is similar to Opponent Modeling Q-learning proposed in [11] because they both make a probabilistic model of opponent's actions. The agent is a Maximum Expected Utility player [7] and in each state chooses action which satisfies:

$$\arg\max_{\alpha_i\in\alpha} \sum_{\beta_j\in\beta} p(\beta_j \mid \alpha_i)u(\alpha_i,\beta_j) \tag{10}$$

In which α is the set of action for agent in current state, β is set of valid actions for opponent, $p(\beta_j|\alpha_i)$ is the conditional probability that the opponent chooses action β_j given that out agent plays α_i, and $u(\alpha_i,\beta_j)$ is the utility for our agent for (α_i,β_j) pair of actions.

A drawback of methods proposed in [11] and [8] is that if opponent's rate of adaptation and learning exceeds that of our agent, then the learning process is not useful. Another problem for both of these methods is that they usually need a long training phase, i.e. the agent should play many games against a particular opponent to make an appropriate and useful model of its behavior. This is not always feasible. In particular humans can learn about their opponents (in general other humans' behavior) in less time.

4. Ingenuity Learning

Here the agent models its opponent behavior based on opponent's ingenuity. In fact what is learned by our agent during a game is opponent's level of ingenuity. In short it can be stated that good moves of opponent will lead to an increase of agent's attitude of opponent's ingenuity, while bad moves of opponent make the ingenuity of opponent less than its current value. What we mean by good move and bad move depends also on current value of opponent's ingenuity from agent's point of view. According to agent's attitude about opponent's ingenuity, the agent will make a model of opponent's behavior whenever necessary. Then using this model the agent will choose its best move in current situation. So the agent is a Maximum Expected Utility player [7] that makes a model of opponent's behavior based on opponent's ingenuity from its own point of view. We call this method "Ingenuity Learning-Maximum Expected Utility" or IL-MEU. The ingenuity of opponent is modeled through parameter k which is in the range [0, 1]. The value of 1 means that the opponent is totally ingenious and plays optimally, so the method acts just like maximin technique and the game tree becomes exactly the minimax tree; the value of 0 means that the opponent has no preference in choosing its actions and plays randomly. In nodes that are related to opponent's actions, based on value of parameter k a possibility distribution [13] on opponent's actions is calculated and the expected utility of that node for our agent is computed. Note that what is obtained as a model of opponent's actions in such nodes is based on parameter k and hence it is not a probability distribution. In [11] and [8] the number of times that the opponent chooses each action is recorded and a probability distribution is obtained using this information. But here we do not record number of times that each action is chosen; rather we make a possibility distribution on opponent's actions based on the value of parameter k.

The value of 1 for parameter k means that the opponent will choose its best action with possibility 1 and will choose other actions with possibility 0. The value of 0 for this parameter indicates that the opponent will choose each of its actions with equal possibility. In fact it means that the opponent behaves randomly.

During the game the value of parameter k is updated according to actions chosen by opponent. As mentioned before, according to parameter k we compute a possibility distribution on opponent's actions. This possibility distribution gives us an expected value for each action of the agent. After choosing action with highest utility, the opponent also makes its action. If the utility gained for agent from this pair of actions is higher than the expected utility, it means that the opponent is weaker than what we expected; so the value of parameter k must be decreased by Δk. On the other hand, suppose that the utility gained from pair of actions is less that what our agent expected. It means that the opponent is more sophisticated than what the agent believes; so the value of parameter k must be increased by Δk.

The model of opponent's behavior is made based on parameter k in nodes of the tree that are opponent's turn to play. Suppose that the opponent can choose from a set of n legal actions in a node of tree. First this set is sorted according to the payoff of each action for opponent (descent). Possibility of each action in sorted list then is computed as:

$$Possibility(b_i) = \frac{1}{i^{\alpha k^{\beta}}} \qquad 1 \le i \le n \tag{11}$$

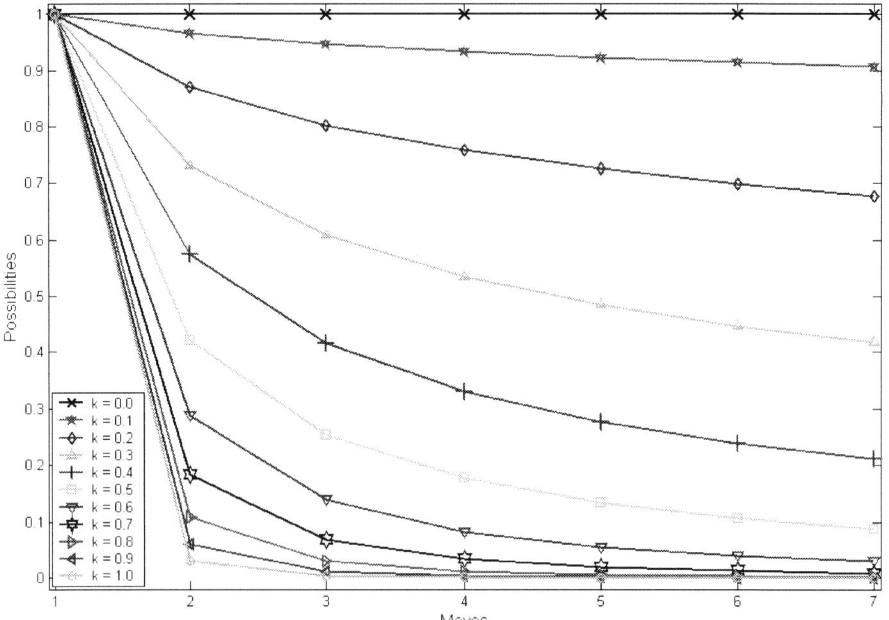

Figure 1. Possibility distributions for seven moves for different values of k

The value of parameters α and β can be in the range [1, 10] to obtain appropriate behavior as described above. In our experiments we use values 5 and 2 for α and β respectively. In Figure 1 possibilities of seven actions for different values of k and mentioned values for α and β are depicted. The only reason to show this figure for the case of seven moves is that in our experiments which will be discussed the maximum number of actions for a player in each state is seven. In general the number of moves in each state of the game has no considerable effect on the performance of proposed idea. The expected utility of choosing action a_j is computed in Eq. 12:

$$EU(a_j) = \frac{\sum_i Possibility(b_i) * u(a_j, b_i)}{\sum_i Possibility(b_i)} \tag{12}$$

In which $u(a_j,b_i)$ is the utility gained by agent after choosing action a_j by agent and b_i by opponent. As Eq. 12 indicates the expected utility of action a_j for our agent is the weighted sum of utilities of different actions of opponent in which weight of each action is the possibility value of choosing that action by opponent. As the agent uses Maximum Expected Utility concept in choosing its action, the action with the highest utility will be chosen:

$$a = \arg\max_{a_j} \{EU(a_j)\} \tag{13}$$

After choosing an action like a by agent, it is opponent's turn to choose its action. Suppose that the opponent chooses an action like b. Then the value of parameter k is updated based on action chosen by opponent and the utility gained by agent:

$$k_{new} = \begin{cases} k_{old} + \Delta k & u(a,b) < EU(a) \\ k_{old} - \Delta k & u(a,b) > EU(a) \\ k_{old} & else \end{cases} \tag{14}$$

In Figure 2 a situation in which the agent must choose from three actions is shown. Suppose that current value of parameter k is 0.6. Numbers in rectangles (leaves) show the utility of that node for our agent. Below each leaf, the possibility value for that leaf is shown. And numbers in ovals show expected utility of each action for agent. As can be seen the recommended action of maximin is different from that of our method. The reason is that the agent has learned its opponent's ingenuity, so it can take advantage of this knowledge.

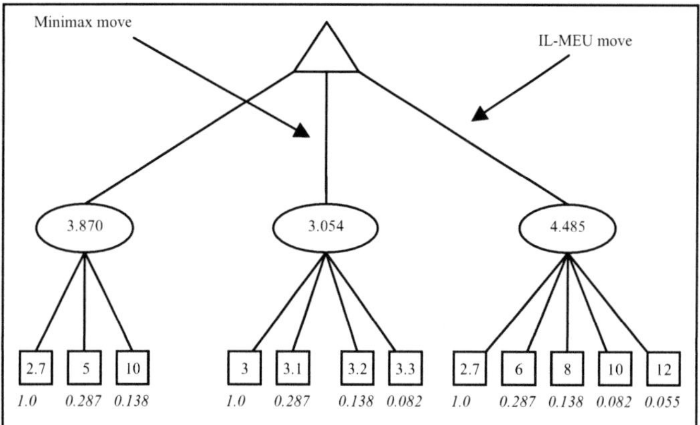

Figure 2. IL-MEU move against minimax move

5. Experimental Results

In order to evaluate the proposed method, the popular game of Connect-4 was chosen as test bed. Connect-4 is a two-player board game in which each player has several round tokens of a specific color (e.g. red and yellow). The board is placed vertically and has seven columns. Each column has six rows to place a token. The game is a turn-taking zero-sum game. A player wins if it could line up four of its own tokens horizontally, vertically or diagonally. Game will end in a draw if the board fills up and no player wins. A situation in which the red (darker) player won the game is depicted in Figure 3. A dashed line in this figure determines the four tokens that caused the winning. In order to evaluate proposed method against different opponents, four kinds of opponents are considered. Each of these opponents has its own weakness. Table 2 shows these four opponents.

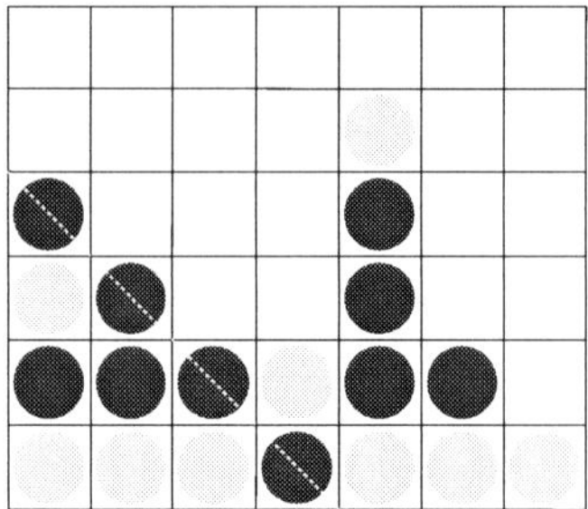

Figure 3. A state of Connect-4 in which Red (darker) player has won the game

In order to demonstrate the effectiveness of proposed method over traditional Maximin technique and method proposed in [8] (Lets call the player which uses this method "Sen player"), a series of 100 games were run against each opponent. In each game, first player to play was chosen randomly. Since three players which are used in our experiments (our proposed player (IL-MEU player), Sen player and minimax player) are deterministic players, they will always do the same sequence of movements in the beginning of games. So in each game the first two moves were played randomly.

The minimax player needs no training phase. IL-MEU player also does not have a special training phase. In particular it can adapt itself during the games in the test phase. But Sen player was trained against each opponent a set of 1000 games before testing against that opponent. The training method was the same as proposed in [8].

One of the major advantages of our proposed method over the method proposed in [8] is in their need for training phase. While the Sen player needs a relatively long period of training, IL-MEU player needs a shorter period of training. In particular in some situations it can adapt itself while playing against an opponent without any special training phase.

Table 2. Opponent types used in experiments

Opponent 1	Minimax player, Cannot detect diagonal threats
Opponent 2	Minimax Player, Chooses from its three best moves randomly
Opponent 3	Minimax player, always chooses its second best move
Random Opponent	Always chooses its action randomly

In our experiments we use 0.1 for Δk. In the set of 100 games for each player, number of wins, number of loses and the average number of moves in winning cases was acquired. The results can be seen in Table 3. The main comparison must be made in the case of average number of moves per game. Because for example all three players can defeat random player in all 100 games, but the period of each game distinguishes the performance of these players. Recall that opponents called Opponent 1, Opponent 2, Opponent 3 and Random Opponent in our experiments were described in Table 2.

Table 3. Results of playing Minimax, Sen and IL-MEU players played against different opponents

	Minimax Player			**Sen Player**			**IL-MEU Player**		
	# of Wins	# of Loses	Avg moves per game	# of Wins	# of Loses	Avg moves per game	# of Wins	# of Loses	Avg moves per game
Opponent 1	74	21	29.78	79	20	27.88	78	14	27.91
Opponent 2	95	5	21.24	96	4	21.72	98	2	18.04
Opponent 3	100	0	15.72	100	0	15.20	100	0	12.47
Random Opponent	100	0	14.80	100	0	14.59	100	0	11.93

The results show that our proposed method performs better than the minimax technique against all kinds of opponents. This preference is especially obvious against opponents 2, 3 and Random. The reason is that the proposed method works well when it is pitted against an opponent with a general weakness rather than a special weakness. The weaknesses of opponents 2, 3 and Random are general. But the weakness of opponent 1 is rather special; it cannot detect only diagonal threats. So the proposed method performs better against opponent 2, 3 and Random.

Compared to method proposed in [8] (Sen player), again our method performs better against opponents 2, 3 and Random while Sen player performs a bit better against opponent 1. The reason is the same as discussed above. Sen player can perform better against opponent 1 because the weakness of opponent 1 is rather special and our proposed method cannot detect such weaknesses easily. On the other hand Opponents 2, 3 and Random have general weaknesses. Sen player cannot perform well against these three opponents despite its long training phase. It means that 1000 games as the training phase was not enough for Sen player to adapt well against these opponents. But our proposed player can perform better because it can detect general weaknesses without any explicit training phase.

Another experiment was performed to evaluate the adaptability of proposed method. This experiment was designed to see if the agent can adapt itself fast enough against changes in opponent's behavior. Here the type of opponent is changed after a predetermined number of games. In this experiment a set of 80 games were used. This set consists of 8 periods; each consists of 10 games against a special opponent. After each period ends, the opponent is completely changed. The types of opponents used in this experiment are those used in previous experiment. The 8 periods and the opponent used in each period are shown in Table 4.

Table 4. Opponents used in 80 games

Game No.	*Opponent Type*
1-10	Opponent 1
11-20	Opponent 2
21-30	Opponent 3
31-40	Random Opponent
41-50	Opponent 1
51-60	Opponent 2
61-70	Opponent 3
71-80	Random Opponent

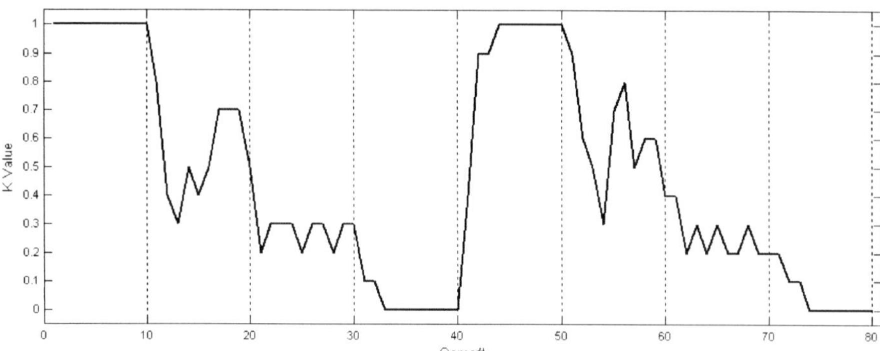

Figure 4. k values in 80 games of second experiment

The value of parameter k which illustrates the agent's belief about its opponent's ingenuity is depicted for 80 games in Figure 4 ($\Delta k = 0.1$). The dashed lines in Figure 4 show the games in which the opponent is changed. Note that the agent can adapt itself well after the opponent was changed every 10 games. This experiment shows one of outstanding properties of this method, i.e. its adaptability. As the agent makes a model of its opponent through learning opponent's ingenuity, it can adapt itself against a special opponent fast. Also if the opponent changes its behavior though a game, the agent can adapt itself again to the new behavior of the opponent in a reasonable time.

6. Conclusions and Future Work

In this paper a new method for opponent modeling in adversarial environments was introduced. The new method works by learning the opponent ingenuity based on its actions and creating a model of opponent behavior base on learned ingenuity.

The new method was tested in the board game of Connect-4 against different opponents. Results showed that this method works well when pitted against opponents with general weaknesses (disabilities) rather than special ones. This can also be inferred considering the idea behind this method. Using this method, the agent cannot detect special weaknesses of its opponent because what it knows about its opponent is just opponent ingenuity. On the other hand the proposed method needs shorter training phase (sometimes no explicit training phase) compared with previous methods. The results also showed that using the proposed method, the agent can adapt itself against changes in opponent's behavior in an acceptable time.

The main drawback of methods which are able to detect special weaknesses of an opponent (e.g. method proposed in [8]) is that they usually need long training phase to make an appropriate model of the opponent. Also they cannot adapt quick enough against changes in opponent behavior. The next step in our work is to use a combination of our method and methods that can detect special weaknesses. In this way the agent can adapt itself in a short time against unknown opponents while making a model of the opponent in order to detect special weaknesses.

Acknowledgements

We'd like to thank Ramin Halavati for his helpful suggestions.

References

[1] Bowling, M. 2003. *Multiagent Learning in the Presence of Agents with Limitations*. Ph.D. thesis. Computer Science Department, Carnegie Mellon University.
[2] Kaelbling, L.; Littman, M.L. and Moore, A.W. 1996. Reinforcement Learning: A Survey. *Journal of Artificial Intelligence Research* 4:237-285.
[3] Littman, M.L. 1994. Markov Games as a Framework for Multi-agent Reinforcement Learning. In *Proceedings of the Eleventh International Conference on Machine Learning*, pp. 157-163. Morgan Kaufman.
[4] Littman, M.L. 2001. Friend-or-Foe Q-Learning in General-Sum Games. In *Proceedings of the Eighteenth International Conference on Machine Learning*, pp. 322-328. Morgan Kaufman.
[5] Morris, P. 1994. *Introduction to Game Theory*. Springer-Verlag.
[6] Press, W.; Teukolsky, S.; Vetterling, W. and Flannery, B. 1992. *Numerical Recipes in C: The Art of Scientific Computing*, Second Edition. Cambridge University Press.
[7] Russell, S. and Norvig, P. 2003. *Artificial Intelligence: A Modern Approach*, Second Edition. Prentice Hall.
[8] Sen, S. and Arora, N. 1997. Learning to Take Risks. *AAAI Workshop on Multiagent Learning*, 59-64.
[9] Strang, Gilbert. 1980. *Linear algebra and its applications*, Second Edition. Academic Press. Orlando, Florida.
[10] Sutton, R.S. and Barto, A.G. 1998. *Reinforcement Learning*. MIT Press
[11] Uther, W. and Veloso, M. 2003. Adversarial Reinforcement Learning. Tech. Report. Carnegie Mellon University. Unpublished.
[12] Watkins, C. and Dayan, P. 1992. Q-Learning. *Machine Learning*, 8(3):279-292.
[13] Zadeh, L.A. 1978. Fuzzy Sets as a Basis for a Theory of Possibility. *FSS* 1, 3-28.

Self-Organization and Autonomic Informatics (I)
H. Czap et al. (Eds.)
IOS Press, 2005

From Bayesian Decision-Makers to Bayesian Agents

Václav Šmídl [1], Jan Přikryl
Institute of Information Theory and Automation
Academy of Sciences of the Czech Republic
Prague, Czech Republic

AbstractBayesian approach to decision making is successfully applied in control theory for design of control strategy. However, it is based on on the assumption that a decision-maker is the only active part of the system. Relaxation of this assumption would allow us to build a framework for design of control strategy in multi-agent systems. In Bayesian framework, all information is represented by probability density functions. Therefore, communication and negotiation of Bayesian agents also needs to be facilitated by probabilities. Recent advances in Bayesian theory make formalization these tasks possible. In this paper, we bring the existing theoretic results together and show their relevance for multi-agent systems. The proposed approach is illustrated on the problem of feedback control of an urban traffic network.

Keywords. Bayesian decision making, multi-agent systems, fully probabilistic design

1. Introduction

In recent years, it becomes obvious that the traditional centralized approach to control of large systems has reached its limits. Decentralization of control and decision making is seen as future direction of research in both academia [13,7] and industry [14]. Many successful applications of so called holonic or multi-agent systems has been published. This paradigm presents a new challenge for designers of these systems, since the traditional methodologies of design became obsolete and no consistent replacement is available [14]. One possible solution of this problem is to extend the existing methodologies to accommodate the distributed setup.

In control applications, we can see an agent as an entity consisting of two principal parts: (i) *autonomous subsystem*, which is responsible for agents ability to act according to its own aims and needs, and (ii) *communication and negotiation subsystem*, which is responsible for exchanging its knowledge with other agents and adjustment of its aims in order to cooperate and thus achieve better overall performance. The autonomous subsystem can be seen as a controller in the traditional sense, hence a number of methodologies for its design is readily available [10]. From this range of theories, we seek a

[1]Correspondence to: Václav Šmídl, Dept. of Adaptive Systems, PO Box 18, CZ-18208 Praha 8, Czech Republic, E-mail: smidl@utia.cas.cz

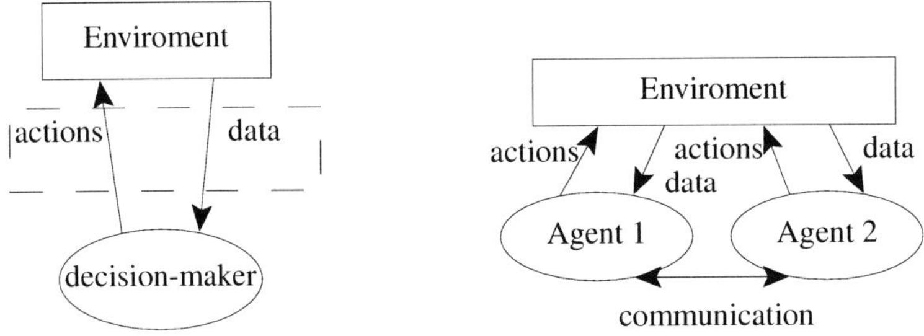

Figure 1. Relation of Bayesian decision making and multi-agent systems.

methodology which is able to embrace not only the autonomous but also the communication and negotiation subsystem. The most promising candidate is the Bayesian theory of decision making, since (i) it is a consistent theory for dealing with uncertainty which is ever present in real environments [5], (ii) the task of agent communication and negotiation can be formalized as decision making problem, and (iii) it is successfully applied in controller design [18] and in design of advanced applications such as advising systems [19].

Traditionally, the decision-maker is assumed to be the only entity that intentionally influences the environment. It consists of a model of its environment, its individual aims, and a pre-determined strategy of decision making. The Bayesian decision-maker is designed by means of the Bayesian theory which results in probabilistic representation of all the components, i.e. model of environment, aims and strategy.

An agent in multi-agent systems is known to influence only a part of the environment, i.e. its *neighbourhood*. The rest of the environment is modelled by other agents, as illustrated in Figure 1. In order to obtain relevant information from distant parts of the environment, an agent relies on communication with other agents in its neighbourhood. If the agents are able to exchange their aims and take them into account, they can cooperate and improve the overall performance of the system. The challenge for Bayesian decision making theory is to formalize communication and negotiation as operations on probability distributions. It was shown that the technique of *fully probabilistic design (FPD)* [17] reduces the task of agent cooperation into *reporting and merging of probability density functions* [1].

In this paper, we review Bayesian decision making in Section 2, and define the Bayesian decision-maker in Section 3. In Section 4, we bring together the latest achievements in application of Bayesian theory to multi-agent systems. The theory is applied to a practical problem of urban traffic control in Section 5.

2. Bayesian Decision Making

Bayesian decision making (DM) is based on the following principle [5]: *Incomplete knowledge and randomness have the same operational consequences for decision making.* Therefore, all unknown quantities are treated as random variables and formulation of the problem and its solution are firmly based within the framework of probability calculus.

This task of decision making can be decomposed into the following sub-tasks.

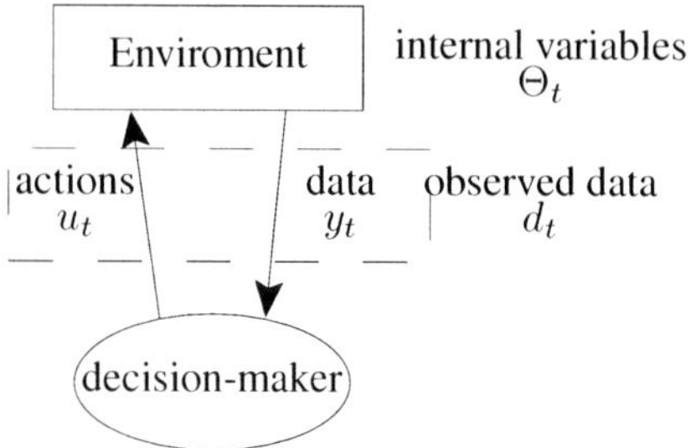

Figure 2. Basic decision making scenario.

Model Parametrization: Each agent must have its own model of its neighbourhood, i.e. part of the environment. Uncertainty in the model is described by parametrized probability density functions.

Learning: Reduces uncertainty in the model of the neighbourhood, using the observed data. In practical terms, parameters of the model are inferred.

Strategy Design: Choose the rule for generating decisions using the parametrized model and given aims.

These tasks will be now described in detail.

2.1. Model Parametrization

The basic scenario of decision making is illustrated in Figure 2. Here, d_t denotes all observable quantities on the environment, i.e. data, y_t, and actions, u_t, $d_t = [y_t', u_t']'$. Θ_t is an unknown parameter of the model of the environment. In Bayesian framework, the *closed loop*—i.e. the environment *and* the decision-maker—is described by the following probability density function:

$$f(d(t), \Theta(t)) =$$

$$\prod_{\tau=1}^{t} f(y_\tau | u_\tau, d(\tau-1), \Theta_\tau) f(\Theta_\tau | u_\tau, d(\tau-1), \Theta_{\tau-1}) f(u_\tau | d(\tau-1)). \quad (1)$$

Here, $f(\cdot)$ denotes probability density function (pdf) of its argument. $d(t)$ denotes the observation history $d(t) = [d_1, \ldots, d_t]$. The model represents the whole trajectory of the system, including inputs u_τ which can be influenced by the decision-maker. The chosen order of conditioning distinguishes the following important pdfs:

observation model $f(y_t | u_t, d(t-1), \Theta_t)$, which models dependency of the observed data on past data $d(t-1) = [d_1, \ldots, d_{t-1}]$, model parameters Θ_t and actions u_t.

internal model $f(\Theta_t | u_t, d(t-1), \Theta_{t-1})$, which models evolution of parameters of the model via data history $d(t-1)$, previous model parameters Θ_{t-1} and the chosen actions u_t.

DM strategy $f(u_t | d(t-1))$, is a probabilistic description of the decision rule.

2.2. Learning via Bayesian filtering

The task of learning is to infer posterior distribution of unknown parameters from the observed data, $f(\Theta_t|d(t))$. This pdf can be computed recursively as follows:

$$f(\Theta_t|u_t, d(t-1)) = \int f(\Theta_t|u_t, d(t-1), \Theta_{t-1}) f(\Theta_{t-1}|d(t-1))\, d\Theta_{t-1} \quad (2)$$

$$f(\Theta_t|d(t)) \propto \frac{f(y_t|u_t, d(t-1), \Theta_t) f(\Theta_t|u_t, d(t-1))}{f(y_t|u_t, d(t-1))}, \quad (3)$$

$$f(y_t|u_t, d(t-1)) = \int f(y_t|u_t, d(t-1), \Theta_t) f(\Theta_t|u_t, d(t-1))\, d\Theta_t. \quad (4)$$

In general, evaluation of the above pdfs is a complicated task, which is often intractable and many approximate techniques must be used [9]. In this text, we are concerned with conceptual issues and we assume that all operation (2)–(4) are tractable.

2.3. Design of DM strategy

In this Section, we review *fully probabilistic design (FPD)* of the DM strategy [17]. This approach is an alternative to the standard stochastic control design, which is formulated as minimization of an expected loss function with respect to decision making strategies [2,6]. The FPD starts with specification of the decision making aim in the form of *ideal pdf* of the closed loop. This ideal pdf—which is the key object of this approach—is constructed in the same form as (1) distinguished by superscript ${}^{\lfloor I}$:

$$f(d(t), \Theta(t)) \rightarrow {}^{\lfloor I}f(d(t), \Theta(t)). \quad (5)$$

Similarly to (1), the ideal distribution is decomposed into ideal observation model, internal model, and ideal DM strategy. Recall, from Section 2.1, that model (1) contains the DM strategy, which can be freely chosen. Therefore, the optimal DM strategy can be found by functional optimization of the following loss function

$$\mathcal{L}(f(u_t|d(t-1)), \mathring{t}) = \mathsf{KL}\left(f(d(\mathring{t}), \Theta(\mathring{t})) \,||\, {}^{\lfloor I}f(d(\mathring{t}), \Theta(\mathring{t}))\right),$$

where $\mathsf{KL}(\cdot,\cdot)$ denotes the Kullback-Leibler divergence between the current (learnt) and the ideal pdf [27], and $\mathring{t} > t$ is the decision making horizon.

The approach has the following special features.

- The KL divergence to an ideal pdf forms a special type of loss function that can be simply tailored both to deterministic and stochastic features of the considered DM problem.
- Minimum of the KL divergence—i.e. the optimal DM strategy—is found in *closed form*:

$$f(u_t|d(t-1)) = {}^{\lfloor I}f(u_t|d(t-1)) \frac{\exp[-\omega(u_t, d(t-1))]}{\gamma(d(t-1))}, \quad (6)$$

where $\omega(\cdot)$ and $\gamma(\cdot)$ are integral functions of all involved pdfs (these are not presented for brevity, see [19] for details). The decisions are then generated using a simplified version of stochastic dynamic programming [4].
- Multiple-objective decision making can be easily achieved using multi-modal ideal distributions [21,12].

3. Bayesian decision-maker

In practise, the task of adaptive decision making is typically solved in two stages [19]: (i) off-line, and (ii) on-line. The off-line stage is dedicated to design of the structure and fixed parameters (such as initial conditions) of the decision-maker. When the structure and fixed parameters are determined, the decision-maker operates autonomously in on-line mode, where it is able to adapt (by adjusting model parameters) to changes in the environment and improve its DM-strategy. Operation needed in both stages are described in this Section.

3.1. *Off-line stage*

In this stage, it is necessary to determine structure of the model (1) and prior distribution of model parameters. These tasks are solved using archives of the observed data as follows.

Model selection: if there is no physically justified model of the environment, this technique test many possible parametrization of the model, and selects one, which is best suited for the observed data. Typically, only a class of models that yields computationally tractable algorithms is examined. A key requirement of tractability is, that the learning operation (2) can be reduced into algebraic operations on finite-dimensional statistics.

Elicitation of prior distributions: The expert knowledge which is not available in the form of pdfs must be converted (often approximately) into probabilistic terms. Moreover, if the available knowledge is not compatible with the chosen model, a suitable approximation (projection into the chosen class) must be found. If there are more sources of prior information available, these must be merged into a single pdf. This operation will be described in detail at the end of this Section.

Design of DM strategy: When the model and ideal distributions are chosen, the optimal DM strategy is given in closed form by the FPD (Section 2). In special cases, the equation (6) can be parametrized by a finite-dimensional parameters, and the implied dynamic programming is reduced into algebraic operations on these parameters.

These tasks are computationally demanding and thus they are traditionally solved off-line, i.e. only once for all available data. This is acceptable, since all expert information is available a priori, and model of the environment is assumed to be constant.

3.2. On-line stage

A typical adaptive decision-maker operates by recursive repetition of the following steps:

1. **read**: the observed data are read from the environment. All the necessary pre-processing and transformation of data is done in this step.
2. **learn**: the observed data are used to increase the knowledge about the environment, namely the sufficient statistics of the model parameters are updated.
3. **adapt**: the decision-maker use the improved knowledge of the system to improve its DM strategy. Specifically, parameters of the DM strategy are re-evaluated using the new sufficient statistics.
4. **decide**: the adapted DM strategy is used to choose an appropriate action. Recall, that the DM-strategy is a pdf. Hence, a realization from this pdf must be selected. Typically, the optimal decision is chosen as expected value of (6).
5. **write**: the chosen action is written into the environment. Similar to the first step, transformation of the results is done in this step.

Note that due to computational constraints, all operations in this stage are defined on finite dimensional parameters or statistics.

3.3. Merging of pdfs

For the task of prior elicitation, we need to define a new probabilistic operation for merging of information from many sources. The merging operation is defined as a mapping of two pdfs into one:

$$f_1(\Theta_t|d(t)), f_2(\Theta_t|d(t)) \stackrel{\text{merge}}{\longrightarrow} \tilde{f}(\Theta_t|d(t)), \tag{7}$$

where f_1 and f_2 are the *source pdfs*, and the $\tilde{f}$ is the *merged pdf*. Many approaches are available, e.g. [16,15,19], with different assumptions and properties. Here, we review results of [24,23] since these have the following properties: (i) defined as optimization problems, with a reasonable loss function, (ii) their results are intuitively appealing and well interpretable, (iii) the optimum is reached for a class of pdfs which is uniquely defined, (iv) is applicable to both discrete and continuous distributions, and (v) algorithmic solutions are available.

We distinguish two kinds of merging:

direct: the source and the merged pdfs are defined on the same variables, such as (7),
indirect: the source distributions are defined on the variable in condition of the merged pdf.

These will be now described in detail.

3.3.1. Direct merging

The task of direct merging can be defined as optimization problem where the optimized function, $\mathcal{L}_M$, is chosen as divergence between the source and the merged pdfs [23] as follows:

$$\mathcal{L}_M(f) = \alpha \mathsf{KL}(f_2||f) + (1-\alpha)\mathsf{KL}(f_1||f). \tag{8}$$

Here, $\mathsf{KL}(\cdot,\cdot)$ denotes the Kullback-Leibler divergence, the weight $\alpha \in \langle 0, 1\rangle$ governs the level of importance of each source, and f is the optimized pdf. The merged pdf $\tilde{f}(d)$ is found by functional minimization of (8):

$$\tilde{f} = \arg\min_{f} \mathcal{L}_M(f). \tag{9}$$

The optimum (9) for merging of distributions of the *same* variable, is found in the form of a probabilistic mixture of the source pdfs:

$$\tilde{f}(d) = \alpha f_2(d) + (1-\alpha) f_1(d). \tag{10}$$

This solution is intuitively appealing and has been proposed using heuristic arguments [15]. Solution of the problem for multivariate distributions with various length of variables is more complicated, the result can not be found in closed form, however an iterative algorithm which asymptotically converge to the optimum is available [22]. This algorithms is feasible for discrete distributions. For continuous variables, complexity of the merged distribution grows with each iteration, and further approximations must be used.

3.3.2. Indirect Merging

The operation of indirect merging is defined as follows:

$$f_1(d_t|d(t-1)), f_2(d_t|d(t-1)), f(\Theta_t|d_t) \overset{\text{merge}}{\longrightarrow} \tilde{f}(\Theta_t|d(t)). \tag{11}$$

This operation can be seen as generalization of the Bayes rule, since it reduces to Bayesian learning (3) if the sources, f_1 and f_2, are empirical densities. Using (10) and generalized Bayesian estimation the problem can be solved as follows [24]:

$$\tilde{f}(\Theta_t|d(t)) \propto f(\Theta_t) \exp\left(-n\int \alpha f_1(d_t|d(t-1)) \ln f(d_t|\Theta_t)\, dd_t\right) \times$$
$$\times \exp\left(-n\int (1-\alpha) f_2(d_t|d(t-1)) \ln f(d_t|\Theta_t)\, dd_t\right)$$

Complexity of this operation is comparable to that of the Bayesian learning (3).

4. Bayesian Agents

The Bayesian agent is an extended Bayesian decision-maker described in previous Section. The additional features are the ability and need of agents to communicate and cooperate. In the Bayesian framework, all knowledge is stored in pdfs. The challenge is to formalize communication and cooperation within the framework of probability calculus. In this Section, we propose a simple probabilistic model of negotiation. For clarity of explanation, we consider only two agents, $A_{[1]}$ and $A_{[2]}$, where agent number is always in subscript in square brackets.

Each agent has the following quantities

Observed data d_t: Naturally, each agent can observe different subset of variables, i.e. $d_{t,[1]}$ and $d_{t,[2]}$, for A_1 and A_2, respectively. The agents can exchange knowledge only in terms of variables that are common for both of them, i.e. $d_{t,[1 \cap 2]}$. Any communication is meaningful only with respect to this subset.

Internal quantities Θ_t: We do not impose any structure of the environment model for the agents, hence, internal quantities $\Theta_{t,[1]}$ and $\Theta_{t,[2]}$ are in general disjoint sets.

Environment Model: $f_{[1]} = f\left(d_{[1]}(t), \Theta_{[1]}(t)\right)$ and $f_{[2]} = f\left(d_{[2]}(t), \Theta_{[2]}(t)\right)$ for each agent.

Ideal distributions: ${}^{\lfloor I}f_{[1]} = {}^{\lfloor I}f\left(d_{[1]}(t), \Theta_{[1]}(t)\right)$ and ${}^{\lfloor I}f_{[2]} = {}^{\lfloor I}f\left(d_{[2]}(t), \Theta_{[2]}(t)\right)$ for each agent.

Negotiation weights: For the purpose of negotiation, we define a scalar quantity $\alpha_{2,[1]} \in \langle 0, 1 \rangle$ denoting the level of belief of agent A_1 in information received from A_2. Analogically, $\alpha_{1,[2]}$ is defined in A_2.

4.1. Communication

The agents can communicate two kinds of information: (i) about the environment, and (ii) about their individual aims. In both cases, the information is stored in the form of pdfs, namely marginal distribution from the environment model for (i), and marginal distribution on ideal pdfs for (ii).

The model of the environment (1) is fully determined by the observation model (Section 2.1) and parameter distribution (3), which is estimated from the observed data $d(t)$. The easiest way how to exchange the information about the environment is to exchange the observed data. The observed data can be seen as a special case of pdf, namely empirical pdf $f\left(d_{[2]}(t)\right)$. Then, the task is formally identical with the task of indirect merging of pdf (11) as described in Section 3.3. The observed data from A_2 are merged with the existing model of A_1 using

$$f_{[1]}, f\left(d_{[2]}(t)\right) \stackrel{\text{merge}}{\longrightarrow} \tilde{f}_{[1]}.$$

and the negotiation weight $\alpha_{2,[1]}$. This weight can be chosen constant or it can be negotiated with the neighbour. When the negotiation is finished, the merged pdf $\tilde{f}_{[1]}$ is then used as the new model of the environment.

The ideal distributions can be communicated and merged in the same way, using direct merging (7). Note that merging of the ideal distributions influences the aim of the agent. The FPD procedure must be performed after each merge in order to recompute the DM strategy. Once again, the result is strongly influenced by the negotiation weights α. These weights can be determined by negotiation strategies.

If the merging operation yields pdfs that are not compatible with the observation model (i.e. can not be reduced into algebraic form), the merged distribution must be projected into the compatible class, as illustrated in Figure 3.

4.2. Negotiation strategies

We distinguish three basic strategies [20]:

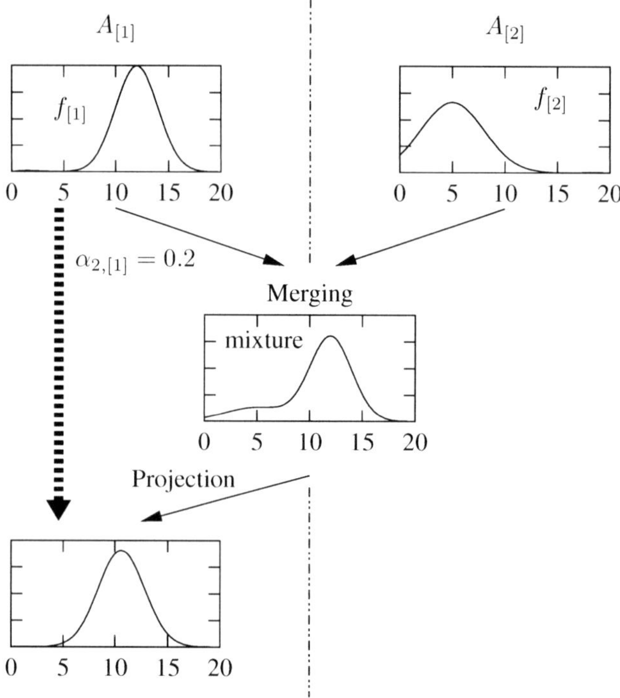

Figure 3. Illustration of merging of two Gaussian distribution. The merged distribution is a mixture of Gaussians for which the operations of learning and design of DM strategy do not have a finite-dimensional parametrization. Thus, the merged distribution is projected into the class of single Gaussians.

- **selfish** — a strategy where each agent freely chooses its own weights. Agent A_1 accepts all information from its neighbour, but it refuse any attempts to change the weight $\alpha_{2,[1]}$ that may be suggested by A_2.
- **hierarchical** — a strategy where the agent have a fixed values of $\alpha_{2,[1]}$, however if the neighbour A_2 is superordinate to A_1, it can assign the value of $\alpha_{2,[1]}$ by communication.
- **cooperative** — a strategy, where both participants have common aim (given by the user using ideal pdfs) to reach an agreement on the negotiation values, i.e. $\alpha_{2,[1]} = \alpha_{1,[2]}$.

4.3. On-line algorithm of Bayesian agents

On-line operation of each Bayesian agent is an extension of the on-line steps of Bayesian decision-maker (Section 3).

1. **read**: the observed data are read from the system (environment). Possible communication (via pdf) from the neighbour is also received in this step. We assume that only one neighbour can communicate in one time step.
2. **learn**: the observed data are used to increase the knowledge about the system (environment).

2a. **merge**: if the communication from the neighbour contains information about the environment, the merge operation is called in order to merge it with the current knowledge. In case of communication of ideal distributions, the FPD procedure is run. Note that this step may be computationally expensive.
3. **adapt**: the decision-maker use the improved knowledge of the environment to improve its DM strategy.
4. **decide**: the adapted DM strategy is used to choose an appropriate action. In multi-agent scenario, the tasks of communication and negotiation are also part of the decision making process. Therefore, in this step, decisions on communication (request communication, negotiate, refuse communication) and negotiation (propose new value of $\alpha_{1,[2]}$) must be also made.
5. **write**: the chosen action is written into the system (environment). If the decision to communicate was made, a message to the neighbour is also written in this step.

Note that acquisition of the observed data is synchronized with communication. In each time step, only one message from the neighbour is received, processed and answered. This allows seamless merging of knowledge from direct observations and from communication. If the periods of data sampling and communication differ, the smaller one is chosen as the period of one step of an agent.

5. Application in Traffic Control

Urban surface transport networks are characterised by high density of streets and a large amount of junctions. In many cities these structures cannot easily accommodate the vast volume of traffic, which results in regular traffic congestions. Efficient traffic control mechanisms are sought that would provide for higher throughput of the urban transport network without changing its topology.

Due to space constraints we cannot present the reader with full introduction to the principles of urban traffic control (UTC). We will just briefly outline terms that will be needed below. More thorough explanation of the UTC methodology exceeds the scope of this paper. Interested readers should refer to any of the existing monographs on this topic, e.g. [28,31].

In most cases, UTC is targeted on *signalled intersections*, where traffic is controlled by *traffic signals*. The sequence of traffic signal settings for an intersection is called a *signal plan*. A signal plan cycle typically consist of several *stages*, where one of the conflicting traffic flows has green and the others have to wait. The lengths of stages, the overall signal plan duration and other parameters are bounded by values reflecting either physical shape of the intersection or other (usually normatively given) rules. An *intersection controller* is an industrial micro-controller that attempts to select the order of stages and to modify stage lengths in such a way that a maximum possible throughput of the intersection is achieved. The ordering of stages may be influenced by public transport vehicles in order to minimise their waiting at an intersection.

Intersection controllers are very often autonomous devices that do not react on traffic conditions at neighbouring intersections. However, in areas with high traffic intensity, intersection controllers may be mutually interconnected in a kind of hierarchical controller that attempts to optimize the throughput of the whole traffic network. Several interesting UTC approaches exist that attempt to solve the traffic control problem using feedback from different traffic detectors [25]. Many agent-based approaches have been implemented as well. For example: (i) agents for setting of the optimal signal plan cycle length [11], and (ii) for distributed coordination of signal plans. The latter are based on game theory [3,8] or Bayesian learning [29]. These applications often use approximate heuristics and long-time statistics to derive the optimal control strategy. Our proposal is to build the strategy adaptively in a collaborative agent-based environment.

In the following text, agents are intersection controllers of some street network. The agents shall agree on the overall traffic signal setting that would minimise time spent by vehicles inside the controlled region and thus maximise the throughput of the network.

5.1. Model

Papageorgiou [30] shows that the total time spent by a vehicle in a controlled micro-region is strongly correlated to queue lengths at signalised intersections of this micro-region. Hence, minimization of waiting queues results in faster vehicle transition and in higher throughput of the network.

We start with a deterministic model describing the behaviour of the traffic at an intersection as a particle flow system [26]:

$$\begin{aligned}\Theta_t &= A\Theta_{t-1} + Bu_{t-1} + F\\ y_t &= C\Theta_t + G\end{aligned}$$

where

$$\Theta_t = \begin{bmatrix}\xi_t\\ O_t\end{bmatrix}$$

is a state vector holding information about waiting queue lengths ξ_t and detected input lane occupancies O_t, u_t is an input variable which represents green settings for a signal plan cycle at this intersection. Matrix A defines transition from an old state to the new one. It is composed from information about waiting queue development, and mutual influence of queues at one lane on other lanes. Matrix B models throughput of the junction, and vector F is composed from the observed incoming traffic intensity. Output vector

$$y_t = \begin{bmatrix}\eta_t\\ O'_t\end{bmatrix}$$

contains information about outgoing traffic intensity η_t and output arm occupancies O'_t. C is a matrix of coefficients transforming waiting queue data into outgoing traffic intensities and vector G models the influence of current incoming traffic and past queues on outgoing traffic.

This model can be transcribed into the following probabilistic internal and observation models:

$$f(\Theta_t|\Theta_{t-1}, u_{t-1}) = \mathcal{N}(A\Theta_{t-1} + Bu_{t-1} + F, Q) \tag{12}$$

$$f(y_t|\Theta_t) = \mathcal{N}(C\Theta_t + G, R) \tag{13}$$

where $\mathcal{N}(\mu, \sigma)$ is a Gaussian probability distribution and Q and R are allowed variances. The internal model (12) describes the probability distribution of queue lengths at an

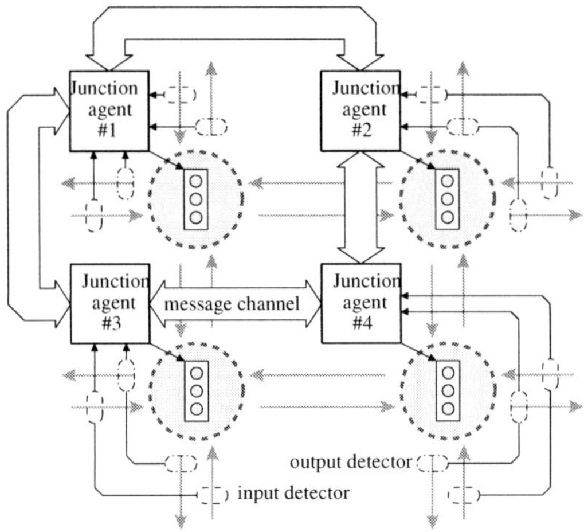

Figure 4. Simple urban traffic network with four controlled junctions and four agents.

intersection at time t given the green settings u_{t-1} and incoming traffic data Θ_{t-1} at time $t-1$. The observation model (13) yields probability distribution of outgoing traffic intensity of the modelled junction at time t, given the queue pdf from the internal model.

5.2. *Ideal distributions*

The global aim of the proposed UTC approach is to minimise waiting queues at every junction. As said in Section 4, agents attempt to reach this aim by exchanging their ideal pdfs, defined on their common data. In our case, agents share information about traffic intensity at particular intersection arms. Hence, the exchanged ideal pdfs specify wishes about intensity of outgoing and incoming traffic.

We propose to model the ideal pdfs as follows:

$$^{\lfloor I}f\left(\xi_t\right) = t\mathcal{N}\left(0, V_\xi, \langle 0, \xi_{\max}\rangle\right), \tag{14}$$

$$^{\lfloor I}f(I_t|\xi_t) = t\mathcal{N}\left(I\left(\xi_t\right), V_I, \langle 0, I_{\max}\rangle\right), \tag{15}$$

$$^{\lfloor I}f(\eta_t|\xi_t) = t\mathcal{N}\left(\eta_{\max}, V_\eta, \langle 0, \eta_{\max}\rangle\right). \tag{16}$$

Here, $t\mathcal{N}\left(0, U, \langle 0, \xi_{\max}\rangle\right)$ denotes a Gaussian distribution with mean value 0 and variance V_ξ, truncated on the interval $\langle 0, \xi_{\max}\rangle$. $\xi_{\max}$ denotes maximal allowed queue length. The ideal (14) favours minimal queue lengths value, since estimates $f\left(\xi_t\right)$ with lower mean value are closer to the ideal than those with larger mean value. The ideal pdf (15) models the agents wishes for input intensities coming from its neighbours. The requested mean value $I\left(\xi_t\right)$ is changing with the current traffic conditions. The variance V_I expresses the "strength" of the request; higher V_I allows higher deviation from ideal $I\left(\cdot\right)$ and leaves the agent more room for adapting to other requests. $I_{\max}$is the maximum possible intensity at the arm (or lane) in concern. In order to communicate these wishes to the neighbours, they must be independent of the internal quantity ξ_t. This can be achieved by marginalization, i.e. $^{\lfloor I}f(I_t|\xi_t) \rightarrow {}^{\lfloor I}f(I_t)$.

Note that output intensities of one intersection are input intensities of its neighbours, i.e. $\eta_{t,[1]} \to I_{t,[2]}$. Hence, the communicated ideals on input intensities $f\left(I_{t,[2]}\right)$ will be merged with ideals on output intensities $f\left(\eta_{t,[1]}\right)$.

5.3. Control cycle

The proposed control cycle of a single agent follows the decomposition from Section 4.3:

1. **read:** The agent reads observed data from the environment and checks for incoming communication from some neighbour.
2. **learn:** Observed data of the agent (measured traffic intensities) are used to increase its knowledge about current traffic conditions, namely pdfs of waiting queue lengths and unobserved intensities of traffic flow.
3. **merge:** If a message from some neighbour arrived, its pdf is merged with the agent's pdfs — either with the current knowledge or with ideal pdfs. In the latter case, FPD procedure that evaluates Eq. (6) is called after the merge in order to reflect the change in ideal aims in the optimal DM strategy.
4. **adapt:** The agent uses the updated knowledge to adapts its DM strategy. Hence, the strategy can be changes in reaction to the changed traffic conditions or in reaction to the message from the neighbour.
5. **decide:** Based on the adapted strategy, the agent decides about its signal plan parameters for the next period. The signal values are typically taken as expected values of the adapted strategy pdf. Decisions whether and what to communicate with agent's neighbour is also made in this moment.
6. **write:** The chosen signal plan is written to the intersection controller. Optionally, communication message is sent to the chosen neighbour.

6. Conclusion

We have presented an application of the Bayesian decision making theory to the area of multi-agent systems. The presented methodology offers clear guidelines and concept of design of multi-agent systems. Since the Bayesian approach formalizes all available knowledge in the form of probability density functions, we had to formalize the key features of agents—i.e. communication and negotiation—using probability calculus. We have shown that the formalization can be achieved using techniques of fully probabilistic design and merging of pdfs.

The work presented in this paper is a conceptual outline of the approach. In spite of the fact that the key techniques are available, many practical issues must be solved before it is ready for real application. The presented application in urban traffic control will be used as testing environment for further research and development of Bayesian agents.

Acknowledgements

This work was supported by grants MŠMT 1M0572 (DAR) and AVČR 1ET 100 750 401 (BADDYR).

References

[1] J. Andrýsek, M. Kárný, and J. Kracík, editors. *Multiple Participant Decision Making*, Adelaide, May 2004. Advanced Knowledge International.
[2] K.J. Astrom. *Introduction to Stochastic Control*. Academic Press, New York, 1970.
[3] Ana L. C. Bazzan. A distributed approach for coordination of traffic signal agents. *Autonomous Agents and Multi-Agent Systems*, 10(1):131–164, January 2005.
[4] R. Bellman. *Introduction to the Mathematical Theory of Control Processes*. Academic Press, New York, 1967.
[5] J.O. Berger. *Statistical Decision Theory and Bayesian Analysis*. Springer-Verlag, New York, 1985.
[6] D.P. Bertsekas. *Dynamic Programming and Optimal Control*. Athena Scientific, Nashua, US, 2001. 2nd edition.
[7] R. Caballero, T. Gomez, M. Luque, F. Miguel, and F. Ruiz. Hierarchical generation of pareto optimal solutions in large-scale multiobjective systems. *Computers and operations research*, 29(11):1537–1558, 2002.
[8] Eduardo Camponogara and Werner Kraus Jr. Distributed learning agents in urban traffic control. In Fernando Moura Pires and Salvador Abreu, editors, *Progress in Artificial Intelligence: Proceedings of the 11th Portuguese Conference on Artificial Intelligence (EPIA 2003)*, volume 2902 of *Lecture Notes in Computer Science*, pages 324–335, Beja, Portugal, December 2003. Springer-Verlag.
[9] Z. Chen. Bayesian filtering: From Kalman filters to particle filters, and beyond. Technical report, Adaptive Syst. Lab., McMaster University, Hamilton, ON, Canada, 2003.
[10] B. Tamer (ed.). *Control Theory*. IEEE Press, New York, 2001.
[11] L. A. García and F. Toledo. An agent for providing the optimum cycle length value in urban traffic areas constrained by soft temporal deadlines. In L. Monostori, J. Váncza, and M. Ali, editors, *Engineering of Intelligent Systems: 14th International Conference on Industrial and Engineering Applications of Artificial Intelligence and Expert Systems, IEA/AIE 2001*, volume 2070 of *Lecture Notes in Computer Science*, pages 592–601, Budapest, Hungary, June 2001. Springer-Verlag.
[12] T.V. Guy, J. Böhm, and M. Kárný. Multiobjective probabilistic mixture control. In IFAC, editor, *IFAC World Congress, Preprints*. IFAC, Prague, 2005. accepted.
[13] Y.Y. Haimes and D. Li. Hierarchical multiobjective analysis for large scale systems: Review and current status. *Automatica*, 24(1):53–69, 1988.
[14] K. H. Hall, R. J. Staron, and P. Vrba. Holonic and agent-based control. In *Proceedings of the 16th IFAC Congress*, 2005.
[15] F.V. Jensen. *Bayesian Networks and Decision Graphs*. Springer-Verlag, New York, 2001.
[16] R. Jiroušek. On experimental system for multidimensional model development MUDIN. *Neural Network World*, (5):513–520, 2003.
[17] M. Kárný. Towards fully probabilistic control design. *Automatica*, 32(12):1719–1722, 1996.
[18] M. Kárný. Tools for computer-aided design of adaptive controllers. *IEE Proceedings — Control Theory and Applications*, 150(6):642, 2003.
[19] M. Kárný, J. Böhm, T. V. Guy, L. Jirsa, I. Nagy, P. Nedoma, and L. Tesař. *Optimized Bayesian Dynamic Advising: Theory and Algorithms*. Springer, London, 2005.
[20] M. Kárný and T.V. Guy. On dynamic decision-making scenarios with multiple participants. In J. Andrýsek, M. Kárný, and J. Kracík, editors, *Multiple Participant Decision Making*, pages 17–28, Adelaide, May 2004. Advanced Knowledge International.
[21] M. Kárný and J. Kracík. A normative probabilistic design of a fair governmental decision strategy. *Journal of Multi-Criteria Decision Analysis*, 10:1–15, 2004.
[22] J. Kracík. Composition of probability density functions - optimizing approach. Technical Report 2099, ÚTIA AV ČR, Praha, 2004.
[23] J. Kracík. On composition of probability density functions. In J. Andrýsek, M. Kárný, and

J. Kracík, editors, *Multiple Participant Decision Making*, volume 9 of *International Series on Advanced Intelligence*, pages 113–121. Advanced Knowledge International, Adelaide, Australia, 2004.

[24] J. Kracík and Kárný M. Merging of data knowledge in Bayesian estimation. In *Proceedings of the Second International Conference on Informatics in Control, Automation and Robotics*, pages 229–232, Barcelona, September 2005.

[25] J. Kratochvílová and I. Nagy. Bibliographic Search for Optimization Methods of Signal Traffic Control. Technical Report 2081, ÚTIA AV ČR, Praha, 2003.

[26] J. Kratochvílová and I. Nagy. Traffic model of a microregion. In IFAC, editor, *IFAC World Congress, Preprints*. IFAC, Prague, 2005. submitted.

[27] S. Kullback and R. Leibler. On information and sufficiency. *Annals of Mathematical Statistics*, 22:79–87, 1951.

[28] Michael Meyer and Eric J. Miller. *Urban Transportation Planning*. McGraw-Hill, 2 edition, December 2000.

[29] Haitao Ou, Weidong Zhang, Wenjing Zhang, and Xiaoming Xu. Urban traffic multi-agent system based on rmm and bayesian learning. In *Proceedings of the American Control Conference*, Chicago, Illinois, June 2000.

[30] M. Papageorgiou. *Applications of Automatic Control Concepts to Traffic Flow Modeling and Control*, volume 50 of *Lecture Notes in Control and Information Sciences*. Springer-Verlag, Berlin, 1983.

[31] Roger P. Roess, Elena S. Prassas, and William R. McShane. *Traffic Engineering*. Prentice Hall, 3 edition, October 2003.

Self-Organization and Autonomic Informatics (I)
H. Czap et al. (Eds.)
IOS Press, 2005

An Energy-Driven Social Behaviour Network Architecture

Mingwei YUAN[a], Ping JIANG[a,b,1] and Julian NEWMAN[c]
[a] *Department of Control Science and Engineering*
Tongji University, Shanghai, China
[b] *Department of Cybernetics and Virtual Systems*
University of Bradford, Bradford, UK
[c] *School of Computing and Mathematical Sciences*
Glasgow Caledonian University, Glasgow, UK

Abstract. The motivation of this paper is to realize an energy-driven self-organising architecture of Social Behaviour Networks(SoBeNet) for the Web application. Internet agents can sense changes in the web environment via virtual web sensors and behavior selection is based on the energy spreading mechanism from the bottom-up paradigm of AI. There is no global coordinator module to control behavior selection and the cooperation between the two agents is implicitly. A social behavior network is formed spontaneously.

Keywords. Energy-driven self-organisation, social behaviour networks, internet agents

1. Introduction

Internet is a highly dynamic and unpredictable environment since it is a complex system composed of multitudinous distributed individuals. Self-organising mechanism of a Virtual Organisation (VO) in the web environment becomes a new study direction of Web Intelligence (WI).

It is well known that there are two branches in the study of AI. One is bottom-up and the other is top-down. They work from different perspectives and tackle problems in different ways. Whilst the top-down approach mimics functionalities of intelligence, the bottom-up approach tries to build structures and mechanisms of neurons. The bottom-up approach [2, 3, 4] dominates research in dynamic distributed AI, e.g. in Robotics and Artificial life, because of their unambiguous and simple interior units as individuals but complex exterior behaviours as a whole. VO through Internet is such a large-scale and dynamic distributed system. It is difficult for such a system to be created, maintained and evolved in a pre-programmed way, yet current mainstream solutions to VO mostly adopt the top-down approach [5], driven by a rigid process

---

[1] Corresponding Author: Dr Ping Jiang, Department of Cybernetics and Virtual Systems, University of Bradford, Bradford, BD7 1DP, UK. Email: p.jiang@bradford.ac.uk

model. This paper argues that a behaviour-based organisation from the bottom-up approach is more appropriate to virtual organisation in the web environment.

The phenomena of self-organisation driven by energy spreading appear in both microworld and macroworld: from atoms to compounds, from cells to organisms, from biological to ecological systems and from planets to galaxies. Sometimes the energy profile cannot be explicitly perceived, but exists in an intangible form such as that in the food chain [1]. There is evidence that energy is the main driving force to achieve self-organisation in the real world. In this paper, we use the energy spreading concept from the perspective of behaviour-based approach to address self-organisation issues for web-based applications.

In this paper, a social behaviour network (SoBeNet) is formed: it is based on an extended behaviour network [6, 7] and composed of autonomous behaviour-based web agents; agents sense the web environment and respond accordingly; the interaction among agents is via the energy spreading. The energy can be acquired from the web environment and from the goals and it can spread among the behaviour modules and self-organisation is driven by activation energy.

2. Related Work

A particularly influential theory about energy models in the analysis of behaviour was proposed by Lorenz [8]. Though Lorenz's original psycho-hydraulic model suffered some criticisms [9,10], in the last few years of the twentieth century, many psychologists have suggested that energy concepts still play a useful theoretical role in behaviour analysis (e.g. [10], [11]). Experiments supported the value of energy models in the analysis of much behaviour such as the nest-building activities of three-spined sticklebacks [12], sleep in humans, and dustbathing in chickens [13].

The researches in psychology motivate us to develop an energy-driven SoBeNet for the web application. The activation spreading based behaviour network proposed by Pattie Maes [14,15] can be adopted to support dynamic behaviour selection which further changes environment energy profile.

The focus of this paper on the social behaviour network as an approach to virtual organisation synthesis differs from empirical studies of social networks in the web [16,17,18,19]. Empirical social network researchers focus on mining user relationship through the information in email, message boards, chat data [18, 19] etc. for discovery of who-to-whom static relationships. The proposed SoBeNet is concerned with formation and execution of dynamic interaction based on goals and changes in the environment. Therefore it focuses more on process rather than static information.

3. The Architecture

A SoBeNet is composed of distributed agents that are simple and they can sense the changes in the environment. Multiple agents are each working on their own individual goals and making decision independently. Every agent has a set of behaviour modules which realize special function. Behaviour modules are activated by energy activation level that is relevant to the goal and present circumstances. No central unit guides energy spreading direction and the interaction among agents happens spontaneously based on energy profiles, which is a self-organisation process.

3.1 The agent architecture

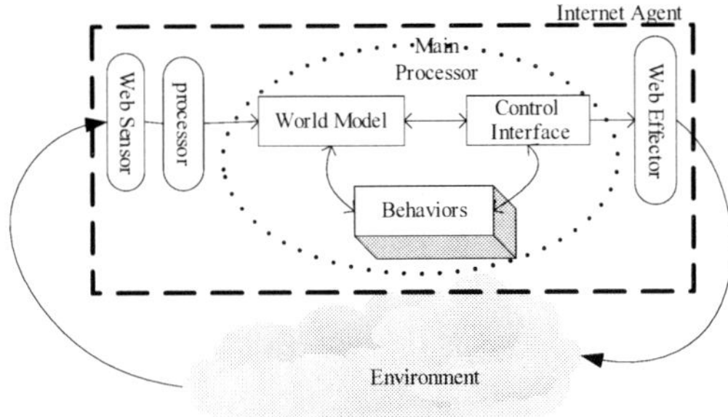

Figure 1: the architecture of the web agent

The architecture of a web agent is shown diagrammatically in Figure 1. Where an agent senses the changes in the web environment through *virtual web sensors*; the *main processor* makes decisions on actions to be taken and that affects the environment through *virtual web effectors*. The design of the agents is based on the behaviour-based paradigm that involves design of local behaviour networks. In this paper, we coordinate agents' actions via a blackboard system which is taken as the *virtual environment* for causal behaviours among agents.

3.2 Blackboard based coordination

The blackboard system provides a multi-agent system an open web environment to support distributed coordination [20, 21]. It functions as the global system context and stores the status variables as common information to all agents. Via the blackboard system, agents sense the environment and share data and knowledge. But agents make decision and take action independently.

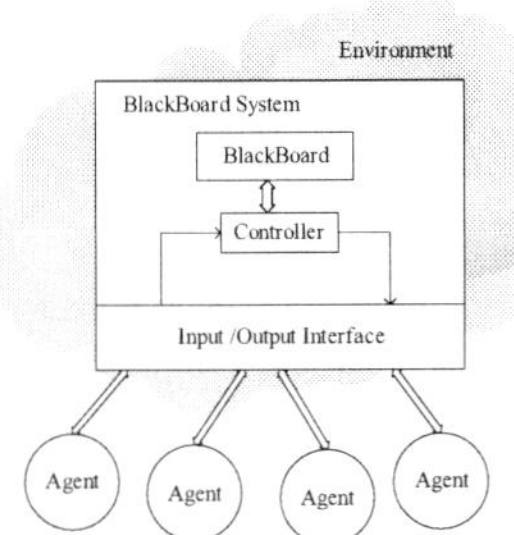

Figure 2: the architecture of the system.

The blackboard system (Figure 2) is composed of Input/Output Interface, Blackboard Controller and Blackboard. The Blackboard is regarded as a shared database and stores the status information. The *Blackboard Controller* is in charge of supervising the blackboard, such as searching, filtering, deleting and altering information. The *Input/Output Interface* provides agents a channel to access the blackboard. An effect caused by *virtual web effectors* of an agent changes the content in the blackboard and other agents may sense this change by *virtual web sensors*. Not all the changes in the blackboard need to be dealt with and the *processor* of agents can filter or preprocess this sensing information based on their local interests [24].

3.3 Energy based behaviour networks

The behaviours could be categorised according to the behaviour matrix in Human Behaviour Representation [22]: reactive behaviours which represent short-term response to short term contingencies, routine behaviours which represent long-term series of actions to achieve a goal and collaborative behaviours, which represent explicit interaction, involve cooperative behaviours and competitive behaviours. In this paper, the cooperation among agents is not explicit but via indirect interaction of local agent behaviours. A reactive behaviour requires that an agent takes actions immediately without thinking, for example, an agent asks the user to shutdown the program once system security exception happens. Reactive behaviours have higher priority than routine behaviours, and hypothesizing here that the energy injected by the precondition is high enough to make reactive behaviours activated once its precondition is satisfied. In the following part we will focus on routine behaviours and discuss a SoBeNet composed of routine behaviours.

The agents in a SoBeNet must be able to select proper behaviours according to the emergence energy for achieving goals and sub-goals. To address this issue, we use activation spreading based behaviour network proposed by Pattie Maes [14, 15]. The term of *activation spreading* stands for a kind of energy propagation to address behaviour selection of agents.

The behaviour network formalism is defined: A behaviour network consists of a tuple (G, M, Π), where G is a set of goals, M is a set of behaviour modules and Π is a set of parameters that controls energy flow. An agent executes a behaviour module in M when its activation energy level exceeds the activation threshold in Π and the expected effects can be reached after the execution.

Energy control parameters could be normalised in [0,1] according to REASM [6] and the parameters in Π is:

$\gamma \in [0,1]$ *activation parameter of module*

$\delta \in [0,1]$ *inhibition parameter of module*

$\beta \in [0,1]$ *inertia parameter of activation*

$\phi \in [0,1]$ *activation parameter of state*

$a \in [0, \hat{a}]$ *activation threshold, with $\hat{a}$ the upper bound for a module's activation*

The energy control parameters have effect on the performance of a behaviour network. How to configure them is another research issue and they could be set by users directly according to their experience or by learning algorithms, such as neural network, to learn from data that are collected from the user profiles or the environment. To focus on illustrating the social behaviour networks architecture, the detailed steps to acquire threshold variables will be not discussed here. But the energy threshold variables of reactive behaviours must be set zero to response to emergencies.

Each behaviour module is a set of attributes stated as the tuple (p_i, b_i, Eff_i, h_i) in which b_i represents a functional behaviour and p_i represents a precondition list to be satisfied in order to activate b_i. $\tau(p_i, s)$ denotes the degree of behaviour executability. Eff_i is a set of effects after the execution of the behaviour b_i. In $Eff_i := \{eff^+, eff^-\}$, eff^+ represents the positive effects to achieve the goal and eff^- represents the negative effects to deviate from the goal, with $eff^+ \cap eff^- = \phi$. Exp(eff^+) and Exp(eff^+) denote the expectation values of eff^+ and eff^- respectively. The value h_i states the activation energy level accumulated by the corresponding behaviour modules from goals and other modules. When the activation energy level exceeds the activation threshold $a_i \in \Pi$, the behaviour could be activated.

3.3.1 Energy spreading

Energy flows along the links between behaviour modules. Suppose there is a link from behaviour A to behaviour B. B is called a successor of A if A has an positive effect (eff^+) that is in the proposition of the precondition of B, and A is called a predecessor of B. There is a conflicting link from A to B if A has a negative effect (eff^-) that is in the proposition of the precondition of B.

Three channels support the energy spreading:

1. Precondition activation

The agent senses the status information in the blackboard via *virtual sensors*. If all p_i in a precondition list of the behaviour module b_i are satisfied, then b_i is executable, but only when its activation energy level exceeds its activation threshold a_i, this behaviour could be activated.

2. Goal activation

Goals inject energy to those behaviours whose eff^+ is in the precondition list of goals and goals extract energy from those behaviours whose eff^- is in the precondition list of goals.

3. Modules activation

An executable behaviour module will reduce its own energy and send it to its successors when the precondition of its successors is unsatisfied. If a behaviour module is not executable, it will reduce its own energy and send it to its predecessor. In a conflicting link, the behaviour lead to a negative effect (eff^-) will be inhibited by its effected behaviour. In some sense, energy from module activation is injected by goals or preconditions indirectly.

3.3.2 Behaviour selection Algorithm in a SoBeNet

The activation energy level calculation formulas from goal activation are expressed as follows:

$$h^T_{kg_i 1} = \gamma \cdot f(\bullet) \cdot Exp(eff^+) \quad (1)$$

$$h^T_{kg_i 2} = -\delta \cdot f(\bullet) \cdot Exp(eff^-) \quad (2)$$

$$h^T_{kg_i 3} = \gamma \cdot \sigma(h^{T-1}_{succ,i}) \cdot Exp(eff^+) \cdot (1 - \tau(p_{succ}, s)) \quad (3)$$

$$h_{kg_i4}^T = -\delta \cdot \sigma(h_{conf,i}^{T-1}) \cdot Exp(eff^-) \cdot (1 - \tau(p_{conf}, s)) \quad (4)$$

Goal-related energy transforms from goals to modules directly or from modules to modules indirectly. T denotes periodicity, $f(\bullet)$ is goal-related latent function which explains static and dynamic characteristics of the goal. $\sigma(\bullet)$ is the transfer function of the modules activation. Equation (1) and (2) denote the activation energy level of a behaviour module that are activated or inhibited by goal g_i directly and Equation (3) and (4) denote the activation energy level of a behaviour module that activated or inhibited by goal g_i indirectly or by other modules whose energy comes from goal activation.

To avoid energy from a goal acting on the same module multiply times, only the strongest path from each goal to a module is taken into account.

$$h_{kg_i}^T = abs\max(h_{kg_i1}^T, h_{kg_i2}^T, h_{kg_i3}^T, h_{kg_i4}^T) \quad (5)$$

It is easy to compute status-related energy transforms from preconditions to modules or among modules referring to aforementioned goal-related module activation level computation. If the sensing statuses are within the precondition of some behaviour k of an agent, the module energy from precondition activation is:

$$h_{kS_j1}^T = \phi \cdot \tau(p_i, s_j) \cdot Exp(eff^+) \quad (6)$$

The precondition activation transforming among modules could be expressed as:

$$h_{kS_j2}^T = \gamma \cdot \sigma(h_{succ,i}^{T-1}) \cdot Exp(eff^+) \cdot (1 - \tau(p_{succ}, s)) \quad (7)$$

$$h_{kS_j3}^T = -\delta \cdot \sigma(h_{conf,i}^{T-1}) \cdot Exp(eff^-) \cdot (1 - \tau(p_{conf}, s)) \quad (8)$$

Only the strongest path from each precondition to a module is taken into account.

$$h_{kS_j}^T = abs\max(h_{kS_j1}^T, h_{kS_j2}^T, h_{kS_j3}^T) \quad (9)$$

$$\tau(p_i, s_j) = \sum_{j=1}^{n} e^{-\alpha d(p_i, s_j)} \quad (10)$$

$\tau(p_i, s_j)$ denotes the degree of behaviour executability. $d(p_i, s_j)$ denotes the minimum virtual distance, which is a measurement of the semantic distance based on domain ontology [24], between precondition p_i and the current status value.

Finally the activation of behaviour k is:

$$h_k^T = \beta(h_{kg}^{T-1} + h_{kS}^{T-1}) + \sum_{i,j}(h_{kg_i}^T + h_{kS_j}^T) \quad (11)$$

Behaviour selection algorithm can be summarised as follows:

1. Compute activation energy level of each module h_k^T (Eq.(11)).

2. Activation and executability of a module are combined by a non-decreasing function into the utility of a module, whereby non-executable competence modules always get the value zero.

3. If the activation energy level h_k of behaviour b_k is larger than the threshold a_i, the behaviour becomes a candidate to be selected. Among the candidate behaviours, the agent chooses the behaviour with the highest activation energy to execute and reset a_i to original value and go to 1.

4. If there is no candidate behaviour, the threshold value a_i is reduced by $x\%$. Go to 1.

4. SoBeNet Example

The formalism of a SoBeNet is driven by different energy spreading directions (see Figure 3). In this section, a job-related application is used to illustrate how a SoBeNet comes into being. In this application there are two kinds of agents: one is job-hunting agent and the other is head-hunting agent.

First, designers need to define each behavior module and its precondition list, add list and delete list explicitly.

The goal of job-hunting agent is Find-new-job and its behavior modules involve:

Search-job:
Precondition list: Job-description
Add list: Candidate-job $\vee$ Search-fail
Delete list:
Activation:

Apply-job:
Precondition list: Job-description $\wedge$ Candidate-job $\wedge$ Information-validity
Add list: Interview-notice $\vee$ Refused-notice
Delete list: Search-fail
Activation:

Modify-job-description:
Precondition list: Search-fail $\vee$ Refused-notice
Add list: Job-description
Delete list:
Activation:

The goal of head-hunting agent is Find-new-Employee and its behavior modules involove:

Publish: (job)
Precondition list: Position-description
Add list:
Delete list:
Activation:

Match: (job-vs-candidate)
Precondition list: Job-description $\wedge$ Position-description $\wedge$ Information-validity
Add list: Interview-notice $\vee$ Refused-notice
Delete list: Validity
Activation:

Second, draw the energy spreading graph according to the possible energy flowing directions, so a SoBeNet is exhibited. The SoBeNet works according to behaviour selection algorithm in section 3.3.2.

The segment of the activation spreading: suppose the current status is S(0)={job-description} and the goal is Find-new-job. The precondition of module Search-job is satisfied and Search-job extracts energy from its precondition (Eq.(6)). Search-job spreads precondition activation to its successors Apply-job and Modify-job-description (Eq.(7)). On the other hand goal Find-new-job injects energy to

Module Apply-job (Eq.(1)). Since the preconditions of Apply-job, or{ Job-description $\wedge$Candidate-job$\wedge$Information-validity }, is unsatisfied, module Apply-job transports energy to its predecessor Module Search-job (Eq.(3) or). Such activation spreading continues and each module update its activation level (Eq.(11)) in every periodicity until there is one behaviour whose activation energy level exceeds the activation threshold according to behaviour selection algorithm.

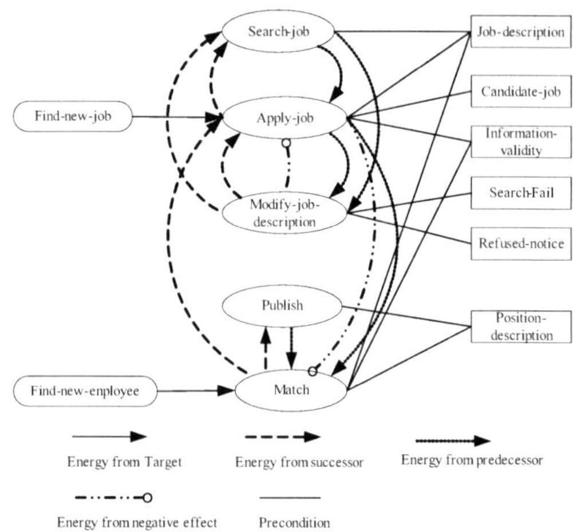

Figure 3: Energy spreading in a social behavior network

The arrow in Figure 3 denotes energy spreading direction. Different behavior modules interact via different links. Two channels energy injects into the SoBeNet: one from the current sensing status on the blackboard and the other from the goals. The rectangle denotes precondition which relies on the information from virtual sensors, the ellipse denotes behaviour module and the round rectangle denotes goal.

Energy spreads through goals, behavior modules and preconditions, which yields agent behavior network. Energy could be transported from one behaviour module such as Apply-job of job-hunting agent to another behavior module such as Match of head-hunting agent. The cooperation between the two agents is implicitly and these interactions reflect sociality. This is the reason why we defined agent behavior network as social behavior network (Figure 3). The agent switches its behavior spontaneously according to behaviour activation energy level and there is no global coordinator module to control behavior selection. Decisions are made during running time, which is of good adaptability to dynamical goals so new behaviour modules could be added into the behaviour network.

5. Conclusions

This paper proposed a self-organisation mechanism for web applications, which is driven by energy spreading. There is no global coordinative mechanism to form a social behaviour network but intangible energy exchange integrates different behaviours of different agents into a social behaviour network. An Extended behaviour network was applied to address the self-organisation problems based on energy spreading, which makes distributed agents aggregate for large-scale, heterogeneous and dynamic cooperation in a web environment.

In a social behaviour network, perception and action are mostly in the form of textual knowledge which can be represented by XML. In order to represent and interpret knowledge on the web, virtual sensors need to be investigated first in the following work.

References

[1] C. Elton and M. Nicholson, The ten-year cycle in numbers of the lynx in Canada, *Journal of Animal Ecology* **11** (1942), 215-244.

[2] R. Brooks, Intelligence without representation, *Artificial Intelligence* **47**(1991), 139-159.

[3] R. Brooks, Artificial life and real robots, *Proceedings of the First European Conference on Artificial Life. F. Varela & P. Bourgine Eds.*, Cambridge, MIT Press, USA 1992.

[4] M. J. Matarić, Behavior-based control: examples from navigation, learning, and group behavior. *Journal of Experimental and Theoretical Artificial Intelligence, special issue on Software Architectures for Physical Agents*, **9**(1997),323-336.

[5] L.M. Camarinha-Matos, H. Afsarmanesh, Elements of a base VE infrastructure, *Computers in Industry*, **51**(2003), 139–163.

[6] K. Dorer, Behavior networks for continuous domains using situation dependent motivations, *Proceedings of the Sixteenth International Conference of Artificial Intelligence* (1999),1233–1238.

[7] K. Dorer, The Freiburg soccer team. In M. Veloso, E. Pagello, and H. Kitano, editors, RoboCup-99: Robot Soccer World Cup III, pages 600–603. Springer-Verlag, Berlin, Heidelberg, New York 2000.

[8] K. Lorenz, über die Bildung des Instinktbegriffes, Naturwissenschaft, **25**(1937): 289-300, 307-318, 324-331.

[9] R. A. Hinde, Energy models of motivation, *Symp. Soc. Exp. Biol.* **14**(1960),199-213.

[10] J.A. Hogan, Animal Behavior, *Foundations of Psychology, Copp Clark Pitman*, Toronto (1996) 138-186.

[11] J A. Hogan, Energy models of motivation: a reconsideration, *Applied Animal Behavior Science* **53**(1997), 89-1005.

[12] K. Nelson, After-effects of courtship in the male three-spined stickleback, *Z. vgl. Physiol*, **50**(1965),569-597.

[13] K. Vestergaard, B. Damm, and D. Abbott, Dustbathing behavior in featherless chicks: a case of central motivational regulation, *Proc. 29th Int. Congress of the International Society of Applied Ethology*(1995), 41-42

[14] P. Maes, How to do the right thing. *Connection Science*, **1**(1989), 291-323.

[15] P. Maes, Situated agents can have goals, *Journal for Robotics and Autonomous Systems* **6**(1990), 49–70.

[16] D. Jensen, J. Neville, Data mining in social networks. *Proceedings of the National Academy of Sciences Symposium on Dynamic Social Network Analysis* (2002).

[17] D. Fisher, P. Dourish, Social and temporal structures in everyday collaboration. *Proc. 2004 CHI*(2004), 551-558.

[18] R. Agrawal, S. Rajagopalan, R. Srikant, Y. Xu, Mining newsgroups using networks arising from social behavior, *Proc 12th International World Wide Web Conference*. ACM, 2003.

[19] V. Tuulos, H. Tirri, Combining topic models and social Networks for chat data mining, *Proceedings of the IEEE/WIC/ACM Conference on Web Intelligence* (WI2004), 206-213.

[20] D. Corkill, Collaborating software: blackboard and multi-agent systems - the future, *Proceedings of the International Lisp Conference*, New York,October,2003

[21] A. Gachet, A new vision for distributed decision support systems. DSIage 2002. Oak Tree Press: Cork, Ireland, 2002.

[22] B.P. Wise, M. McDonald, L.M. Reuss, J. Aronson, ATM human behavior modeling approach study, http://www.asc.nasa.gov/aatt/research.html#69

[23] R.C. Bolles, Theory of motivation, Harper and Row, New York, 1967.

[24] P. Jiang, Y.H. Peng, Q. Mair, Concept mining for distributed alliance in multi-agent based virtual enterprises, 2004 IEEE Conference on Cybernetics and Intelligent Systems (2004), Singapore.

Self-Organization and Autonomic Informatics (I)
H. Czap et al. (Eds.)
IOS Press, 2005

A Variable Resolution Virtual Sensor in Social Behaviour Networks

Ping JIANG[a,c], Yonghong PENG[a], Quentin MAIR[b], Mingwei YUAN[c]
[a]*School of Informatics, University of Bradford, UK*
[b]*School of Computing and Mathematical Sciences, Glasgow Caledonian University, UK*
[c]*Department of Information and Control Engineering, Tongji University, China*

Abstract. In today's hyper-competitive business environments virtual organisations are becoming highly dynamic and unpredictable. Individuals may want to work together across organisation boundaries but do not have much prior knowledge of others. The semantic web and its associated new standards appear very promising as candidates to support a new generation of virtual organisations. In this paper a behaviour based organisation, Social Behaviour Networks, is proposed. In order to sense the changes on the web this paper focuses on a virtual sensor for allocating tasks amongst agents based on the announcements of tasks and capabilities of agents in DAML (DARPA Agent Markup Language). Due to the autonomy of agents the announcements are often vague and in a very high dimensional space. The ontology can provide useful information for achieving variable-resolution sensing from an individual agent's perspective and reducing the dimensions of the virtual space. The variable-resolution virtual sensors are based on hierarchical clustering analysis to reveal the level of similarity of announcements in the web.

Keywords. Virtual organisation, clustering, matchmaking, social behaviour network

Introduction

In the application areas of virtual organisations (VOs) distributed cooperation and integration have become more common and more important. There are two main approaches to implementing a VO: transaction-oriented layer-based and distributed agent based [1]. The transaction-oriented layer-based approach usually addresses the definition of goals and subgoals in a top-down manner. A protocol for collaboration is defined beforehand by a consortium or a standards body as a reference model, meta-data and process model; this is then deployed customising a cooperation layer of an existing IT infrastructure. As a result the process models which define cooperation are generally pre-compiled and remain static during the life-cycle of the system.

In today's hyper-competitive business environments we are beginning to see VOs which are highly dynamic and unpredictable. A key point to address in the organisation of VOs is how to organise independent individuals, with limited understanding of one another, through the exchange of local information and the integration of local processes. This can be addressed by the distributed agent based approach [2][3][4]. Whilst this approach has attracted research interest it is still far from having widespread deployment in terms of dynamic and uncertain environments, e.g. in the web. To support a self-organisational and self-motivated VO in a web environment it is possible to study VOs from the perspective of behaviours; that is as a Social Behaviour Network (SoBeNet). This involves:

- A general representation of knowledge in terms of goals and competences in a machine interpretable way, i.e. a virtual sensor space;
- A concept mining methodology to discover the context of a VO, i.e. virtual sensors;
- A mechanism to support goal driven process execution i.e. virtual controllers;
- A learning and adjustment mechanism to respond the changes in a domain, i.e. virtual adaptive machines.

In a SoBeNet we define the virtual sensor space of an organisation using semantic web technologies, thereby representing the environment in a machine interpretable way. Virtual sensors are constructed for each individual to sense changes in the web and to activate appropriate primitive behaviours accordingly. In fact a virtual sensor is a knowledge discovering and clustering algorithm classifying web content based on its interests. The selection of the primitive behaviours will depend on the decision of a virtual controller dependent on the activation energies of behaviours. Activation energy is a variable which models the probability of a behaviour occurring. Activation energies may be changed by goals, current circumstances and trust in the candidates.

Semantic web technologies are a family of techniques and languages built upon the extensible markup language XML. This hierarchy of languages (XML, RDF, RDFS, DAML/OWL) allows knowledge to be represented in the world wide web in a way accessible both to humans and machines. So far most existing reasoners (http://www.daml.org/reasoning/) in the semantic web are symbolic rule based systems. These have been successful in applications with structural information processing, often required in the semantic web. However this approach has exhibited some limitations in a dynamic or uncertain environment. For example it is too constrained to be able to deal with exceptions to rules or to exploit fuzzy, approximate, or heuristic fragments of knowledge. Using current techniques the quality of service discovery and binding largely relies on how well humans have compiled service advertisements and interface specifications [5]. In this paper DAML+OIL is used to define the virtual sensor space as a shared vocabulary of terms in a community. Instead of logical reasoning, data mining techniques (that were primarily applied for data classification) are used to sense changes in the virtual space. From the principle of IPDI (Increasing Precision but Decreasing Intelligence) [6] of an intelligent system the sensing resolutions required by different agents are determined by their roles in an organisation. A variable-resolution virtual sensor, based on hierarchical clustering analysis of ontologies, is proposed to cope with the ambiguity and uncertainty of the linguistic announcements from an individual agent's perspective. Therefore an ontology description can provide not only a rule base for logic reasoners but also a numerical scale about textual terms in the semantic web environment. Data mining techniques can be extended for concept-mining purposes and can enhance intelligent capabilities that previously were limited by rule-based reasoning.

1. Semantics and Numeric Representation of a Virtual Sensor Space

In the SoBeNet each actor plays two roles in a system: in a micro-community it could be an initiator who is searching for qualified candidates to complete a desired task; in another community it could be a participant who is advertising its capability for getting a commitment from other actors. This means that a complex organisational network can be established by the expansion of actors' partnerships across micro-communities as shown in Fig.1.

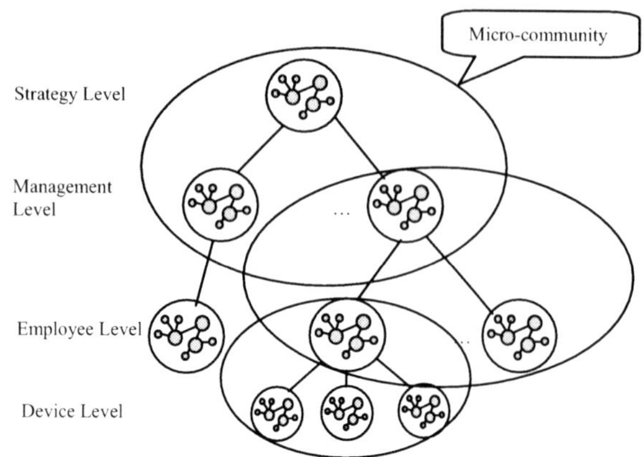

Fig.1 Organisational dimension of a SoBeNet

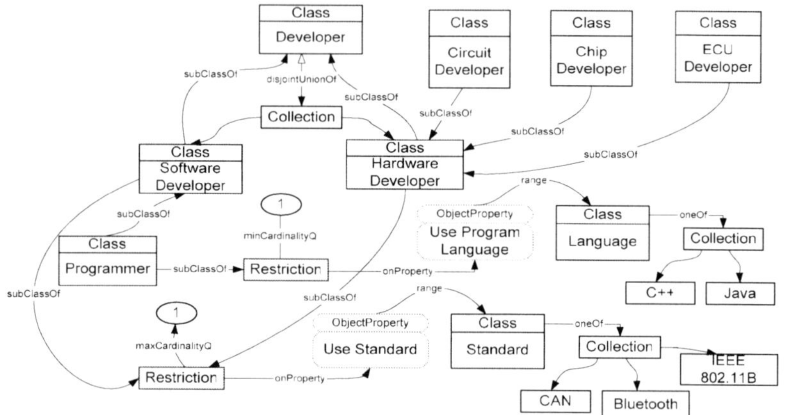

Fig.2 Ontology of a community for electric device development

Suppose an actor a_j, belongs to two communities, as a task initiator in $C_I = \{a_{I1} \quad a_{I2} \quad \dots \quad a_{IL}\}$ and as a task participant in $C_P = \{a_{P1} \quad a_{P2} \quad \dots \quad a_{PM}\}$ respectively. Each actor is capable of decomposing a given task $t_{input}(C_p)$ from C_p into a series of subtasks of $t_{output}(C_I) = [t_{o1} \quad t_{o2} \quad \cdots]$ as a demand for partner seeking in C_I.

Within a community $C(C_I$ or $C_p)$, in order to facilitate cooperation and understanding amongst agents, the semantics of commonly used concepts and terminologies needs to be defined. Suppose that the ontology in a community C can be defined as:

$$\Omega_C \equiv R(e_1, e_2, ..., e_N) \quad (1)$$

where $e_1 ... e_N$ are entities (concepts, terminologies, properties) of a virtual space and $R(\cdot)$ is the relationship among the entities. A simple example of an ontology defined using DAML-OIL for Electric Device Development is shown in Fig. 2, where the entities $e_1 ... e_{14}$ are {*Developer, Software Developer, Hardware Developer, Programmer, Language, Standard, C++, Java, CAN, Bluetooth, IEEE 802.11b, ECU developer, Chip Developer, Circuit Developer*}.

Sharing the same ontology, an initiator a_{Ii} advertises a commitment request of a task and participants a_{Pj}, j=1...M advertise their capability for a contract in a community. The advertisements of both initiator and participant are instances of the local ontology Ω_C, both of which are XML documents forming a virtual sensor space. As an example a hardware developer, as a participant, can announce himself/herself as "A hardware developer using the CAN standard for a given task". As shown in Fig.3 he/she is seeking a suitable job from the community of the Electric Device Development team. In the same way an initiator can announce a task for finding qualified partners e.g. for "a programmer with knowledge and experience of Bluetooth and C++" as shown in Fig.4. Partnership seeking then becomes a process of matchmaking between the desired task and the capability descriptions. A feature vector can be defined as a numerical representation of an advertisement/Web Document:

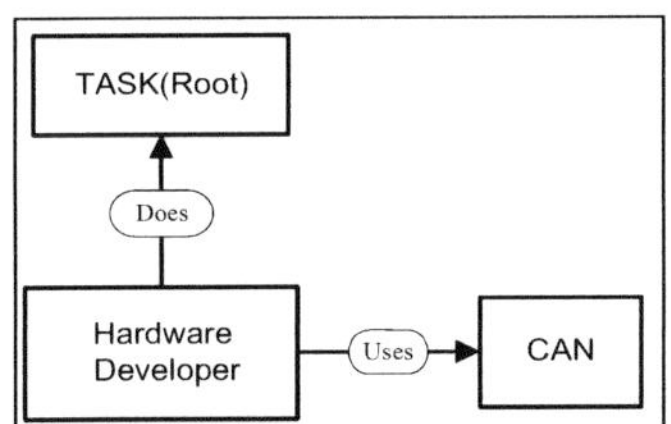

Fig.3 Advertisement of a participant

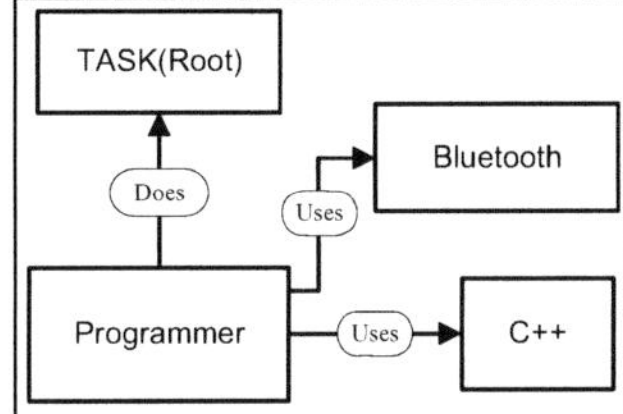

Fig.4 Advertisement of an initiator

$$V(t)=[s_1, s_2, \ldots, s_N]^T \quad (2)$$

The component s_i of $V(t)$ has a one-to-one correspondence to the entity e_i in the ontology definition (1). The $s_i \in [0,1]$ is the semantic closeness, inverse of a semantic distance, between e_i and the root(task):

$$s_i = \begin{cases} e^{-\alpha DIS(e_i, root)} & \text{if } e_i \text{ appeared} \\ 0 & \text{if } e_i \text{ not appeared} \end{cases} \quad (3)$$

where the $Dis(e_i, root)$ is the semantic distance between the entity e_i and the root calculated from an advertisement e.g. Fig.3 for a participant and Fig.4 for an initiator. The α is a steepness measure [7].

In order to deal with the automation of knowledge extraction, semantic distances or similarities between concepts have been researched in recent years. This work includes semantic web matchmaking [5] and conceptual clustering of database schema [8]. Semantic distance can be used to characterise similarities or dissimilarities between two concepts, terminologies or objects. Usually a distance between two objects in a graph is the shortest path between them. The path description greatly depends on the viewpoint of observations. Different types of semantic distances were proposed in [8]: visual distance, hierarchical distance and cohesive distance etc. For example a visual distance is defined from the observation that two objects semantically linked by a relationship are very often graphically close. In fact anyone can define their own semantic distance from a domain perspective.

The semantic representation of an advertisement in the virtual space goes through two steps. Firstly an advertisement in natural language is represented as a semantic graph using the ontology definition. Then, based on the semantic distance, the virtual sensor can sense advertisements in a form of a numerical feature vector. This gives information both about how many entities are related to a task and about the kind of relationships between the entities involved.

2. Concept Clustering with Variable Resolution

In the aforementioned VO an initiator senses the web and looks for the best-matched partner based on the capability announced by participants. Suppose that an actor a_i is an initiator in a community C_I and is seeking for partnerships in a community C_I with a desired capability of:

$$X_d(t_{oj})=[x_{d1}, x_{d2},\ldots,x_{dN}]^T, \quad x_{di}\in[0,1],\ i=1\ldots N, j=1\ldots K \tag{4}$$

The feature vector $X_d(t_{oj})$ specifies what kind of capabilities and contents are required for undertaking a subtask t_{oj}, where x_{di} is between 0 and 1.

At the same time, participants a_j, $j=1\ldots M$, in community C_I advertise their knowledge and capabilities for finding a suitable task from initiators:

$$W(a_j)=[w_1, w_2,\ldots,w_N]^T, \quad w_i\in[0,1],\ i=1\ldots N, j=1\ldots M \tag{5}$$

where a higher weight w_i indicates a specific participant a_j has a stronger ability in the area of concept e_i.

Partnership creation then depends on the similarity between the demand of an initiator and the capability of participants. A straightforward matchmaking method is to compare the demand vector in (4) directly with the capability vectors in (5) to find the best-matched partner. However this full space comparison is problematic. The announcements of tasks by initiators and capabilities by actors are from the perspective of distributed individuals. It is infeasible to restrict both sides to use exactly consistent terminologies. Sometimes, in feature vectors, there are a few dimensions on which numerical distances are far from one another even though the essences of them are very close. The reason is that the entities in the $(e_1, e_2, \ldots, e_N)$ space are highly correlated by their semantic relationships, but the comparison in the full vector space assumes them irrelevant. For example someone who advertised that he/she could do a job of "software development" but who did not explicitly say that he/she could program using "C++", may fail to get a job which required "C++ programming". In fact "Software Developer" and "C++" have a tight semantic relation as shown in the ontology of Fig. 2. A virtual sensor with the aid of the ontology should be able to deal with uncertainties and ambiguity from the initiator's view. According to the principle of IPDI[6], different roles in an organisation may have different precision to concepts. For instance a strategy-maker in Fig.1 requires more intelligence but less precision than an employee. A variable-resolution sensor could sense the difference and discover the meaning behind terminologies from the role's perspective using a proper resolution. In fact a variable-resolution virtual sensor is a method to deal with semantic redundancy within a feature vector based on semantic relations. It is implemented by the hierarchical clustering analysis of ontologies:

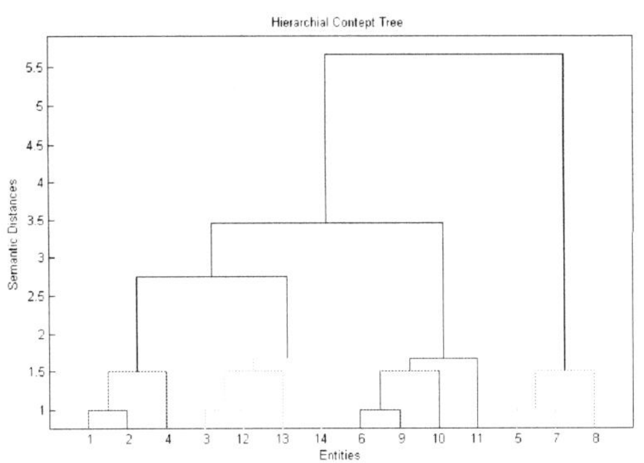

Fig. 5 Hierarchical concept tree.

Step 1: Extract semantic distances between all entities of $\{e_1, e_2, \ldots, e_N\}$ from ontology.

$$D = \begin{bmatrix} 0 & d(1,2) & \cdots & d(1,N) \\ d(2,1) & 0 & \cdots & d(2,N) \\ & & \cdots & \\ d(N,1) & d(N,2) & \cdots & 0 \end{bmatrix} . \qquad (6)$$

where each component of $d(i,j)$ is a semantic distance between the concept e_i and the concept e_j, for example, using visual distance [8] that measures the shortest path between e_i and e_j in a DAML-OIL diagram, e.g. Fig.2.

Step 2: Apply a data clustering method to reveal the hierarchical relationship between all the entities in terms of ontology distances.

Step 3: For the purpose of matchmaking, a resolution is set by a contract initiator in terms of a given task. The feature vectors of demands and capabilities are then compressed to a short feature vector with an appropriate dimension to reflect the desired resolution.

Step 4: Apply matchmaking of the short feature vectors. The participant which has the highest similarity will be selected as the partner for the given task.

Step 2 plays the key role in achieving an appropriate decision. Hierarchical clustering analysis (HCA) [9] is employed in this step. HCA can create a hierarchical tree as illustrated in Fig.5 for the ontology of Fig.2, which represents the hierarchical relationship between all entities. A bottom-up method is employed to generate this tree from the semantic distances between the entities in the ontology. Initially each entity is considered as one independent cluster at the bottom level and at each step the most similar two clusters are merged to form a new cluster at a higher-level. This procedure is repeated upwards. The end result is a binary tree. In terms of decreasing resolution (increasing Semantic Distance) the merged concepts reflect a view of "increasing intelligence".

3. Virtual Sensors in the SoBeNet

A schematic view of the SoBeNet approach using the proposed virtual sensors is shown in Fig.6. After sampling and preprocessing of the web documents a feature vector is generated. The virtual sensor will then transfer the feature vector into a short feature vector

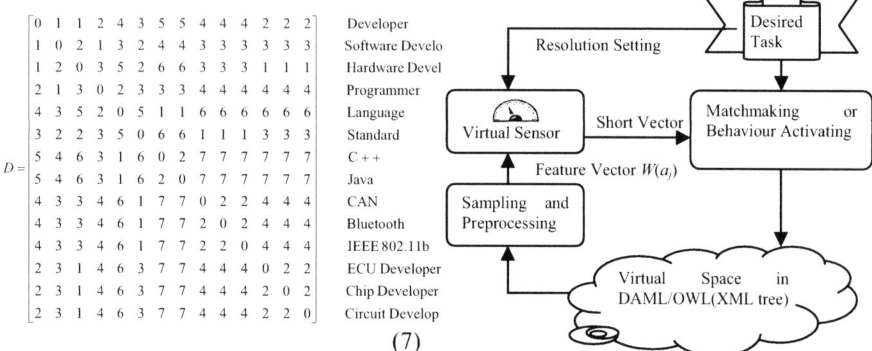

$$D=\begin{bmatrix} 0&1&1&2&4&3&5&5&4&4&4&2&2&2\\ 1&0&2&1&3&2&4&4&3&3&3&3&3&3\\ 1&2&0&3&5&2&6&6&3&3&3&1&1&1\\ 2&1&3&0&2&3&3&3&4&4&4&4&4&4\\ 4&3&5&2&0&5&1&1&6&6&6&6&6&6\\ 3&2&2&3&5&0&6&6&1&1&1&3&3&3\\ 5&4&6&3&1&6&0&2&7&7&7&7&7&7\\ 5&4&6&3&1&6&2&0&7&7&7&7&7&7\\ 4&3&3&4&6&1&7&7&0&2&2&4&4&4\\ 4&3&3&4&6&1&7&7&2&0&2&4&4&4\\ 4&3&3&4&6&1&7&7&2&2&0&4&4&4\\ 2&3&1&4&6&3&7&7&4&4&4&0&2&2\\ 2&3&1&4&6&3&7&7&4&4&4&2&0&2\\ 2&3&1&4&6&3&7&7&4&4&4&2&2&0 \end{bmatrix} \quad (7)$$

Fig. 6 SoBeNet with a variable-resolution sensor

based on the resolution setting of the user. This is achieved as follows:

Firstly the semantic distance matrix can be extracted from Fig.2 according to the definition of the visual distance [8], as shown in Eq.7.

Based on this 14×14 matrix a hierarchical tree is then obtained to reveal the relationship among all the entities as shown in Fig.5. For a given resolution a series of clustered concepts on an appropriate level can be merged to form a short feature vector. For example a very high resolution less than distance 1 requires a full space comparison, i.e. N=14. Reducing resolution means allowing fuzzy matchmaking and using higher level concepts. For a resolution of distance 1, i.e. level 1 in Fig.5, entity pairs e_1&e_2, and e_3&e_{12}, e_6&e_9 and e_5&e_7 have similar meanings and are merged to new concepts. As result a total of 10 concepts are extracted, i.e. N=10; these include the following entities respectively: $V=\{(e_1\&e_2),e_4,(e_3\&e_{12}),e_{13},e_{14},(e_6\&e_9),e_{10},e_{11},(e_5\&e_7),e_8\}$.

If an initiator decreases resolution further, e.g. with distances less than 2 (level 3 in Fig.5), a different set of concepts can be extracted with N=4 and the new short vector is:

$$V=\{((e_1\&e_2)\&e_4),(((e_3\&e_{12})\&e_{13})\&e_{14}),(((e_6\&e_9)\&e_{10})\&e_{11}), ((e_5\&e_7)\&e_8)\} \quad (8)$$

From Eq. (8), we can understand the meanings of these merged concepts. $((e_1\&e_2)\&e_4)$ and $(((e_3\&e_{12})\&e_{13})\&e_{14})$ represent Software Developer Related and Hardware Developer Related respectively while $(((e_6\&e_9)\&e_{10})\&e_{11})$ and $((e_5\&e_7)\&e_8)$ represent Standard Related and Computer Language Related respectively.

The elements of the short feature vector are calculated by a weighted accumulation of the low level entities to reveal relevance to a higher-level concept

$$e_{l,i} = \omega \cdot e_{l-1,i1} + (1-\omega) \cdot e_{l-1,i2}. \quad (9)$$

where $e_{l,i}$ is the revealed concept on the l-th level; $e_{l-1,i1}$ and $e_{l-1,i2}$ denote its two children; ω is a given weight. In this example, ω=0.5.

Suppose that a contract initiator from the automobile industry is seeking partners in a community for Telematic Control Unit (TCU) development. The initiator has decomposed the task into 6 subtasks:

Task	C++ software work for Bluetooth application	TCU programming for integration of Bluetooth, CAN, IEEE.802.11b	Bluetooth communication chip design	Bluetooth board design based on the designed chip	TCU board design for connecting CAN, Bluetooth, and IEEE 802.11b	Integration of hardware work with software work
Feature Vector $T(i)$, $i=1...6$, $\alpha=7/max(Dis)$	[0, 0, 0, $e^{-\alpha}$, 0, 0, $e^{-2\alpha}$, 0, 0, $e^{-2\alpha}$, 0, 0, 0, 0]	[0, 0, 0, $e^{-\alpha}$, 0, $e^{-2\alpha}$, 0, 0, $e^{-3\alpha}$, $e^{-3\alpha}$, $e^{-3\alpha}$, 0, 0, 0]	[0, 0, 0, 0, 0, 0, 0, 0, 0, $e^{-2\alpha}$, 0, 0, $e^{-\alpha}$, 0]	[0, 0, 0, 0, 0, 0, 0, 0, 0, $e^{-2\alpha}$, 0, 0, $e^{-\alpha}$, $e^{-\alpha}$]	[0, 0, 0, 0, 0, 0, 0, 0, $e^{-2\alpha}$, $e^{-2\alpha}$, $e^{-2\alpha}$, 0, $e^{-\alpha}$, $e^{-\alpha}$]	[$e^{-\alpha}$, $e^{-2\alpha}$, $e^{-2\alpha}$, 0, $e^{-3\alpha}$, $e^{-3\alpha}$, 0, 0, $e^{-4\alpha}$, $e^{-4\alpha}$, $e^{-4\alpha}$, 0, 0, 0]

There are three participants in the community:

Participant Capabilities	Actor1: hardware development work	Actor2: software development	Actor3: software and hardware integration
Feature Vector $A(i)$, $i=1...3, \alpha=7/max(Dis)$	[0,0,0, 0,0,0,0,0,0, 0,0, $e^{-\alpha}$, $e^{-\alpha}$, $e^{-\alpha}$]	[0, $e^{-\alpha}$, 0,0,0,0, $e^{-2\alpha}$, $e^{-2\alpha}$, 0, 0,0,0,0,0]	[$e^{-\alpha}$, 0, 0, 0,0, $e^{-2\alpha}$, 0,0, 0,0,0, $e^{-2\alpha}$, $e^{-2\alpha}$, $e^{-2\alpha}$]

Considering a level-3 matchmaking in Eq.(8), The 14 dimensional task vectors can be compressed to 4 dimensional short vectors using Eq. (9):

$T(1)$ = [0.1839, 0, 0.0338, 0.0338]; $T(2)$ = [0.1839, 0, 0.0605, 0];
$T(3)$= [0, 0.0920, 0.0338, 0]; $T(4)$= [0, 0.2759, 0.0338, 0];
$T(5)$= [0, 0.2759, 0.1184, 0]; $T(6)$= [0.1258, 0.0169, 0.0222, 0.0124].

Similarly, the short vectors of the capabilities can be obtained for the 3 participants:
$A(1)$= [0, 0.3219, 0, 0];$A(2)$= [0.0920, 0, 0, 0.1015];$A(3)$= [0.0920, 0.1184, 0.0169, 0].

The extracted short vectors can be easily verified by their implied meaning, e.g. A(1) means an actor with hardware development ability.

In order to sense the best-matched partners for all tasks, the correlation matrix can be calculated in terms of actors and tasks:

$$C = A * T' = \begin{bmatrix} 0 & 0 & 0.0296 & 0.0888 & 0.0888 & 0.0054 \\ 0.0204 & 0.0169 & 0 & 0 & 0 & 0.0128 \\ 0.0175 & 0.0179 & 0.0115 & 0.0332 & 0.0347 & 0.0139 \end{bmatrix}$$

According to this the results of allocation of the tasks can be determined below: T1->Actor2; T2->Actor3; T3->Actor1;T4->Actor1; T5->Actor1; T6->Actor3.

4. Conclusions

This paper presented a virtual sensor for distributed alliances in a virtual organisation on the semantic web. An agent could be aware of the capability of other agents by using semantic web technology e.g. through his/her advertisement in DAML-OIL which constitutes the virtual space. The virtual sensor with variable resolution is constructed by ontology definition and concept-mining. It is able to deal with ambiguities of announcements that are due to the diversity of views of distributed and autonomic individuals. The case study has demonstrated that the proposed virtual sensor is capable of avoiding inconsistencies of announcements and the high-dimensional difficulties of correlation analysis. The proposed method can also be used in other web-based applications, e.g., job hunting, investment opportunity seeking, and student placement, etc.

References

[1] L. M. Camarinha-Matos, H. Afsarmanesh, Elements of a base VE infrastructure, *Computers in Industry* **51**(2003), 139–163.

[2] T. J. Norman, A. Preece, *et al.*, Agent-based formation of virtual organisations, *Knowledge-Based Systems*, **17**(2004), 103–111.

[3] C. V. Goldman and J. S. Rosenschein, Evolutionary patterns of agent organizations, *IEEE Trans. Systems, Man, And Cybernetics—Part A*, **32**(2002), 135-148.

[4] R. Subbu, A. C. Sanderson, Network-based distributed planning using coevolutionary agents: architecture and evaluation, *IEEE Trans. Systems, Man, And Cybernetics—Part A*, **34**(2004), 257-269.

[5] K. Sycara, M. Klusch, S. Widoff, and J. Lu, Dynamic service matchmaking among agents in open information environments, *ACM SIGMOD Record*, **28**(1999), 47—53.

[6] G.N. Saridis, Analytical formulation of the principle of increasing precision with decreasing intelligence for intelligent machines, *Automatica*, **25**(1989), 461-467.

[7] J. Williams, N. Steele, Difference, distance and similarity as a basis for fuzzy decision support based on prototypical decision classes, *Fuzzy sets and systems*, **131**(2002), 35-46.

[8] J. Akoka, I. C. Wattiau, Entity-relationship and object-oriented model automatic clustering, *Data & Knowledge Engineering*, **20**(1996), 87-117.

[9] L. Kaufaman and P.J. Rousseeuw, Finding groups in data: an introduction to cluster analysis. New York: John Wiley & Sons, 1990.

Part II

Self-Organization/Adaptation of Multi-Agent Systems

Developing Self-Organized Architecture Solution according to Model Driven Generative Domain Engineering

LIANG Hai-Hua[a, 1] and ZHU Miao-Liang[a]

[a] *Department of Computer Science and Technology, Zhejiang University Hangzhou 310027, China*

Tel: +86-13606618140, E-mail: lhhwanmit@hotmail.com, http://cs.zju.edu.cn

Abstract: The development of multi-agent based solution for outdoor mobile robot navigation is a complex multi-level process. Model Driven Generative Domain Engineering is one domain engineering method aim to developing optimized, reusable architectures, components and aspects for application engineering. According to MDGDE, we designed a set of event-driven agents, which enable the robot to initiate action adaptive to the dynamical changes in the environment. This paper describes our approach as well as its motivations and our practice.

Keywords: Generative programming, Agent, MDA

1. Introduction

Designing autonomous mobile robots is a long term project in our lab. The purpose is to design a machine which is able to do environmental navigation on its own without a human directly manipulating. The robot must be able to perceive its surroundings through different kinds of sensors and initiate appropriate actions in that environment through actuators to achieve its designed goals.

The robot architecture is a key issue in the design of a mobile robot. Architecture defines the principles involving organizing hardware and software function modules, integration methods and supporting tools [1]. In this paper, we describe a domain engineering method called Model Driven Generative Domain Engineering (MDGDE) used to design and implement domain specific optimized architecture and reusable assets for mobile robots.

This paper is organized as follows. Section 1 argues that agent make great contribution to model complex software system and vice versa. Section 2 presents the MDGDE method. Section 3 introduces the application of MDGDE to analyze the agent domain and design Feature model. Architecture design is also illustrated in section 3 in detail. Section 4 discusses related works. Finally, section 5 presents the experimental results and section 6 concludes this paper.

2. Agent Means Much to Software Engineering and Vice Versa

MAS (multi agent system) provide a new viewpoint for Software Engineering. Adopting an agent-oriented approach to software engineering means decomposing the problem into multiple, interacting, autonomous components (agents) that have particular objectives to achieve.

The key abstraction agents, interactions and organizations. M. Wooldridge defined agent follows [2]: an agent is an encapsulated computer system that is situated in some environment, and that is capable of flexible, autonomous action in that environment in order to meet its design objectives.

We can divide system into object, feature and agents. Objects encapsulate state, action. But they are passive because they can determine themselves when and how they are called. Features are powerful for us to analyze system. But they do not describe the process. Agents have their own plan, objective, action. Agent is one choice to model some kind of complex system. When we combine these three means, we make magic works.

3. Model Driven Generative Domain Engineering Method

Model driven Generative Domain Engineering (MDGDE) is improved on the base of Generative programming (GP) [3] for the purpose of benefiting from the new emerging software engineering methods, such as Model Driven Architecture (MDA) [4] and Aspect-oriented software development (AOSD) [5]; besides, MDGDE propose the guidelines of combining Object-Oriented (OO) methods into domain engineering. The foundation of Generative programming is Domain Engineering (DE). MDA is a framework for software development defined by the Object Management Group (OMG), Key to MDA is the importance of models in the software development process [4] [6].

Software is the knowledge storage medium, while the software development process can be viewed as a knowledge acquisition activity [7]. MDGDE is divided into 3 phases: domain analysis, domain design and domain implementation, as in Figure 1.

The purpose of domain analysis in MDGDE is : Choosing and scoping the domain under studying; collecting domain information, then unified them into a consistent domain model.Domain scoping defines the interested domain, stakeholder and the target of stakeholders. Domain scope can be defined using exited applications or counterexamples in the domain. Building domain metamodel involves seven steps:identify domain's stakeholders, and their viewpoints and requirements; make the list of features and high level requirements; building the feature model; model some special subsystems as agents, define their objective, plan, and action (for our case); identify the system objective and main business process and activity; identify the context of system, divide the system into subsystem and module; identify the shared features between modules; the shared features would be modeled as aspects; make object-oriented analysis about the results; identify use-case and calss; design DSL; describe metamodel using DSL.

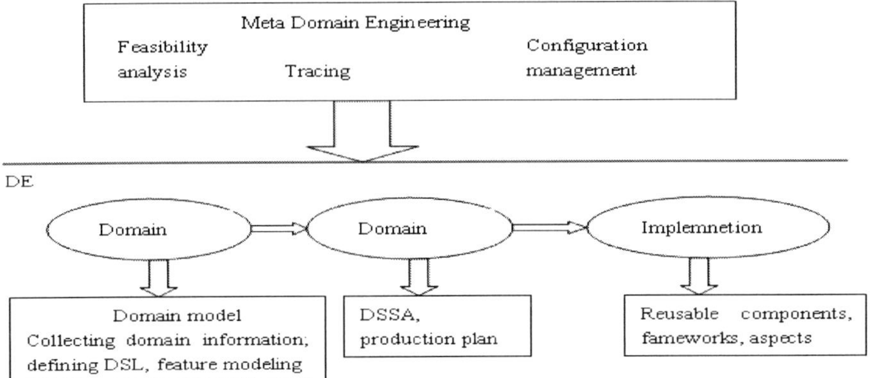

Figure 1.MDGDE development procedure

The purpose of domain design in MDGDE is to develop domain specific architectures for the applications in the domain, including the guideline for how to apply the architectures; and map from PIM to PSM, design components to implement common feature and variable feature.

The purpose of designing DSSA is to design one or one set of architecture satisfying the requirement of all stakeholders [8], involving divided system into components and interaction between components [9]. Also the rules about how to apply or select architecture are defined.

Designing DSSA involves 3 steps: Identify modules; Define the relations between modules; Choose architecture style. The definition of architecture style includes: A set of module type; The relations between the modules, represented as topological layout; A set of semantics constraint; A set of interaction mechanism (such as subroutine calling, blackboard) which determine how the modules to go with each other.

Architecture style enables the perfect idea about architecture design to be preserved and communicated. We can find many architecture style definition and the rules about how to choose an appropriate style in [10].

Design of common features includes design of universal component, class and aspect. Because common features may evolve too, emphases of common features design are hierarchy and interfaces of components which should satisfy the Demeter and open-closed principle. We found mixin [11] very useful in our practice of design hierarchy of classes. In our case, the agent communication classes is designed utilizing mixin. The benefit is that we need not have to define one communication class for every agent type and reduce the cost of dynamic binding.

The PIM (platform independent model) in MDGDE comprises architecture and platform independent part of component design, as shown in Figure 2. Mapping from PIM to PSM can be accomplished manually, or through some tools and some rules. Eclipse+BasicMTL of INRIA are such one choice.

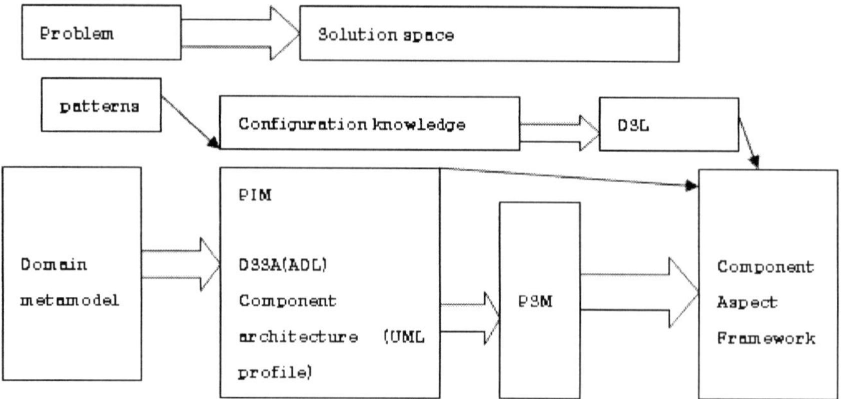

Figure 2. The solutions space and problem space in MDGDE

The main jobs of Domain implementation are design the detail of the components, frameworks and aspects, implement the architectures, component and aspect as what the production plan prescribed and develop configuration DSL. The process of Domain Engineering is a domain too (called meta domain). (shown in Figure 1).

4. Analyzing the Mobile Robot Domain and Designing Reusable Assets

Each agent comprises 3 parts: kernel, IO ports and switch. The kernel contain specific function library or knowledge base, as shown in Figure 3. IO ports enable agent to communicate with other agents. Switch enable agents make decision based on system state, choose function and IO port, evaluate the input data, and output the result as event to Bulletinboard, as shown in Figure 3.

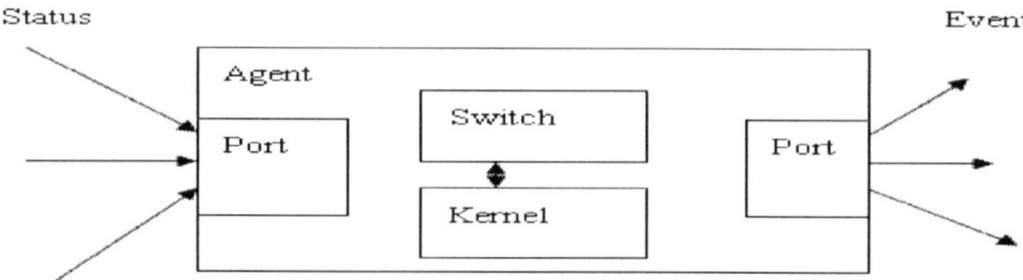

Figure 3. A simple agent

According their responsibility, agents are classified as follows:

Sensory agents get the environment information as input, and outputs description of the environment to Fusion, including: RFCCD (Road Following CCD), STCCD (Stereo CCD) and Drive; Processing agents process the information from other agents: Fusion (integrates information from the sensory agents), GPL (Global Path Planning), LPL (Local Path Planning), Bulletinboard (will discuss in detail later), Localization etc.

Bulletinboard agent is the information center, the primary function is: (1) Receives the event from agents and evaluates; (2) Deduces the system status; (3) Broadcasts the status to agents through ROBIX in order to control and organize the agents [12]. Thus, the robot architecture is dynamically reorganized to adapt the dynamic environments. As a high level monitor, Bulletinboard collects event and evaluates the current system status. The current system status is the key for activity and cooperation between agents. So the evolutions must be accurate and fast enough. The internal structure of the Bulletinboard agent composed of a real-time reasoning system based on DES (Discrete Event System) model and events Monitor, as shown in Figure 4.

4.1 Feature Modeling and Architecture Design

We identified three kinds of features in MDGDE: Aspect feature: this kind of feature capture crosscutting concerns, we implement aspect features using AspectC++ [13] [14]. Log is an example of aspect feature; Concrete feature: this kind of feature will mapped to an implement component. For example, algorithm feature of Fusion agent will mapped to some kind of concrete function library; Abstract feature: this kind of feature has no direct implementation. They will be implemented by combination of components and aspects. Reliability is an abstract feature.

For the mobile robot, we define a layered architecture [9]. The lower layer provides service to the upper ones. There exist three main classes of control architectures developed for mobile robots. One is based on functional decomposition; the other is based on behavioral decomposition [15] [16] [17]; and the third is the hybrid architecture that combines the former two [18] [19] [20].

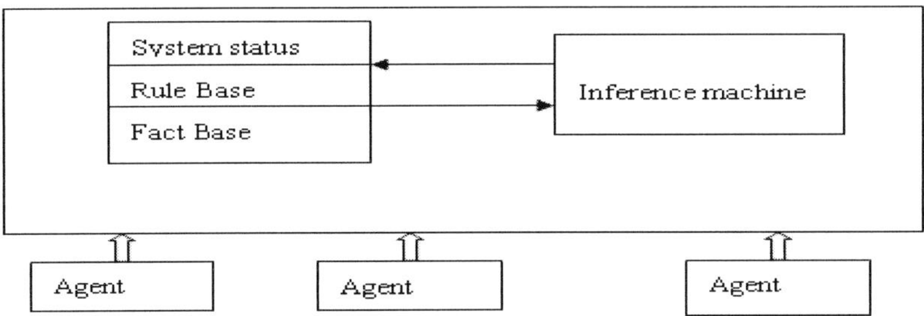

Figure 4. Bulletinboard agent

A robot reorganizes different sets of agents to achieve its goal. For example, When the state of robot is RS{road straight}, robot will organize 5 agent to achieve the task: Bulletinboard agent get the system state is "road straight"; RFCCD, LADAR, STCCD, LPL, Drive and Location agents will be organized to accomplish the move of robot. Whenever LADAR agent reports there are suspected obstacles again to Bulletinboard; Bulletinboard agent deduces the new system status RO {road driving with obstacle}, then broadcasts this new status; some agents get awaked, such as Fusion, it requests outputs from RFCCD and LADAR to make a unique description about the environments and sends processing results to LPL; some agent will change their interior function, such as LPL will change to compute the path of sheering obstacle;

thus relatively safe driving is achieved. These two processes are depicted in figure 5.a and figure5.b.

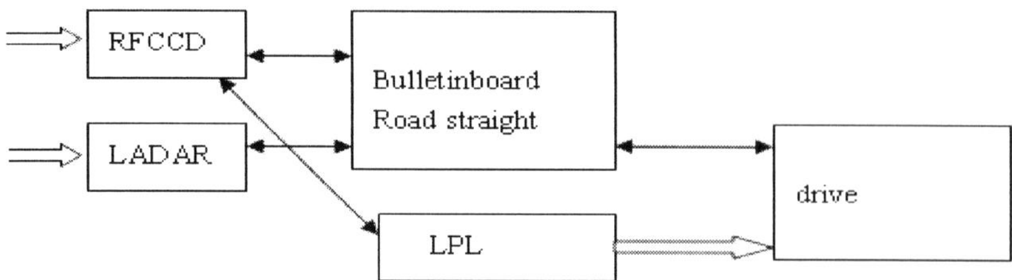

Figure 5.a.Architecure when status is RS

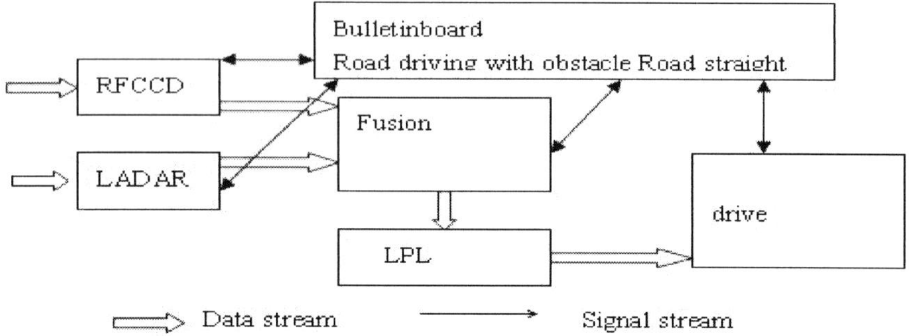

Figure 5.b.Architecture when status is RO

There exist two different information/control streams in the system, one is data stream, and another is signal steam. Through separation of data stream and signal stream, adaptive capacity of the robot is improved notably.

5. Related Works

Kinny et al. developed the first methodologies for the development of BDI agents based on OO technologies [21]. The agent methodology distinguishes between the external viewpoint and the internal viewpoint. External view is characterized by two models: Agent Model and Interaction model. Internally view comprise of Beliefs, Desires and Intentions, which represent, respectively, their informational, motivational and deliberative states. MESSAGE (Methodology for Engineering Systems of Software Agents) [22] is a methodology which builds upon best practice methods in current software engineering such as for instance UML for the analysis and design of agent-based systems. MESSAGE presents five analysis models: Organization Model, Goal/Task Model, Agent/Role Model, The Domain (Information) Model, The

Interaction Model. Tropos [23] is another good example of a agent-oriented software development methodology that is based on object-oriented techniques. In particular, Tropos relies on UML and offers processes for the application of UML mainly for the development of BDI agents and the agent platform JACK. The Tropos specification makes use of the following types of models: Actor and Dependency Model, Goal and Plan models, Capability diagram, Agent interaction diagrams:

6. Experimental Results

We have test the architecture solution in this paper through simulation and make outdoor field experiments for both on road and off road navigation under different weather conditions. Experimental results showed that the self- organized architecture solution enable the mobile robot to move on structured roads with clear marks, semi-structured roads paved with gravels and moderate undulation cross-country roads with sparse vegetation. The system architecture is dynamically reorganized according to the changes in the environment. Through MDGDE methods, we developed reusable assets that accelerated our development process and improved the reliability of the system. This is a notable gain from this project.

7. Conclusion

MDGDE is based on GP, UML profile and AOSD, combined with design patterns and OO methods, provide good support to software reuse. Domain engineering provided reusable DSSA, component, framework and aspects, also including other products, such as document, design principle. That is a kind of knowledge reuse. And in this paper, we developed a self-organized multi-agent based architecture solution for outdoor mobile robot and gained satisfied results. We hope that we can make the MDGDE more systematic, and practice this method to make more intelligent robots.

References

[1] Alami, R., Herrb, M., Morisset, B., Chatila, R., Ingrand, F., Moutarlier, P., Fleury, S., Khatib, M., Simeon, T., 2000. Around the lab in 40 days [indoor robot navigation]. Proceedings of IEEE International Conference on Robotics and Automation, San Francisco, CA, 3:88-94.

[2] M. Wooldridge (1997) "Agent-based software engineering" IEE Proc. on Software Engineering, 144 (1) 26-37.

[3] Krzysztof Czarnecki, Ulrich W.Eisenecker.Generative Programming: Methods, Tools, and Application:Addison-Wesley, 2000

[4] Anneke Kleppe, Jos Warmer, Wim Bast.MDA Explained: The Model Driven Architecture: Practice and Promise: Addison Wesley, 2003

[5] Robert E. Filman, Tzilla Elrad, Siobhán Clarke, Mehmet Aksit, Aspect-Oriented Software Development, Addison Wesley Professional, 2004

[6] Jim Arlow, Ila Neustadt. Enterprise Patterns and MDA: Building Better Software with Archetype Patterns and UML: Addison-Wesley, 2003

[7] Philip G.The Laws of Software Process: A New Model for the Production and Management of Software, Auerbach Publications, 2004

[8] Eric Evans.Domain-Driven Design: Tackling Complexity in the Heart of Software: Addison-Wesley, 2003

[9] Daniel J. Duffy, Datasim Education BV, Amsterdam, Netherlands, Domain Architectures:Models and Architectures for UML Applications, John wiley & sons 2004

[10] Len Bass, Paul Clements, Rick Kazman.Software Architecture in Practice, Second Edition:Addison-Wesley, 2000

[11] Yannis Smaragdakis, Don Batory, Mixin layers: an object-oriented implementation technique for refinements and collaboration-based designs, ACM Transactions on Software Engineering and Methodology (TOSEM), Volume 11 Issue 2

[12] Zhu, M., Zhang, X., Wang, X., Tang, W., 2000. Computer integration system of autonomous intelligent robot self-organization structure IRASO. Pattern Recognition and Artificial Intelligence, 13(1):36–41 (in Chinese).

[13] Daniel Lohmann, Georg Blaschke, and Olaf Spinczyk, "Generic Advice: On the Combination of AOP with Generative Programming in AspectC++", In Proceedings of GPCE'04, October 24th-28th, 2004, Vancouver, Canada.

[14] Andreas Gal, Wolfgang Schröder-Preikschat, Olaf Spinczyk, "AspectC++: Language Proposal and Prototype Implementation", Accepted at the OOPSLA 2001 Workshop on Advanced Separation of Concerns in Object-Oriented Systems, Tampa, Florida, October 14th, 2001

[15] Rosenblatt, J. K., Payton, D. W., 1989. Fine-grained alternative to the subsumption architecture for mobile robot control. Proceedings of IJCNN International Joint Conference on Neural Networks, Washington, DC, p.317–323.

[16] Sowmya, A., 1992. Real-time reactive model for mobile robot architecture. Proceedings of SPIE Conference on Applications of Artificial Intelligence X: Machine Vision and Robotics, Orlando, FL, 1708:713–721.

[17] Watanabe, M., Onoguchi, K., Kweon, I., Kuno, Y., 1992. Architecture of behavior-based mobile robot in dynamic environment. Proceedings of IEEE International Conference on Robotics and Automation, Nice, Fr, 3:2711–2718.

[18] Pons, N., Delaplace, S., Rabit, J., 1993. Mobile robot architecture dedicated to asynchronous events management. Proceedings of the 8th International Conference on Applications of Artificial Intelligence in Engineering, Toulouse, Fr, 2:547–560.

[19] Ollero, A., Mandow, A., Munoz, V., Gomez De Gabriel, J., 1994. Control architecture for mobile robot operation and navigation. Robotics and Computer-Integrated Manufacturing, 11(4):259–269.

[20] Low, K.H., Leow, W.K., Ang Jr., M.H., 2002. A hybrid mobile robot architecture with integrated planning and control. Proceedings of the 1st International Joint Conference on Autonomous Agents and Multiagent Systems, Bologna, Italy, p219–226.

[21] Kinny, D., Georgeff, M., Rao, A.: A Methodology and Modeling Technique for Systems of BDI Agents, in Proceedings of the Seventh European Workshop on Modelling Autonomous Agents in a Multi-Agent World (MAAMAW 96), LNAI 1038, Springer, 1996

[22] Giovanni Caire, Wim Coulier, Francisco Garijo, Jorge Gomez, Juan Pavon, Philippe Massonet, Francisco Leal, Paulo Chainho, Paul Kearney, Jamie Stark, Richard Evans, Agent Oriented Analysis using MESSAGE/UML, Proceedings AOSE 2001, Springer 2001.

[23] Paolo Bresciani, Paolo Giorgini, Fausto Giunchiglia, John Mylopoulos, and Anna Perini.TROPOS: An Agent-Oriented Software Development Methodology. Journal of Autonomous Agents and Multi-Agent Systems. 2003. Kluwer Academic Publishers, 2004.

Self-Organization and Autonomic Informatics (I)
H. Czap et al. (Eds.)
IOS Press, 2005

A Realistic Simulation Testbed for Studying Game Playing in Robotic Soccer

Mahdi Asadpour [1], Mansour Jamzad
Sharif University of Technology
Department of Computer Engineering
Tehran, Iran

Abstract. The main goal of any testbed is to facilitate the trial and evaluation of ideas that have promise in the real world. In fact, in a real platform several physical or hardware restrictions exist and maybe arise for an agent or robot. These restrictions affect the performance of implemented algorithms. To the best of our knowledge, at present, only RoboCup Soccer Server is a common testbed for simulating a soccer game, but it can not support the above mentioned physical limits on real robot. In order to overcome these problems, we designed and implemented a realistic simulation testbed called SharifCE. In addition, this testbed allows the user to define a probability for a fault occurred on any internal part of the robot. That is done by a fault injection procedure. Our experimental results convinced us that SharifCE Testbed is quite appropriate for Multiagent Systems (MAS); because not only it is similar to a real platform, but also it supports the necessities such as Movement, Communication, Supervision, Cooperation and Learning. As an experimental result, a practical implementation of this testbed is presented.

Keywords. SharifCE Testbed, Realistic Simulation, Multiagent Systems, Soccer Simulation, Fault Injection

1. Introduction

It is known that the main goal of any testbed is to facilitate the trial and evaluation of ideas that have promise in the real world [1]. Although many robotics issues can only be studied in the real world situation, but there are many issues that can be studied in simulation. The main work that has been done in this field is RoboCup Soccer Server which is one of RoboCup competitions leagues. Although this soccer server has many advantages that has made it very popular, but it is very abstract and can not support real physical robots with all restrictions and constrains imposed on them.

SharifCE Testbed is a simulation robotic soccer designed to simulate real autonomous robots. The researchers can study game playing and other related algorithms (e.g. multiagent systems) by implementing their algorithms on this testbed.

SharifCE Testbed is constructed under Webots software (http://www.cyberbotics.com) that provides us with a rapid prototyping environment for modeling, programming and simulating mobile robots [3].

[1] Correspondence to: Mahdi Asadpour, AI and Robotics Lab, Computer Engineering Dept., Sharif University of Tech., Azadi ave., Tehran, Iran. Tel.: +98 21 6616 4114; E-mail: asadpur@ce.sharif.edu

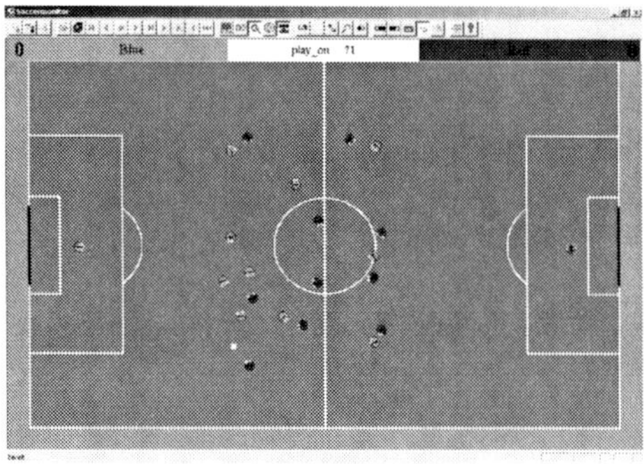

Figure 1. A typical view of SServer.

We highly recommend the reader to make a look at the movies and programs source code of this simulated testbed in the web page of this paper[1].

The rest of the paper is organized as follows: In section two, we overview the related work, simulation robotic soccer. In section three, we present the structure and features of SharifCE Testbed. Then, in the remaining sections we discuss about the suitability of this testbed for MAS algorithms and describe the implementation of a practical work. Finally we will conclude our discussion in the last section.

2. Overview

2.1. Need for Simulation

Although the final aim is real robotics and one may need to study a problem in its ideal situation by controlling each part of the system, but because of non-avoidable facts such as friction and atmosphere, . . . such demands become intangible in real world situation.

There are many situations that we prefer to perform simulations prior to investigations with real robots. This is because simulations are *easier to setup*, *less expensive*, *faster* and *more convenient to use*, thus allowing a better design exploration. In addition, simulation often *runs faster* than real robots while all required parameters can easily be displayed on a screen [3].

2.2. RoboCup Soccer Server

A particularly good simulator for this purpose is the "soccerserver" developed by Noda [5] that is pictured in Figure 1. This simulator is realistic in many ways: the players' vision is limited; the players can communicate by posting to a blackboard that is visible to all players; all players are controlled by separate processes; each player has 10 teammates and 11 opponents; each player has limited stamina; actions and sensors are noisy; and playing occurs in real time. The simulator provides a domain and supports users who wish to build their own agents. Furthermore, teams of agents can be evaluated by playing against each other, or perhaps against standard teams [1].

[1] http://ce.sharif.edu/~asadpur/sharifce.html

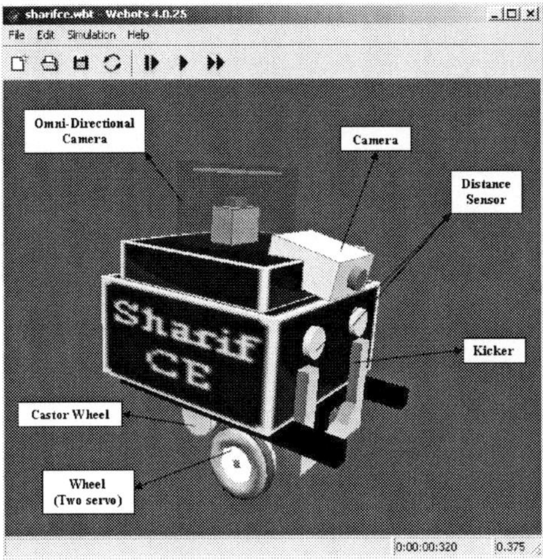

Figure 2. SharifCE robot model with details in Webots window.

A Client in Soccer Server (SServer) can extract teammates and opponents state from information transferred by a **Server** and other **Clients** through standard commands which are documented in the Server Manual [6].

3. SharifCE Testbed

We list the structure and features of SharifCE Testbed in the following, but the detail materials about how to model and simulate this testbed with Webots is described in [9] and [11].

3.1. SharifCE Soccer Robot

A robot for this testbed should be generic, that is it should have the necessary requirements of a robotic soccer as:

- Movement toward rear and front or left and right,
- Strike the ball for shooting or passing,
- Recognize the position of each object of SharifCE Testbed by the robot itself or to be informed by another agent.

The robot that is modeled here is based on SharifCE robot[2] which won the first place in Middle Size RoboCup competitions, Stockholm, 1998 [7]. However for generalization, we added several additional capabilities to that robot. Figure 2 shows a visualization of this simulated robot. The equipments used in this robot are summarized in the following.

1. Two wheels with two Servo Motors for each wheel. One for moving toward rear and front and another for a steering mechanism to rotate toward left and right.

[2]The new version of ARVAND robot

2. A Kicker, for stroking (shooting, passing) the ball: it can shoot the ball with various strokes by setting the force and velocity of kicking.
3. A Castor Wheel that is positioned in the rear of the button of robot to equilibrate or balance it.
4. A Normal Camera and an Omni-directional Camera that captures 360 degrees field of view of environment. In Webots you can define a camera node and set its fields such as Field Of View (FOV) and as a result, Webots generates images of scene that is in FOV of robot. For generating 360 degree images, we use three separate cameras in one point; each camera has 120 degree (about 2.094 radian) FOV and has 120 degree gap with both two immediate cameras. Therefore, if we combine these three images, we can make a 360 degree view.
5. The robot can rotate around the ball by calling a defined function [3], this function computes the velocity and the angle of each wheel in regards to the position of ball. In result, the robot will rotate around the ball while the ball still is in control of robot [7].

6. Emitter and Receiver for each robot. Each robot can broadcast its messages by emitter node and another robot can receive these messages by its own receiver node and vice versa.
7. We defined and programmed a Supervisor object. This supervisor gathers some series of information from the position of each robot and the ball. This information is saved in a buffer. Each robot can receive this buffer from the supervisor.
8. Two Distance Sensors in front of robot.

3.2. Soccer Field

We modeled a soccer field similar to Middle Size RoboCup (Figure 3). This soccer field has two goals, blue and yellow, white game lines, and also one orange soccer ball. Whereas the out lines are presented in new soccer RoboCup rules, but because of difficulties that arise in contrast, we defined an obstacle above the out lines to avoid the ball from going out of playing field. However, we can simply delete these obstacles: the supervisor can be programmed to place the ball on out line after it goes out.

3.3. Fault Injection

The destructive nature of a crash and long error latency make it difficult to identify the causes of failures in the operational environment. It is particularly hard to recreate a failure scenario for a large, complex system. Engineers most often use low-cost, simulation-based fault injection to evaluate the dependability of a system that is in the conceptual and design phases. It assumes that errors or failures occur according to predetermined distribution. We inject faults into the system in order to

- Identify dependability bottlenecks,
- Study system behavior in the presence of faults,
- Determine the coverage of error detection and recovery mechanisms, and

[3] with name arvand()

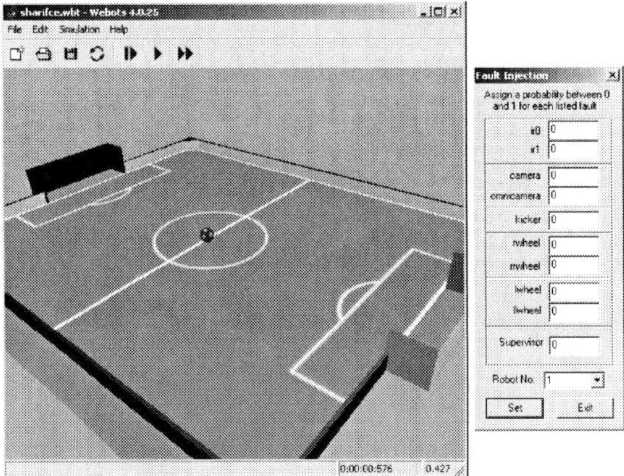

Figure 3. Soccer field: blue and yellow goals with a ball. Fault Injection dialog box is included.

- Evaluate the effectiveness of fault tolerance mechanisms and performance loss [4].

In SharifCE Testbed, the users can inject some faults or noises which possibly might occur in robot in real environment. This capability helps researchers to study or test their embedded system precisely and more realistically. At this stage for the simplicity, we assume that the fault might occur only in the robot but not in other component of testbed. Faults in SharifCE Testbed are allowed to occur in the following components of the robot: motors power supply, servo motors, kicker, cameras, distance sensors, emitter and receiver.

At the first time when we run SharifCE Testbed simulation, it asks the user via a dialog box (Figure 3) to assign a probability (between 0 and 1) for all faults listed there. The value 0 means a fault never occurs, and the value 1 means total failure for that device in the robot. Any value in (0,1) shows the demolition chance of that device. In practice, we call a random number generator function to produce the probability of *Success* and *Failure* for each device.

3.4. Game

To set up a game, we use three copies of SharifCE robot and one copy with some modifications as a goalie. Two wheels similar to the front wheels are placed in the rear of goalie of blue team. This mechanism gives goalie the potency of rotating around its geometrical center and moving across goal line [8]. The goalie is shown in Figure 4. For opponent team simple soccer robots are used (yellow color).

In addition as mentioned in earlier sections, the robots can communicate with each other by using the supervisor buffer information or by using its own receiver to receive the emitted messages from other robots.

4. Comparison of SharifCE Testbed with SServer

As mentioned before (section 2), in SServer, each robot or agent is graphically shown with a shape e.g circle that is a very simple and abstract model of a real robot. This

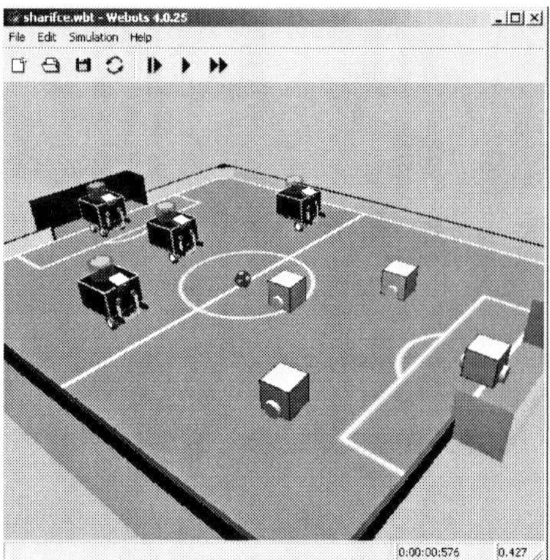

Figure 4. Game field: two teams of robots opponent each other.

platform is fairly noise-free, even the robots do not crash at all. Physical restrictions (e.g. weight, friction, volume, bounding) and system faults are not supported in this Soccer Server. All these factors do exists in a real robot. SharifCE Testbed not only supports all capabilities of SServer, but it also covers its weaknesses. In this testbed, the ball and robots have friction, bounding object for obstacle avoidance purposes and also they may face with faults injected randomly. These criteria are summarized in Table 1.

Table 1. Comparison

Capabilities	SServer	SharifCE Testbed
Move, shoot	X	X
Real-like scale		X
Physical properties (weight, friction, ...)		X
Capable of using Supervisor	X	X
Damaging others		X
Missing/Losing ball		X
Robot customization		X
Object detection	X	X

5. Applications for Multiagent Systems

Soccer is a rich domain for studying multiagent issues. Teams of players must work together in order to put the ball in the opposing goal while at the same time defending their own. Learning is essential in this task since the dynamics of the system can change as the opponents' behaviors change. Some low-level skills, such as dribbling, are entirely individual in nature, others, such as passing and receiving passes, are necessitated by the multiagent nature of the domain [2].

Since occasionally MAS algorithms are studied in robotic soccer area, we have designed this testbed in such a way that it is suitable and applicable for MAS, too. Moreover, researchers may want to test their works on MAS issues in a testbed more close to reality and measure their performance with respect to restrictions that are related to physical limitations. These are some of the main reasons that we surely recommend SharifCE Testbed.

5.1. Agreement with MAS Taxonomy

In the following, we will describe the agreement of SharifCE Testbed for each group of MAS taxonomy. The taxonomy is derived from P.Stone and M.Veloso classification [1].

In both of below categories, the robots can be *Homogeneous*[4] or *Heterogeneous*[5].

- **Communicating agents:** In this group, the agents can communicate to each other directly (with themselves) or indirectly (with a supervisor).
  Since the robots in SharifCE Testbed have emitter and receiver, they can communicate directly. A robot emits the messages that indicates its position and its experiences of soccer field and game playing, etc, to every other robot. There can be a predefined protocol for the style of transferred information, so that teammates are able to process and recognize them. This schema can be used in fields of *Cooperation* and *Learning*.
  On the other hand, the supervisor in SharifCE Testbed can be used as a connector between each robot because supervisor can inform each robot from others and from the ball position. The supervisor can behave like a game advisor; it can be programmed to supervise robots actions and get reward for good working or punishment for bad working regarding their result. This schema is useful for studying in *Supervised Learning*, and *Reinforcement Learning*.
- **Non-communicating agents:** In this group, the agents can not communicate to each other directly (with themselves) or indirectly (with a supervisor). The robots can only conclude and decide from knowledge that is gathered by themselves.
  Without using supervisor, the robots should learn facts by themselves.
  On the other hand, when using a supervisor, the supervisor can only work as a teacher; get reward and punishment for each action. For disabling receiver and emitter of each robot, the operator must assign the value 1 to related parts of robot in Fault Injection Dialog (see section 3).

6. A Practical Implementation

We have implemented several practical works on SharifCE Testbed. In the following one such work is being described that is very simple to study and implement in SServer, though, but is very difficult for real robotics because of limitations in controlling the passed ball and so on.

---

[4] all of the agents have the same internal structure including goals, domain knowledge, and possible actions [1].

[5] Agents might be heterogeneous in any of a number of ways, from having different goals to having different domain models and actions [1].

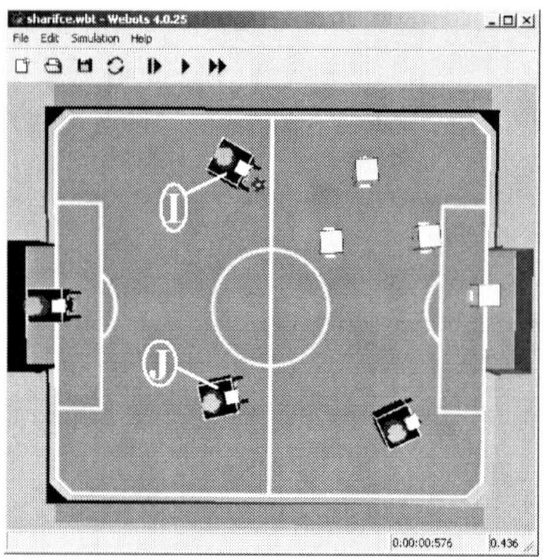

Figure 5. Top view: practical multiagent cooperation of $Robot_i$ and $Robot_j$ to do a better kick to opponent goal.

Assume one robot of the blue team wants to make a decision: Kick the ball to opponent goal or Pass it to a teammate? What decision is better? We solve this problem in three steps using an Artificial Intelligence Problem Solving methods [10].

Initial state: Two teams (three players plus a goalie in each team), blue and yellow, are ready in testbed. The ball is in control of $robot_i$ in blue team. The snapshot of initial state is shown in Figure 5.

Goal state: A decision should be made between shooting the ball to opponent goal without considering the position of opponent team robots, or pass the ball to a teammate.

Algorithm: We define two criteria to select the better action:

1. D_i, that is calculated by this formula:

$$d_{ij} : \text{Distance from } robot_i \text{ to } robot_j \tag{1}$$

$$D_i = \sqrt{d_{i1}^2 + d_{i2}^2 + d_{i3}^2} \tag{2}$$

   This criterion means that whatever the defender robots are far from $robot_i$, then D_i will be bigger. Thus if $D_i \geq D_j$ then $robot_i$ is in better situation than $robot_j$.

2. M_i, is calculated as follows:

$$M_i = \min(|d_{i1}|, |d_{i2}|, |d_{i3}|) \tag{3}$$

   It means that if the closest opponent robot from $robot_i$ is farther from other teammates of $robot_i$, thus $robot_i$ is in better situation that others.

For combining both of the above criteria, we define F_i as $F_i = M_i \times D_i$. Now, if F is larger for one robot, thus that robot is in better situation than others. $Robot_i$ calculates F_1, F_2 and F_3 for itself and other teammates. If F of $robot_i(F_i)$ is the largest, then $robot_i$ will have the best opportunity to kick the ball to opponent goal. Otherwise, the $robot_i$ will pass the ball to the one that has the best fitness regarding the value of F. In case of pass, the $robot_i$ will broadcast the calculated Fs too by emitter to $robot_j$ that has the best situation. Next, $robot_j$ will receive values by its receiver. $Robot_j$ does not need to recalculate Fs twice, because it **learns** from $robot_i$ and approximately they are unchanged. Finally, $robot_j$ kicks the ball to opponent goal.

7. Conclusion

We introduced SharifCE Testbed that is constructed with Webots software as a realistic simulation testbed for robotic soccer. This testbed includes two teams of robots, ball and a soccer field. It provides the physical restrictions for robots. The capabilities such as, Move forward and backward or towards left or right, Kicking the ball, Sensing the front obstacles, Capturing images in the camera field of view, Receiving and Emitting messages, Interacting with supervisor, can be disable as a fault that is injected to the system.

The comparison of SharifCE Testbed versus other related work(s) convinced us of its superiorities. Our practical implementation of this testbed showed that it is a perfect testbed for studying Multiagent Systems. Our testbed is more suited for situations in which we want to define the agents to be as close as possible to real robots and the real situations occur during a soccer game.

References

[1] P. Stone and M. Veloso, *Multiagent Systems: A Survey from a Machine Learning Perspective*, Journal of Autonomous Robots, Vol. 8, No. 3, pp 345-383, 2000.

[2] P. Stone and M. Veloso, *Towards Collaborative and Adversarial Learning: A Case Study in Robotic Soccer*, International Journal of Human-Computer Systems (IJHCS), 1997.

[3] Olivier Michel, *WebotsTM: Professional Mobile Robot Simulation*, Journal of Advanced Robotics Systems, Vol. 1, No. 1, pp 39-42, 2004.

[4] Mei-Chen Hsueh et al., *Fault Injection Techniques and Tools*, IEEE Computer Society Press, Vol. 30, pp 75-82, April 1997.

[5] Itsuki Noda, *Soccer server : a simulator of robocup*, Proceedings of AI symposium '95, pp 29Ű34. Japanese Society for Artificial Intelligence, December 1995.

[6] Mao Chen et al., *Users Manual RoboCup Soccer Server version 7.07*, 2001.

[7] Mansour Jamzad et al., *ARVAND: a Soccer Player Robot*, AI Magazine, Vol. 21, No.3, pp 47-51, Fall 2000.

[8] Mansour Jamzad et al., *A Goal Keeper for Middle Size RoboCup*, Lecture Notes in Artificial Intelligence: RoboCup-2000: Robot Soccer World Cup IV, pp 583-586, Springer 2001.

[9] Mahdi Asadpour, *Simulating ARVAND Middle Size Soccer Player Robot with Webots Software*, BS thesis, Sharif University of Technology, Department of Computer Engineering, Feb 2005.

[10] S. Russell and P. Norvig, *Artificial Inteligence: A Modern Approach*, Prentice Hall, 1995.

[11] http://www.cyberbotics.com

[12] http://www.robocup.org

Self-Organization and Autonomic Informatics (I)
H. Czap et al. (Eds.)
IOS Press, 2005

Knowledge Management to Support Situation-aware Risk Management in Autonomous, Self-managing Agents

Martin Lorenz [a] Jan D. Gehrke [a] Joachim Hammer [b] Hagen Langer [a] Ingo J. Timm [a]
[a] *University of Bremen, {mlo|jgehrke|hlanger|i.timm}@tzi.de*
[b] *University of Florida, jhammer@cise.ufl.edu*

Abstract. We present a novel approach to enable decision-making in a highly distributed multiagent environment where individual agents need to act in an autonomous fashion. Our architecture framework integrates risk management, knowledge management, and agent deliberation to enable sophisticated, autonomous decision-making. Instead of a centralized knowledge repository, our approach supports a highly distributed knowledge base in which each agent manages a fraction of the knowledge needed by the entire system. Our approach also addresses the fact that the desired knowledge is often highly dynamic, context-sensitive, incomplete, or uncertain. Thus risk management becomes an integral component which enables context-based, situation-aware decision making, which in turn supports autonomous, self-managing behavior of the agents. A prototype system demonstrating the feasibility of our approach is being developed as part of an ongoing funded research project.

Keywords. Intelligent Agents, Multiagent Systems, Knowledge Management, Risk Management, Decision Support, Logistics

1. Introduction

In this paper we describe a new approach to enable robust decision-making in a highly distributed, multiagent environment where agents need to act in an autonomous fashion. Our application is the logistics domain where autonomous agents are seen as a promising and effective approach to represent the different planning, scheduling, and controlling processes in an enterprise. For example, we can envision a scenario in which agents are used to represent real-world entities such as truck and container, abstract objects such as weather or traffic service, or even human decision makers, such as the ramp agent at a loading dock. In this scenario, the agents need to make decisions about which containers to transport, what the fastest route to a specific destination is given current road or weather conditions, or what to do with goods damaged during unloading, for example. We believe this kind of autonomous, decentralized decision-making can help make the operational processes more efficient, cost-effective, and allow the participating enterprise to stay competitive. It is also a major improvement over traditional centralized approaches in which individual agents are ill-equipped to deal quickly with sudden events since control usually resides with the entities that are removed from the scene of the event and thus have only delayed access to the relevant information.

Enabling this type of autonomous decision-making is challenging given the potentially large number of agents that could be involved, the dynamic and sometimes even competitive environment in which the agents operate. In principle, enabling a technical system, e.g., an autonomous agent, to make decisions that are designed to impact real-world entities delegates the assessment of consequences of the decisions to the agent. To the same extent as the future is perceived as decision-dependent, any decision to be made by the agent must be regarded as risky [18, p.77]. In the context of this work we define risk as uncertainty about the future state of the world which implies that any decision by the agent might turn out wrong. The goal of risk management (RM) is to attempt to optimize the agent's decisions in the presence of incomplete, imprecise, or debatable information by reducing the uncertainty about future events.

Knowledge management (KM) is an important means to achieve this. Here KM is regarded in terms of a knowledge consumer role presupposing knowledge providers but more complex interactions are possible. Our approach to KM aims at finding a rational basis to obtain needed information and to choose an appropriate provider. Furthermore, the agent is challenged by the fact that the knowledge that is needed is often highly dynamic, context-sensitive, incomplete, or uncertain. Thus the integration of risk and knowledge management enables context-based, situation-aware decision-making, which in turn supports autonomous, self-managing behavior of the agents.

Current agent architectures are not designed to model this complex decision-making process which requires agents to process knowledge about internal structures and organizations, show awareness of other agents and communicate or even cooperate with them, and perceive changes in their environment. A common approach in the agent community is to discriminate the steps `percept`, `reason`, and `do` as a basis for decision-making (deliberation cycle). In more sophisticated approaches, logical reasoning behavior is described. For example, in the *BDI* (belief, desire, intention) approach as introduced in [21], the strategic layer of agents may be modeled within desires, operational aspects within beliefs, and tactical features within intentions or plans. The BDI approach also attempts to closely mimic human decision-making ([3]) and represents the dominant approach for modeling intelligent behavior within the agent research community [6].

The major shortcoming of current agent deliberation cycles is the relatively simple discovery and evaluation of alternatives. The standard approach to creating consistent subsets (goals) for action selection is not sufficient for dynamic environments, as the agent must often conduct multi-criteria optimization, which may also be based on competing goals. Hence an important challenge for this project is to augment the agent's deliberation cycle with the ability to identify and assess the underlying risks that are associated with the options that determine the next course of action. If necessary, the agent must be able to augment its knowledge base with missing or updated knowledge, for example, from other agents, to be able to properly assess and evaluate the feasible options.

In the remainder of the paper we introduce our architecture framework, which integrates risk management, knowledge management, and agent deliberation to enable sophisticated, autonomous decision-making.

2. Framework

Our framework is depicted in Fig. 1. It includes explicit *risk* and *knowledge* management, termed decision-support in the figure, which may work in an inter-leaved fashion to augment the deliberation cycle of the agent. Generally speaking, we use risk management to identify and assess the risks associated with one or more options, and knowledge management to acquire missing knowledge, for example, to improve risk assessment or to generate additional options. Our decision-support system can be integrated into any intelligent agent that utilizes some form of deliberation with separate option generation and selection phases.

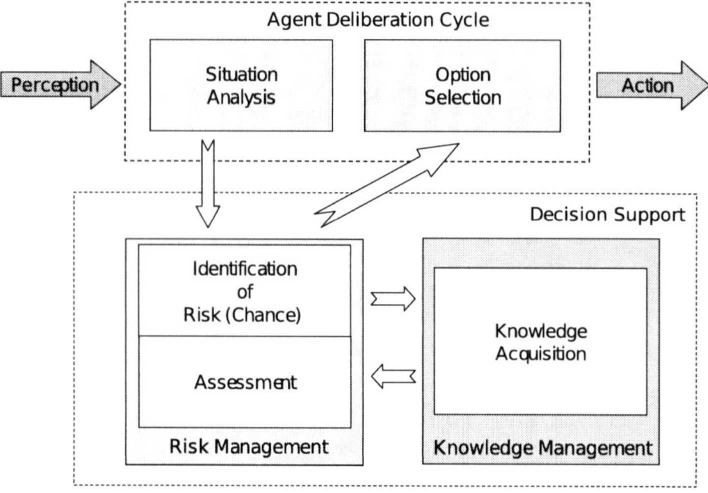

Figure 1. Conceptual overview of the framework depicting the interaction between agent deliberation, risk management, and knowledge management.

We realize that not all deliberation and subsequent option selection involves sophisticated risk management. In fact, many important actions are the result of a trained response (e.g., to avoid imminent danger). However, in this paper, we are focusing on agents in complex decision situations, such as the ramp agent wondering whether it is safe to start unloading paper rolls from a ship given the possible threat of a rain storm. We will elaborate on the use of risk and knowledge management in this scenario later in Sec. 5.

Starting with the deliberation cycle at the top of Fig. 1, we assume that some perceptions are leading to a situation, where the agent has to decide on its next action. Before making a decision, the agent invokes risk management to help with the assessment of the option(s) (e.g., unloading the paper rolls immediately or delaying it until the next morning). We are envisioning that all components have access to a common repository or knowledge base (not shown in the figure) containing the options currently under review. For the remainder of the paper, we will use the term "beliefs" to refer to this knowledge base. By invoking the risk management module, the agent also passes along a pointer to the option(s) currently under deliberation as input.

The first step in risk management is the identification of potential risks associated with each option. For example, in our scenario, a risk of unloading immediately could be that the rolls get wet if it starts raining. Each identified risk must be evaluated to assess the magnitude of the risk and its probability of occurrence. In the ideal case, the agent has sufficient knowledge to arrive at a meaningful risk assessment. Upon completion, the result of the assessment is returned to the deliberation process which uses the information to aid in the selection of the best possible option.

Due to incomplete or uncertain knowledge (e.g., weather information has only a limited life-span and must be updated frequently), risk management may be unable to decide on risk. The exact approach for estimating both is described in more detail in Sec. 3. This triggers knowledge management to acquire the missing information or detailed information on the current situation — including alternative actions. Knowledge acquisition may retrieve knowledge from other agents (e.g., weather service) or directly from external sources/sensors (e.g., a barometer).

A central component of our approach is the representation of *decision-support parameters* which govern the RM and KM processes as well as the interactions between them. For example, when RM invokes KM to acquire missing knowledge to help assessment of risk, it communicates the *importance* of obtaining the missing knowledge to KM. This helps KM to select the proper strategy, which could be to obtain weather information from a free Internet service in case the importance is low to trading or even purchasing weather information from a reputable broker in case the perceived importance is high. Another parameter used by KM is *availability* which expresses the probability that an item of knowledge is available from any known source at this time. Availability of knowledge is based on prior experiences and used by KM, for example, in deciding which knowledge items should be acquired (in case there are choices). So far we have identified a total of eight parameters for RM and KM which will be described in more details in the following sections.

As we mentioned above, RM and KM are closely intertwined and can be invoked multiple times during a single decision-support cycle. For example, during knowledge acquisition, there could be a need to decide between different knowledge brokers both offering similar information. Based on the importance of the intended knowledge (importance parameter), the perceived trust of each knowledge broker (confidence parameter), the cost of the information (cost parameter), as well as the perceived value of the offered information compared to what is expected (similarity parameter), the RM module can be invoked by KM to help assess the "risk" of using one broker over the other.

In the following sections we describe the RM and KM modules including the shared decision-support parameters in more detail. A short scenario illustrating the use of our framework in a logistic environment is presented following the overviews in Sec. 5.

3. Risk Management

As stated in the introduction we associate uncertainty with risk. Thus the acquisition of facts that can reduce uncertainty is one strategy to handle risk. In this section we present an approach to assess the the amount of uncertainty and a strategy to reduce it by invoking knowledge management.

Risk management is a continuous process that will trigger further deliberation as soon as a fact is added to the knowledge base, which makes the situation risky. As already mentioned in the introduction, risk arises whenever a subsequent decision must be based on incomplete knowledge and thus might turn out wrong.

Our concept of risk management is heavily depending on knowledge. Therefore it can only function in close collaboration with a knowledge management infrastructure. In the following we will describe the mechanisms of this collaboration and subsequently describe the core task of knowledge-based risk assessment.

3.1. Interplay Between RM and KM

The correlation between risk and knowledge management is at least threefold. First of all knowledge of risk is one part of an agent's beliefs. Thus it can be communicated by our approach of knowledge management.

Secondly, an agent can use its knowledge of the world to identify risks. From that point of view—the *knowledge-based risk identification* view—knowledge is needed when the agent wants to reason about the possible risks it will face. Without knowledge risks degrade to a *threat*, i.e., to some incalculable future state of the world the agent will tumble into.

And thirdly, the act of communicating knowledge is in itself a risk to the agent because it can fail in various ways and of course the agent has to incorporate the possibility of false information into its reasoning about the value of intended information items.

As sketched out in Fig. 1 and mentioned earlier, the first element of the risk management process is to evaluate each incoming perception as to whether or not it adds evidence to the beliefs which would make the next decision risky.

3.2. Risk Identification

The initial task and most important prerequisite for successful risk management is its ability to identify risk and evaluate its potential consequence. Risk identification in an autonomous knowledge-based system can be achieved by matching fractions of the beliefs with situation patterns.

In the situation analysis phase of an agents deliberation cycle (see Fig. 1) incoming perceptions are integrated with the current beliefs B. Subsequently the agent generates a set of options $\Omega = \bigcup_{i \in \{1..n\}} \{O_i\}\ |\Omega| = n$ that are accessible given the current situation (for details and a formal specification of this process we refer to recent work by Timm [25]). Following the formalization each option O_i contains a desire and a plan to achieve it. Furthermore, the agent is able to generate an assumption on a future state of affairs based on its beliefs which is again part of its beliefs[1] (referenced as B^+). For risk identification a generalization B^{O_i} of B^+ is created, which contains only those beliefs that might be affected or are required by the plan. Risk identification will then work on B^{O_i} and the option itself to search for incidents that may impact the execution of O_i.

Following the approach presented by Lattner et al. [16] we define a risk pattern as a formal description of a situation where certain occurrences may be dangerous for the agent. A risk pattern $\mathscr{P} = \langle S, \chi \rangle$ is defined by a situation description sentence S and a gravity value χ. χ is a value for the possible outcome of the incident described by that pattern. A risk pattern is marked as identified if S subsumes B^{O_i}.

Consider the following simplified example:

[1] For the sake of simplicity we omit all modal or temporal operators in the following.

$$\langle\,(\exists \mathtt{x},\mathtt{y},\mathtt{z}\ \mathtt{agent}(\mathtt{x}) \wedge \mathtt{carries}(\mathtt{x},\mathtt{y}) \wedge \mathtt{water}(\mathtt{z}) \wedge \mathtt{damage}(\mathtt{z},\mathtt{y})), 0.8\rangle$$

This pattern expresses the fact that water damages the load an agent carries.This situation is associated with a gravity of 0.8.

In the next step of risk assessment the agent evaluates all evidences $E = \{E_1, .., E_j\}$ (i.e., beliefs) that lead to a match of the risk pattern according to the degree of uncertainty it has about each evidence E_j. The agent's attempt to acquire additional knowledge is then triggered by a threshold on the combined gravity and uncertainty.

3.3. Risk Parameters

To order the acquisition of new knowledge from KM, risk management assigns a parameter *importance* to every item k it requests.

Importance $\mathrm{Imp}(\mathcal{A}, k, t)$ denotes the importance of an agent $\mathcal{A}$'s intention to obtain the item of knowledge k at time t. It is a measure for k to contribute to the achievement of strategic goals. The parameter value ranges from 0 (irrelevant) to 1 (maximum significance).

To determine $\mathrm{Imp}(\mathcal{A}, k, t)$, the risk identification process interprets every perception in E as a belief k that supports a given risk hypothesis. Every belief k is associated with two probability values which denote the subjective confidence the agent has. In this we follow the basic idea of the Dempster-Shafer theory of evidence (c.f. [8] or [14]).

Support and plausibility *Support* $Supp(k)$ for a hypothesis k indicates the probability mass given to sets of evidence that are enclosed by it. In other words, it gives the amount of belief that directly supports a given hypothesis. *Plausibility* $Pl(k)$ is 1 minus the masses given to sets of events whose intersection with the hypothesis results in an empty set. Again, in other words, it gives an upper bound on the belief that the hypothesis could possibly happen, i.e., it "could possibly happen" up to that value, because there was no evidence that would contradict that hypothesis. Hence $Supp(k) = 0 \equiv Pl(k) = 1$ denotes total ignorance concerning k.

Based on this support values the agent can express its need for new evidences. More precisely, it calculates the difference $\psi_i = Pl(k_i) - Supp(k_i)$ for all relevant facts in its set of beliefs. Relevant are those facts k_i that are present in a risk pattern i.e., in E.

Ignorance factor The *ignorance factor* $\psi = \sum \psi_i$ denotes the agents lack of crisp knowledge to be able to soundly evaluate the risk in question. Together with a *gravity value* χ these define the Imp parameter for knowledge management. A first approach to derive Imp from ψ and χ is given by $\mathrm{Imp} = \chi * \psi$.

Gravity value χ is part of an agents knowledge base. It is tied to the risk pattern and provides a value for the possible damage caused by the incidence of that risk. Motivation for this approach is taken from the classical formalization of risk as function of the occurrence probability of an incident and the severity of its consequence. As a part of the knowledge base χ is in itself subject to knowledge management and can therefore be communicated. The gravity value is considered to be an a-priori set constant here. We will consider a dynamic gravity function which includes several factors like e.g. occurrence probability, amount of loss/damage, damage classification, etc. in an ad-hoc manner.

Based on a threshold for the ignorance factor ψ, risk management decides whether the evidence that is already present in the beliefs is sufficiently crisp to assess the risk which was identified. If ψ is below the threshold, the option O_i is annotated with a risk value composed from all gravity values assigned to it in E and returned to option selection (see Fig. 1). If the ignorance is too high to give a concise risk estimation for O_i, new evidence is necessary to support or contradict the risk hypothesis. Therefore a request consisting of the belief that needs to be updated and the importance derived from ignorance and gravity is sent to KM.

4. Knowledge Management

If the risk management component identifies the need for additional information, knowledge management is invoked as depicted in Fig. 1.

Our approach to knowledge management consists of three main components: **conceptual knowledge**, **roles**, and **parameters**. The conceptual knowledge is represented as an OWL ontology. For the purpose of our logistic application domain this ontology includes a representation of the transportation or production network, the basic types of agents and their properties (e.g., for a vehicle, its average and maximum speed, the types of routes in the network it can use, and its load capacity), and the properties of 'inactive' objects, such as highways, depots, etc.

4.1. Roles

In contrast to previous approaches to agent-based knowledge management [28] we do not presuppose a one-to-one correspondence between agents and knowledge management functions, such as *providing* knowledge or *brokering* knowledge. In our approach these functions are implemented as *roles*. A knowledge management role includes certain reasoning capabilities, a visibility function on an agent's beliefs, a deliberation pattern (i.e., a plan how to accomplish the KM task), and a communication behavior with interacting roles. The aim of KM roles is to provide a formal description of KM tasks that eases the development of agents and reduces computational complexity by means of a minimum set of processed knowledge and applied reasoning capabilities.

One agent can assume different roles and may change them over time. The minimum role model includes the roles of a *provider* offering information and a *consumer* being in need of information. The next extension would be a *broker* mediating between the two [29]. In [15] we introduced an extended role model of eight roles. It incorporates for instance a *translator* between different knowledge representation formalisms.

4.2. Parameters in Knowledge Acquisition

The deliberation pattern of roles is primarily determined by decision parameters as a rational basis to allow for a design of logical foundations, and for analysis and verification of decision processes. Each role has a set of parameters characterizing aspects of decisions to make. In case of the knowledge consumer role, i.e., the rational process of selecting an item of knowledge from a set of possible items and their providers, we introduce a set of four decision parameters. Together with *importance* (see Sec. 3.3) these are correlated and balanced by an agent during a knowledge transfer process in order to reach a rational decision. The parameter set is considered necessary but not definitely complete. Except for *similarity*, all parameters are agent-specific, i.e., they need not to be objective.

Availability $\text{Avail}(\mathcal{A}, k, t)$ is the probability supposed by agent $\mathcal{A}$ of an item of knowledge k to be available from any source at time t in principle. Due to its independence of specific sources the parameter does not help chose an appropriate source. *Availability* is consulted to select the most intended item of knowledge.[2]

Cost $\text{Cost}(\mathcal{A}, \mathcal{B}, k, t)$ determines the costs resulting from the knowledge transfer of k between consumer $\mathcal{A}$ and provider $\mathcal{B}$ at time t. This includes costs arising in the communication process and possible costs to obtain k as payment to the knowledge provider.

Confidence $\text{Conf}(\mathcal{A}, \mathcal{B}, k, t)$ describes the confidence of the knowledge-consuming agent $\mathcal{A}$ at time t that knowledge-providing agent $\mathcal{B}$ will answer the request for the intended item of knowledge k correctly. The parameter value ranges from -1 to 1. -1 means $\mathcal{A}$ feels certain that $\mathcal{B}$ is lying or just has incorrect beliefs, whereas a confidence of 1 corresponds to absolute confidence in $\mathcal{B}$'s answer. 0 stands for neutral confidence, i.e., agent $\mathcal{A}$ has no clue whether $\mathcal{B}$'s answer will be rather right or wrong.

Similarity $\text{Sim}(k, k')$ denotes the similarity of two items of knowledge k and k' to compare the intended answer with the one actually obtained. The value ranges from 0 (no similarity) to 1 (exact match). The obtained item k' may differ in terms of integrity and accuracy. Integrity concerns missing knowledge, whereas accuracy concerns deviations, e.g., spatial, temporal, or precision of measurement. The consumer needs to evaluate kind and scale of a potential deviation in order to plan and execute appropriate actions (intentions) to finally get the knowledge needed. Similarity should be calculated based on information provided by the ontology, e.g., information on deviations and partonomies of spatial concepts.

The parameters discussed above are closely interrelated and determine the impact of the knowledge transfer on the behavior of the consuming agent. The first decision is what knowledge item should be obtained at first. Here the agent tests each item's *importance* (specified by RM) and *availability*. The availability estimation may be based on background knowledge and experiences made. Next the provider of the intended knowledge has to be selected. Influencing decision parameters are *cost* of transfer and *confidence* in the provider. When the transfer is finished the providers answer to the send query is evaluated using the *similarity* function. If similarity is low *confidence* in the provider will decrease or increase otherwise. Thus, successful knowledge transactions with a particular provider agent will strengthen the connection between the involved agents and increase the likelihood of future transactions.

If *similarity* and *confidence* are too low in relation to the intended item's *importance* the agent may try to get an answer from another agent provided that there is enough time and money left. The obtained knowledge item is added to the agent's beliefs if it is considered credible and valuable. After this KM informs RM of the belief update (cf. Fig. 1) enabling a more precise risk assessment.

[2]Figure a case of an information item being extremely useful (in terms of importance) but with little or no chance of obtaining it, e.g., the winning numbers of tomorrow's lottery drawing.

5. Case Study

We now illustrate the RM and KM concepts introduced in the previous two sections using a simple case study which expands on the logistics examples used so far. Let us assume the existence of a ramp agent at a dock who is responsible for unloading a shipment of paper rolls which must be brought to a warehouse on the dock before being transported to their customers by truck.

What makes this simple scenario interesting for our decision-support framework is the fact that despite their weight (e.g., a roll typically weighs between 1,000 and 2,000 lbs.), paper rolls are very sensitive to shock, temperature changes, and moisture, and thus require special handling and care during loading and transport.

For example, water damage caused by rain during the unloading process or by excessive moisture in the storage rooms (e.g., as a result of a sudden temperature change) can render all affected rolls unusable. Considering that the cost of a roll ranges from € 1300 to € 2000, that delivery schedules are specified down to the desired hour of the day, and that rolls have to be subjected to weather several times during the unloading process our scenario represents a very challenging decision problem that requires careful planning and on-the-fly re-planning by the ramp agent. For example, in case of a severe weather threat, the ramp agent must decide quickly whether unloading should be interrupted risking delays in the delivery of rolls to customers as well as additional docking fees for the ship, or if it should continue, risking damage to some of the rolls.

For the rest of this case study, we will focus on the ramp agent and its decision to continue the unloading process. Let us assume the ramp agent is represented as an intelligent agent $\mathcal{A}$[3]. Let us further assume that $B(\mathcal{A}, t)$ and $I(\mathcal{A}, t)$ represent *beliefs* and *intentions* of agent $\mathcal{A}$ at time t. For simplicity, we assume that B and I as well as items of knowledge (k, k', etc.) are represented as (sets of) definite clauses. For example, on the day of this fictitious example, $\mathcal{A}$ has intention I_1, "take paper rolls to the warehouse for quality control and redistribution to the delivery trucks." Another intention I_2 could be to "identify all rolls that do not meet the quality requirements of the recipient before the rolls leave the warehouse." These intentions could have been formed by the agent's situation analysis given one or more desires. In addition, at the same time t, $\mathcal{A}$'s beliefs contain items k_1, "each ship carries approximately 5,000 paper rolls" and k_2, "our fork lift crew can unload a new roll every 60 sec.", as well as weather-related items such as k_3, "it is late spring and weather conditions change quickly".

The agent beliefs k_4 = "Heavy rain within 20 minutes" with $Supp(k_4) = 0.2$ whereas $Pl(k_4) = 0.95$ such that $\psi = 0.75$ which is far to high given that contact with water will cause a total loss of approx. € 1000 (χ might be 0.9 in that case). So the agent requests a current short term weather forecast with Imp(≈ 0.83). Recall that Imp(k) is the importance RM assigns to a knowledge item k. RM invokes KM with the request to obtain knowledge item k containing temperature readings, air pressure, wind speed, etc. for the next 12 hours for the specific geographic location of the port.

Obtaining the needed weather information k is the task of knowledge management. It involves the ramp agent $\mathcal{A}$ in the role of a *knowledge consumer* and two weather service agents, $\mathcal{A}_{w1}$ and $\mathcal{A}_{w2}$, in the role of *knowledge providers*. We assume for simplicity that $\mathcal{A}$ has prior experience about knowledge providers for weather information. Oth-

[3] In more complex scenarios requiring interactions between the shipment and the personnel on the dock, individual paper rolls may also be represented as agents.

erwise, $\mathcal{A}$ would have to consult a *knowledge broker* (a different type of role) first. $\mathcal{A}$ intends to add k to its beliefs $B(\mathcal{A})$, where k is a fully instantiated clause subsumed by another clause q, the *query*. We further assume that k is not already included in $B(\mathcal{A})$ and that k cannot be inferred from $B(\mathcal{A})$, given $\mathcal{A}$'s current inferential abilities. In a knowledge transfer, k' is part of an informative communicative act directed to $\mathcal{A}$ by some other agent, for example, $\mathcal{A}_{w1}$.

In communicating with $\mathcal{A}_{w1}$ and $\mathcal{A}_{w2}$, $\mathcal{A}$ learns that there are two weather information packages k_{w1} and k_{w2} available with different level of detail and at different cost to the agent. In order to judge the usefulness of each package (e.g., a package may include incomplete or inaccurate data), in addition, to the importance value obtained from RM, $\mathcal{A}$ uses the availability, confidence, and cost measures described in Sec. 4 to decide which of the two weather providers should be used. In our scenario, $\mathcal{A}_{w1}$ may be a free Internet weather service with unknown reputation and whose information contains relatively little detail for small regions such as the port where the unloading is going on. On the other hand $\mathcal{A}_{w2}$ may be the weather service at the nearby airfield providing highly detailed weather information for the desired area. $\mathcal{A}_{w2}$, which is known for its reliable data and high availability, requires a fee.

Despite the fact that weather agent $\mathcal{A}_{w2}$ requires a fee, $\mathcal{A}$ decides to obtain k_{w2} given the high value for importance (provided by RM), as well as confidence (based on prior experience with the two weather services). Availability of information does not enter into the decision-making process at this time since the information was readily available from both services.

When the requested information from $\mathcal{A}_{w2}$ arrives, agent $\mathcal{A}$ computes the similarity measure between the intended information k and the actually obtained information k_{w2}. $\text{Sim}(k, k_{w2})$ is then used to update the confidence measure for $\mathcal{A}_{w2}$ using the following rule of thumb: the greater the similarity, the more confidence one has in the service of the agent and vice versa. In addition to k_{w2}, $\text{Sim}(k, k_{w2})$ will be returned to RM, which uses it to evaluate the quality of the acquired information. Presumably, if the similarity is small, RM may decide to continue to request additional information.

Risk management uses the new information k_{w2} to compute the risk values as follows. k_{w2} adds new evidence to the beliefs thus $Supp(k_4)$ and $Pl(k_4)$ can be recalculated. The new evidence reduces $\mathcal{A}$'s ignorance concerning k_4 such that now a risk value can be assigned to the option for unloading paper rolls and it is forwarded to option selection.

Finally the agent decides to postpone the unloading, because the new evidence k_{w2} has increased the support of k_4 "heavy rain" to 0.7. The option to delay and avoid rain damage to the rolls has been favored over the option to proceed and avoid any late fees.

In this case study we showed an example for a straightforward invocation of KM by RM. However, other scenarios are possible. RM could be called by KM as well to determine the risk of acquiring new knowledge, e.g., lying or ill-informed knowledge providers.

6. Related Work

In the multiagent literature a variety of decision-making strategies has been described. Most multiagent systems however employ rather simple decision strategies and concentrate on phenomena emerging from the collaboration and communication mechanisms. An interesting theoretical approach to decision mechanisms in collectives and complex systems is presented by Tumer and Wolpert[27].

One of the most important concept for deliberative decision-making in autonomous agents has been developed by Rao and Georgeff[21] based on the belief-desire-intention theory of human rational action by Michael Bratman [3]. Kakas and Moraitis consider argumentation depending on the particular context that the agent finds himself[11].

Previous research on MAS in the logistics domain has put a strong emphasis on price negotiations and auctions. In these approaches the inter-agent communication often reduces to bidding (cf., e.g., [31]), or the internal structure is defined by a set of equations (e.g., [2]). Scholz et al. [22] apply MAS to shop floor logistics in a dynamic production scenario. It aims at flexible and optimal scheduling of production plans in a heterogeneous shop floor environment. Hofmann et al. [9] aim at replacing conventional tracking and tracing in the logistics domain based on sending (i.e. pushing) EDIFACT messages by an agent-based pull mechanism. Smirnov et al. [24] present a prototype of a multiagent community implementation and a constraint-based protocol designed for the agents' negotiation in a collaborative environment.

According to Dieng-Kuntz and Matta [5] "Knowledge Management [...] aims at capturing explicit and tacit knowledge [...] in order to facilitate its access, sharing out and reuse." This rather organization-centered view can be applied to information technology (IT) as supplementary technology (cf. e.g. [19]) as well as to KM within pure IT driven systems like the autonomous logistics scenario we proposed in this paper. Agent-based or agent-mediated knowledge management (cf. [28], [29]) is a relatively young but currently very active field of research. Van Elst et al. [29] give a comprehensive overview of approaches, that use agent concepts for knowledge management. They hereby distinguish three areas: single agent systems, homogeneous MAS and heterogeneous or society-oriented MAS. Single agent approaches to KM usually are personal assistants like the well-known seminal works by Maes and Henry Lieberman [17], the anticipatory knowledge mediator "Watson" [4], and others. [20] explore how cognitive agents can be used to design systems that implement their vision of knowledge management and that in particular support the knowledge management processes in social, organizational and individual dimension. Our proposed approach employs deliberative agents for which Timm [25] introduces a formal model.

Serafini and Tamilin [23] use a P2P architecture to define a sound and complete algorithm for global subsumptions based on local knowledge. Borgida and Serafini [1] investigate the issue of integration of information from multiple sources in a cooperative information system.

Risk assessment as scientific topic is basically known from management science, finance, environmentalism and health care. Therefore a number of methodologies for organizational risk identification and management can be found in the literature [7,26]. Risk identification is thereby described as the ongoing risk management task of identifying the significant risks to the success of an endeavor. All proposed techniques are of organizational nature, i.e. checklists of risks and their factors, brainstorming of risks and their factors, cross functional teams, interviews with stakeholders and domain experts, etc. In the later literature much attention is paid to software engineering risk management (cf., e.g., [13]) which tends to adapt existing methodologies to the special needs of software development projects.

An upcoming field is the development of computer-based tools to assist in the risk management process. Zoysa and Russell [32] give a thorough overview on "computerized knowledge-based methodologies [...] to capture and reuse risk-related knowledge". An additional interesting approach which fits in this category is proposed by Kim [12].

Explicit, knowledge-based risk identification based solely on autonomously acquired data (in contrast to specific software-guided user input), i.e., a fully automated knowledge-based risk management system, has not yet been proposed in the considered literature.

7. Status and Conclusion

We have described our conceptual framework for enabling autonomous decision-making in agents. Our approach, which integrates risk and knowledge management, allows an agent to evaluate decisions/options based on the likelihood of certain beliefs that the agent uses as supporting evidence. In case, the supporting evidence is weak, i.e., below a certain threshold, or even missing, knowledge management attempts to provide the missing information. Instead of a centralized knowledge repository, our role-based approach supports a highly distributed knowledge base in which each agent manages a fraction of the knowledge needed by the entire system. Our approach also addresses the fact that the desired knowledge is often highly dynamic, context-sensitive, incomplete, or uncertain. Thus the integration of risk and knowledge management enable context-based, situation-aware decision-making, which in turn supports autonomous, self-managing behavior of the agents.

7.1. Benefits and Contributions

The approach described in this paper has the following three important benefits: (1) Our approach augments agent deliberation with sophisticated decision-making capabilities not found in current architectures. (2) By using risk management to also support the acquisition of knowledge, our approach is better equipped to manage the highly dynamic, context-sensitive, and uncertain information needed to make autonomous decisions in realistic environments. This is of particular importance, since we do not presuppose benevolent behaviour. (3) Our role-based knowledge management enables the distribution of knowledge and knowledge management functionality which eliminates the need for a centralized knowledge repository. On the other hand, it provides the necessary flexibility to allow existing KM infrastructure to co-exist with our approach.

As such we expect that our project will not only contribute to a better understanding of the use of autonomous agents in the logistic domain but also provide new theories and algorithms for the efficient management of risk and knowledge in large-scale multiagent systems. Other important contributions include the development of a formal representation that is powerful enough to represent agents, their roles, and the underlying decision-support mechanism, as well as an efficient implementation of agents to allow experimental validation of the accuracy, precision, and promptness of autonomous decision-making in complex and dynamic environments.

7.2. Plans for Prototype Development

We have conducted an initial feasibility study of the concepts proposed here using a simplified model of our logistic scenario. We are currently developing a proof-of-concept prototype system to help validate our approach.

Specifically, we are developing a distributed multiagent system based on the FIPA compliant agent platform JADE.[4] This platform is aimed to be a testbed for various applications of autonomous agents in logistic scenarios. Basically those scenarios consist of a number of active objects modeled as agents and a traffic network of nodes and edges. For example, agents could model packages to be shipped as well as trucks, that want to maximize their utilization. Nodes may be logistic sources and sinks, or traffic junctions. Edges represent roads, railroads, waterways and the like that connect nodes. Manually or stochastically triggered *world events*, e.g., a traffic jam or a breakdown of a truck, force agents to reconsider their plans.

The case study in Sec. 5 serves as our starting point for describing knowledge management needs in the logistics domain.

In order to make use of such a scenario in our prototype, we defined a *logistics ontology* which forms a common ground for all KM-related tasks within the simulated world. Currently, the agents in our prototype are already able to exchange knowledge in a rudimentary way. The proposed decision-support system will be realized as part of an agent's behavior. Knowledge management roles follow from agent communication that will be specified as FIPA interaction protocols for each interacting role pair.

We will use the prototype as a means for validating whether (1) our set of decision parameters is complete and minimal; (2) the assessment of risk can enhance the deliberation cycle; (3) our approach to distributed knowledge management is robust; and (4) the use of roles will reduce the computational costs of reasoning within the agents. Our long term objective is to evaluate possibilities and limitations of autonomy in logistics.

8. Acknowledgments

This research is funded by the German Research Foundation (DFG) as the Collaborative Research Centre 637 "Autonomous Cooperating Logistic Processes: A Paradigm Shift and its Limitations" (SFB 637). Additional Information can be found at `http://www.sfb637.uni-bremen.de/`.

We would like to thank the reviewers for valuable hints and suggestions which helped to improve this paper.

References

[1] A. Borgida and L. Serafini. Distributed description logics: Directed domain correspondences in federated information sources. In R. Meersman and Z. Tari, editors, *On the Move to Meaningful Internet Systems 2002: Confederated International Conferences CoopIS, DOA, and ODBASE 2002*, volume 2519 of *Lecture Notes in Computer Science*, pages 36–53. Springer-Verlag, 2002.

[4] `http://jade.tilab.com/`

[2] A. Bos, M. M. de Weerdt, C. Witteveen, J. Tonino, and J. M. Valk. A dynamic systems framework for multi-agent experiments. In *European Summer School on Logic, Language, and Information*. Foundations and Applications of Collective Agent Based Systems Workshop, 1999.

[3] M. E. Bratman. *Intentions, Plans, and Practical Reason*. Harvard University Press, Cambridge, MA, USA, 1987.

[4] J. Budzik and K. Hammond. Watson: Anticipating and contextualizing information needs. In *Proceedings of the Sixty-second Annual Meeting of the American Society for Information Science*, 1999.

[5] R. Dieng-Kuntz and N. Matta, editors. *Knowledge Management and Organizational Memories*. Kluwer Academic Publishers, July 2002.

[6] M. d'Inverno, M. Luck, M. P. Georgeff, D. Kinny, and M. Wooldridge. The dMARS architecture: A specification of the distributed multi-agent reasoning system. *Autonomous Agents and Multi-Agent Systems*, 9(1–2):5–53, 2004.

[7] S. Figlewski and R. M. Levich, editors. *Risk management: the state of the art*. Kluwer Academic Press, 2002.

[8] J. Y. Halpern. *Reasoning about uncertainty*. The MIT Press, 2003.

[9] O. Hofmann, D. Deschner, S. Reinheimer, and F. Bodendorf. Agent-supported information retrieval in the logistics chain. In *HICSS '99: Proceedings of the Thirty-second Annual Hawaii International Conference on System Sciences*, volume 8, page 8028. IEEE Computer Society, 1999.

[10] M. C. Horsch and D. Poole. An anytime algorithm for decision making under uncertainty. In G. F. Cooper and S. Moral, editors, *UAI '98: Proceedings of the Fourteenth Conference on Uncertainty in Artificial Intelligence, July 24-26, 1998, University of Wisconsin Business School, Madison, Wisconsin, USA*, pages 246–255. Morgan Kaufmann, 1998.

[11] A. Kakas and P. Moraitis. Argumentation based decision making for autonomous agents. In *AAMAS '03: Proceedings of the second international joint conference on Autonomous agents and multiagent systems*, pages 883–890, New York, NY, USA, 2003. ACM Press.

[12] D. Kim. Structural risk minimization on decision trees using an evolutionary multiobjective optimization. In M. Keijzer, U.-M. O'Reilly, S. M. Lucas, E. Costa, and T. Soule, editors, *Genetic Programming. 7th European Conference, EuroGP 2004, Proceedings*, volume 3003 of *Lecture Notes in Computer Science*, pages 338–348. EvoNet, Springer-Verlag, 5-7 Apr. 2004.

[13] J. Kontio. *Software Engineering Risk Management: A Method, Improvement Framework, and Empirical Evaluation*. PhD thesis, Helsinki University of Technology, Sept. 2001.

[14] R. Kruse, E. Schwecke, and J. Heinsohn. *Uncertainty and Vagueness in Knowledge Based Systems*. Springer-Verlag, 1991.

[15] H. Langer, J. D. Gehrke, J. Hammer, M. Lorenz, I. J. Timm, and O. Herzog. Emerging knowledge management in distributed environments. In *Workshop on Agent-Mediated Knowledge Management (AMKM 2005) at the Fourth International Joint Conference on Autonomous Agents & MultiAgent Systems (AAMAS 2005)*, 2005. To appear.

[16] A. D. Lattner, I. J. Timm, M. Lorenz, and O. Herzog. Knowledge-based risk assessment for intelligent vehicles. In *International Conference on Integration of Knowledge Intensive Multi-Agent Systems (KIMAS '05): Modeling, Evolution and Engineering*, Waltham, MA, USA, Apr. 2005.

[17] H. Lieberman. Letizia: An agent that assists web browsing. In C. S. Mellish, editor, *Proceedings of the Fourteenth International Joint Conference on Artificial Intelligence (IJCAI-95)*, pages 924–929, Montreal, Quebec, Canada, 1995. Morgan Kaufmann Publishers Inc.

[18] N. Luhmann. *Soziologie des Risikos*. Walter de Gruyter, 2003.

[19] H. Maurer. Important aspects of knowledge management. In R. Klein, H.-W. Six, and L. Wegner, editors, *Computer Science in Perspective: Essays Dedicated to Thomas Ottmann*, volume 2598 of *Lecture Notes in Computer Science*, pages 245–254. Springer-Verlag, July 2003.

[20] T. Nabeth, A. A. Angehrn, and C. Roda. Enhancing knowledge management systems with cognitive agents. *Systémes d'Information et Management*, 8(2), 2003.

[21] A. S. Rao and M. P. Georgeff. Decision procedures for BDI logics. *Journal of Logic and Computation*, 8(3):293–342, 1998.

[22] T. Scholz, I. J. Timm, and P.-O. Woelk. Emerging capabilities in intelligent agents for flexible production control. In B. Katalinic, editor, *Proceedings of the International Workshop on Emergent Synthesis (IWES 2004)*, 2004.

[23] L. Serafini and A. Tamilin. Distributed reasoning services for multiple ontologies. Technical Report DIT-04-029, Informatica e Telecomunicazioni, University of Trento, 2004.

[24] A. Smirnov, M. Pashkin, N. Chilov, and T. Levashova. Multi-agent knowledge logistics system "KSNet": Implementation and case study for coalition operations. In V. M. et al., editor, *CEEMAS 2003*, pages 292–302, 2003.

[25] I. J. Timm. *Dynamisches Konfliktmanagement als Verhaltenssteuerung Intelligenter Agenten*. PhD thesis, Universität Bremen, Bremen, Germany, Apr. 2004.

[26] M. J. S. Timothy McDaniels, editor. *Risk analysis and society : an interdisciplinary characterization of the field*. Cambridge University Press, 2004.

[27] K. Tumer and D. Wolpert, editors. *Collectives and the design of complex systems*. Springer, 2004.

[28] L. van Elst, V. Dignum, and A. Abecker, editors. *Agent-Mediated Knowledge Management: International Symposium AMKM 2003, Stanford, CA, USA, March 24-26, Revised and Invited Papers*, volume 2926 of *Lecture Notes in Computer Science*. Springer-Verlag, 2004.

[29] L. van Elst, V. Dignum, and A. Abecker. Towards agent-mediated knowledge management. In *Agent-Mediated Knowledge Management: International Symposium AMKM 2003, Stanford, CA, USA, March 24-26, Revised and Invited Papers* [28], pages 1–30.

[30] J. Vassileva, S. Breban, and M. C. Horsch. Agent reasoning mechanism for long-term coalitions based on decision making and trust. *Computational Intelligence*, 18(4):583–595, 2002.

[31] S. Zice, L. Zhengping, Q. Runtao, and S. Mansoor. Agent-based logistics coordination and collaboration. Technical Report SIMTech (AT/01/011/LCI), Singapore Insitute of Manufacturing Technology, 2001.

[32] S. D. Zoysa and A. D. Russell. Knowledge-based risk identification in infrastructure projects. *Canadian Journal of Civil Engineering*, 30(3):511–522, 2003.

Self-Organization and Autonomic Informatics (I)
H. Czap et al. (Eds.)
IOS Press, 2005

Ontology-based Multi-agent Systems Support Human Disease Study and Control

Maja HADZIC and Elizabeth CHANG
Curtin University of Technology, School of Information Systems, Perth, Western Australia

Abstract. The medical milieu is an open environment characterized by a variety of distributed, heterogeneous and autonomous information resources. Coordination, cooperation and exchange of information is important to the medical community. Efficient storage and acquisition of medical knowledge requires structured and standardized organization of data. We design a new ontology, called Generic Human Disease Ontology (GHDO), for the representation of knowledge regarding human diseases. The concepts of the GHDO ontology are organized into the following four 'dimensions': Disease Types, Symptoms, Causes and Treatments. We align and merge existing ontologies against the four dimensions of GHDO. The designed ontology makes our query system suitable for different user categories. The process of problem decomposition into smaller sub-problems within a multi-agent system becomes much easier as well. We also design a multi-agent system framework over different information resources. The multi-agent system uses the common GHDO ontology for query formulation, information retrieval and information integration. This intelligent dynamic system provides opportunities to collect information from multiple information resources, to share data efficiently and to integrate and manage scientific results in a timely manner.

Keywords. Human disease ontologies, multi-agent systems, ontology-based multi-agent systems, intelligent information retrieval.

1. Introduction

In the past, biomedicine was purely a laboratory-based science. However, in recent years, biomedicine has also been transformed into an information science. These changes from laboratory-based information science represent a dramatic shift in biomedical sciences. New generations of biomedicine scientists are both computational scientists and laboratory scientists [1].

Traditionally, medical knowledge has resided within the heads of experienced scientists and doctors, who have devoted much study to become experts in their particular domain of study. This approach worked well in the past when the flow of data was not so great as to overwhelm the expert. However, this situation is rapidly changing. New modern techniques are providing huge, rapidly accumulating amounts of information.

Good predictions can only be made against a substantial collection of existing medical knowledge. We use term "knowledge base" to represent the collection of existing medical knowledge. The bigger the knowledge base the better the predictions that can be made. However, the size of the existing knowledge base is too large for any human to assimilate. Therefore, predictions are only being made against a small subset of the available knowledge. Unfortunately, in most cases some important information is being neglected. There is a need to create systems that can use the existing knowledge base for various purposes such as knowledge acquisition and intelligent information retrieval. Such systems could play a crucial role in filtering the flood of data to the point where human experts could again apply their knowledge sensibly.

In order to make all these data really useful, one needs tools that will access and retrieve exactly the information one needs. The available online information needs to be intelligently queried. We have chosen application domain of human disease research and control to investigate the structure and content of information resources. A characteristic of the domain is that trusted databases exist but their schemas are often poorly documented, if at all. The management of resources and services is important in the biomedical community and associated smaller communities of people committed to a common goal. The network of biomedical databases forms a loose federation of autonomous, distributed, heterogeneous data repositories ready for information integration.

This paper is structured as following:

- In the Section 2, we provide a short overview of our research aim.
- In the Section 3, we discuss assumptions regarding human diseases taken during process of the system construction.
- In the Section 4, we describe how we designed Generic Human Disease Ontology (GHDO).
- In the Section 5, four different types of agents within the multi-agent system are identified and described.
- Section 6 describes how this multi-agent system operates on the basis of the GHDO ontology.
- In the Section 7, we discuss related work in the biomedical domain.
- Finally, in the Section 8, we conclude and provide our final remarks.

2. Aim of the research

We aim to develop a methodology to access, extract and manipulate information from various information resources. Ontologies may be seen as shared formal conceptualization of domain knowledge. Therefore, ontologies constitute an essential resource for enabling interoperation in an open environment such as the internet. We focus on the task of formalizing and combining the knowledge regarding human diseases into a single coherent unifying framework.

The multi-agent system makes use of this ontology for the purpose of intelligent and dynamical information retrieval. Within the multi-agent system, ontology is used at the different levels:

- Firstly, ontology is used to *locate and retrieve* requested information. Information content within an information resource can be described using an ontology. Only then, an agent committed to this ontology is able to 'understand' the information contained within these resources and is able to exactly locate and retrieve the information requested by a user.
- Secondly, ontology is used to enable cooperatively working agents to *communicate* with each other during the process of the information retrieval. Use of ontology permits coherent communication between the different agents and facilitates sharing of the information among different agents.
- Thirdly, ontology is used to *analyze and manipulate* the retrieved information. In this way, the redundant and/or inconsistent information is removed. Only relevant information is selected, assembled together and presented to the user.
- Fourthly, ontology is used to *present* the retrieved information to the user in a meaningful way. The information is presented to the user in a way that makes it easier for the researcher, physician or patient to have an overview of the requested knowledge

regarding human disease of interest. Moreover, the inherited organisation of ontologies adds a taxonomical context to search results, making it easier for the users to spot conceptual relationships in data.

In our work, we introduce an ontology-based multi-agent model for the information retrieval and representation of biomedical knowledge related to human diseases. The ontology is realized in multi-agent system designed to aid medical researchers, physicians and patients in retrieving relevant information regarding human diseases. But we believe that the way we approach our problem is applicable to other knowledge domains as well.

3. Assumptions

We assume that each disease is uniquely described by its different types, symptoms, causes and treatments. By characterizing all of those four dimensions, it is possible to specify a particular disease. This is shown in Figure 1.

It is impossible for two different diseases to have same characteristics describing their disease types, symptoms, causes and treatments. At least one of the mentioned four components needs to be different in the case of different diseases.

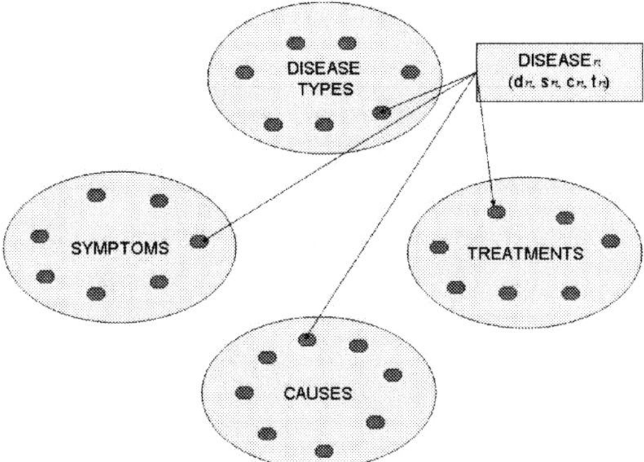

Figure 1. A disease is uniquely specified by its four components

Let d, s, c and t describe disease types, symptoms, causes and treatments of a diseases respectively. For the two diseases Disease$_k$ and Disease$_l$ we have that:

If Disease$_k$ = Disease$_l$, then
$(d_k = d_l) \wedge (s_k = s_l) \wedge (c_k = c_l) \wedge (t_k = t_l)$

If Disease$_k$ ≠ Disease$_l$, then
$(d_k = d_l) \vee (s_k = s_l) \vee (c_k = c_l) \vee (t_k = t_l)$

This means that in the case of two different diseases, at least one of the four components (disease types, symptoms, causes and treatments) needs to be different.

On the basis of this assumption, we develop an ontology to cover the information of all the four 'dimensions'.

4. Human Disease Ontology

We have identified two main user categories of the system:

1. medical researches that are mainly interested in *causes* of a disease, and
2. physicians and patients that are faced with a situation of a disease, and are mainly interested in disease *symptoms* and *treatments*.

Having these two users categories in mind and on the basis of the assumption described in the previous section, we construct Generic Human Disease Ontology (GHDO) [2]. The GHDO has four main branches:

1. *disease types*, describing different types of a disease;
2. *phenotype*, describing disease symptoms;
3. *causes* responsible for that disease that can be genetic and/or environmental;
4. *treatments*, providing an overview of all treatments possible for a particular disease;

Top-level hierarchy of the GHDO is illustrated in Figure 2.

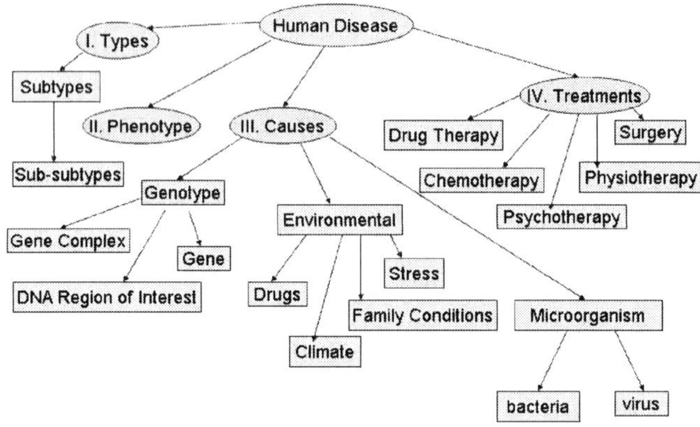

Figure 2. Top-level hierarchy of the GHDO

The information presented in this figure state that a disease may have different types that also further may be divided into subtypes and sub-subtypes. For each disease, there is a corresponding phenotype (or observable characteristics of an ill individual), namely symptoms of a disease. Each disease is caused by cause(s) which can be genetic (genotype), environmental or a disease may be caused by a microorganism. Genetic causes can be a mutated gene, a complex of genes or DNA region of interest. DNA region of interest is a region in the DNA sequence that potentially contains a gene responsible for the disease. This region needs to be further examined in order to correctly locate the mutated gene. Environmental causes of a disease can be stress, climate, drugs or family conditions. Microorganisms that may cause a disease may be virus or bacteria. Possible treatments for a disease can be drug therapy, chemotherapy, surgery, psychotherapy or physiotherapy.

Researchers in the medical ontology-design field have developed different terminologies and ontologies in many different areas of medical domain. In order to obtain some uniformity across different ontologies, definitions from other published and consensual ontologies can be reused [3]. The four different branches (sub-ontologies) of the GHDO ontology can serve as a reference point against which the concepts from the existing ontologies can be reorganized, aligned and merged. We can use other ontologies such as LinkBase [4], TAMBIS ontology [5] and UMLS [6]. The way that these concepts are organized within the existing ontologies is not suitable to be used by our system. LinkBase contains more then one million concepts and only concepts suitable for and needed by our information system need to be selected. If a gene is mutated, it may result in non-functional protein. Presence of a non-functional protein in human body can cause a disease. This is the reason why some bimolecular ontologies, such as TAMBIS ontology that represents general knowledge in regard to proteins and associated genes, would be suitable to cover the ontology part in regard to genetical disease causes. UMLS terminology can be used to validate the selected concepts. So, we use terminology from existing ontologies but select and organize the concepts in a way that can be used by our system.

5. The multi-agent system architecture

The organization of the different types of agents within our information system is presented in Figure 3. This is a distributed multi-agent system with the agents as its processing nodes. All the agents within this information system are dependent on each other with the respect to the same goal. To be able to achieve this goal, they are performing different tasks and cooperatively working on different levels within this multi-agent system.

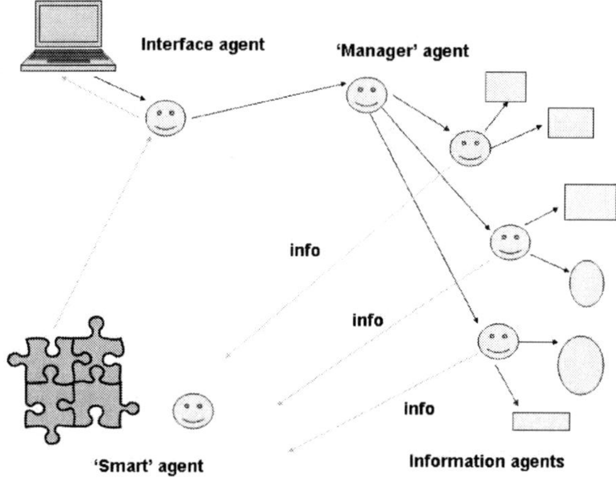

Figure 3. Interface, 'Manager', Information and 'Smart' agents

In our multi-agent system architecture, we distinguish following agent types:

- *Interface* agents to assist the user in forming queries as well as to present the retrieved and assembled information back to the user. Interface agents communicate user's request to the 'Manager' agents.
- *'Manager'* agents decompose the overall task into smaller tasks and assign these subtasks to the various Information agents.

- *Information* agents retrieve the requested information from a wide range of biomedical databases. Each information agent may have a set of databases assigned to it. The information agents send the retrieved information to the 'Smart' agents.
- *'Smart'* agents analyze this information, assemble it correctly and send to the Interface agent directing it back to the user as an answer to his/her query.

Because the data transfer is important and the common goal of agents is to answer the query in the most efficient way, communication within this system needs to be synchronized. There are three different parties in communication within this information system: user-agent, agent-information resource, agent-agent. The Interface agents communicate to the user during the process of the query initialization. The Information agents are communicating with the information resources, and have to move between different information resources to access various data. Communication between agents, such as communication between 'Smart' and Information agents, allows agents to share information, to coordinate their activities and enables them to work collaboratively.

6. Use of ontologies within multi-agent system

The predominant approach has been to use modal logic for reasoning about multi-agent systems. The idea is to develop logics that can be used to characterize the mental states of agents as they act and interact. The two main problems to be addressed in development of a logical formalism for multi-agent systems to represent intentional notions are the syntactic one and a semantic one [7]. We approach those problems from ontological point of view. The syntactic problem can be solved by use of common ontology specification language. All agents within this multi-agent system need to agree and commit to this language. Only when communicating using a common language, the agents are able to interact, freely share and combine their tasks and results in the most accurate and efficient way. How the semantic problem is approached is described in following of this section. Important fact is that ontologies that represent medical knowledge as well as user's queries about this knowledge, can be derived from a single generic ontology (Generic Human Disease Ontology, GHDO) [8].

We adopted the principle of double articulation for formalizing ontologies [9]: specifying of ontology base and specifying of ontology commitments. We firstly define a set of (binary, even) conceptual relationships in the ontology base. The other domain knowledge and its formal semantics are specified in the commitment layer.

The *ontology base* holds binary conceptual relations, called lexons, which are intuitive statements or facts within a specific real world domain. A lexon is formally described as $< \gamma$, term1, role, co − role, term2 $>$, where γ is an abstract context identifier used to group lexons that are intuitively related to each other within a conceptualization of a domain. We can write this stetement also in the form of $\mathcal{R}$[t1, t2] to represent the binary relationship between terms t1 and t2. We take following example:

- *<human diseases, disease, has, is of, type>* of the form $< \gamma$, term1, role, co − role, term2 $>$

This means that in the context of human diseases *'disease has type'* and *'type is of disease'*.

This can be represented by two binary relationships of the form $\mathcal{R}$[t1, t2]: *has [disease, type]* and *is of [type, disease]*.

The *commitment layer* contains a set of interpretations, also known as commitments. Each commitment provides a specific interpretation to a subset of lexons in the ontology base by means of rules and constraints. As examples we take following commitments:

- 'each disease has zero-or-more type'
- 'each disease shows at least one phenotype'
- 'each disease is caused by at least one cause'

Definition 1:
We define **GHDO ontology** as composed of the following two layers:

- an *ontology base* holds intuitive conceptualizations of a domain. In the ontology base, we describe the binary relationships $\mathcal{R}$ between terms t1 and t2: $\mathcal{R}$ [t1, t2].
- a layer of *ontological commitments*, where each commitment operates on the ontology base and holds a set of domain rules. We represent a commitment as $\mathcal{C}$.

As a result of user's query, the overall problem to be solved is constructed as Specific Human Disease Ontology (SHDO) template from GHDO by Interface agents. Retrieving and adding of relevant information upon this SHDO template results in Specific Human Disease Ontology (SHDO). This is shown in Figure 4.

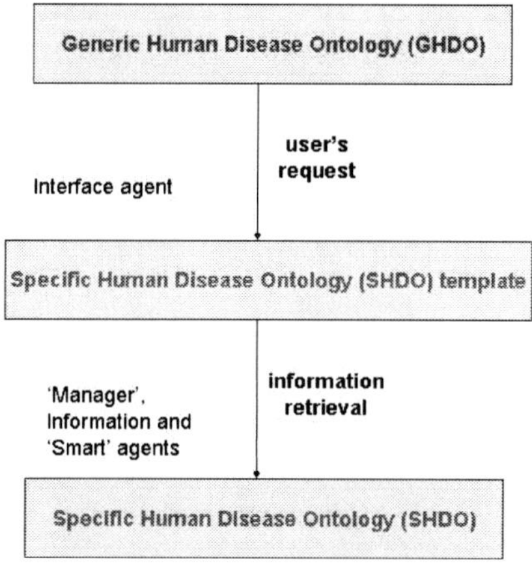

Figure 4. GHDO, SHDO template and SHDO

We will define the formation function for the purpose of defining mechanism by which a user's query is constructed.

Definition 2:
We define **formation function** τ that operates on GHDO and maps ontology base relations $\mathcal{R}$ between terms t12 and t2 to true or false values, and ontology commitments $\mathcal{C}$ to true or false values.

We say that the specific ontology templates Specific Human Disease Ontology (SHDO) templates are formed from generic ontology GHDO by formation function τ,

$\tau : [(\mathcal{R} \rightarrow \{\text{true, false}\}) \wedge (C \rightarrow \{\text{true, false}\})]$.

A user may only be interested in information regarding the human disease knowledge that can be represented by a part of the GHDO. For example, a user may only be interested in Symptoms of a particular disease and it is not needed to represent the information regarding Treatments, Causes and Disease types of that disease. Relationships within the ontology base and commitments within the commitment layer of those three sub-ontologies need then to be mapped to false values, while the relationships within the ontology base and commitments within the commitment layer of the Symptoms sub-ontology need to be mapped to true values.

Definition 3:
The ontologies of an individual agent, have two main components:

- internal stable component, noted as $\mathcal{A}O_i$. For example, domain knowledge or knowledge regarding tasks this agent performs.
- external variable component, noted as $\mathcal{A}O_e$. This component is dependable on the user's query.

Agent ontology is then written as $\mathcal{A}O$, where $\mathcal{A}O = \mathcal{A}O_i \cup \mathcal{A}O_e$.

From the moment of query specification to answering this query, we differentiate four different stages. The whole process is shown in Figure 5.

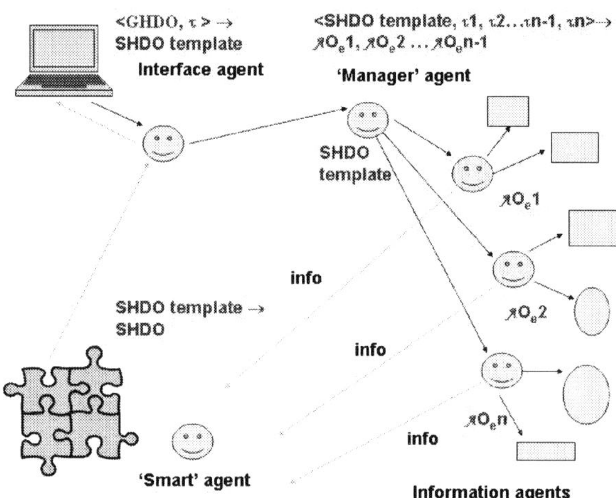

Figure 5. Mechanism of answering user's query

1. Specific Human Disease Ontology template (SHDO template). Upon a user's request, the Specific Human Disease Ontology template is formed from Generic Human Disease Ontology (GHDO) by function formation τ,

$\tau : [(\mathcal{R} \rightarrow \{\text{true, false}\}) \wedge (C \rightarrow \{\text{true, false}\})]$.

We represent this as < **GHDO, τ** >.

The *Interface* agent gives the resulting SHDO template over to the 'Manager' agent.

2. Task sharing. SHDO template is decomposed into smaller sub-problems by *'Manager'* agents. This kind of decomposition is hierarchical so that sub-problems are further decomposed into smaller sub-sub-problems and so on. The SHDO template is first decomposed into its four sub-ontologies (disease types, symptoms, causes and treatments). These sub-ontologies are further decomposed into smaller sub-sub-ontologies. The goal of this problem decomposition is to reach a stage where sub-problems are of an appropriate granularity so that they may be solved by individual Information agents. A task assigned to an Information agent can be composed of more atomic actions. The grain size of sub-problems is important, and decomposition can continue by Information agents until the sub-problems represent atomic actions that cannot be decomposed any further. The different levels of decompositions will often represent different levels of problem abstraction. Each of these different levels in the problem solving hierarchy represents the problem at the progressively lower level of abstraction.

We show an example when a user is interested in symptoms (phenotype) and causes of some disease. The query is structured as SHDO template by Interface agent. This template is composed of two sub-ontologies: Symptoms and Causes sub-ontologies. The SHDO template is decomposed by 'Manager' agent and smaller sub-sub-ontologies are assigned as tasks to individual Information agents, as shown in Figure 6. The Information agents may continue the process of decomposition until a state is reached described only by atomic actions. The Information agents then perform those atomic actions. The process of problem decomposition and task assignment assumes that the agents must have the appropriate expertise to do this. They must have knowledge of the task structure and must know how the task is put together. For example, 'Manager' agents needs to know which Information agents are suitable for performing a specific task so it knows how and to who to assign different tasks. Also, Information agent needs to know how and where to perform atomic actions of the overall task that was assigned to him. This is the reason why ontology is used to represent domain knowledge as well as the task structure in our system.

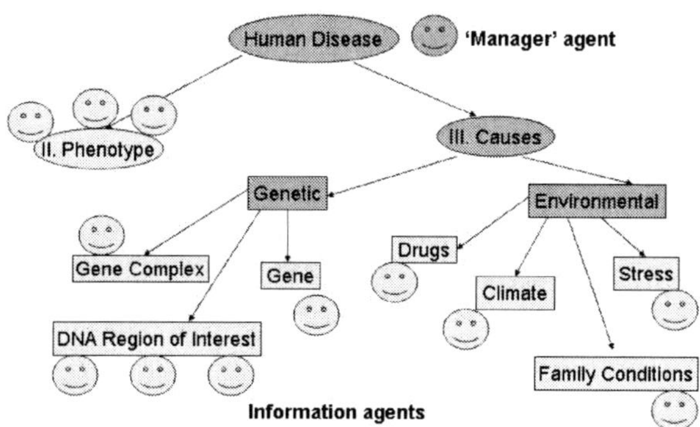

Figure 6. 'Manager' agent assigns task to different Information agents

In this step, SHDO template needs to be partitioned into smaller tasks, $\mathcal{A}O_e$s. $\mathcal{A}O_e$ represents task assigned to an individual Information agent. $\mathcal{A}O_e$ is constructed from the SHDO template using the formation function τ_a which maps relationships and commitments of the SHDO template to true and false values,

$\tau_a : [(\mathcal{R}_{\text{SHDO template}} \rightarrow \{\text{true, false}\}) \wedge (C_{\text{SHDO template}} \rightarrow \{\text{true, false}\})]$.

If there are n Information agents, then n of such $\mathcal{A}O_e$ are formed so that the whole information content within SHJDO template is covered, or

$\mathcal{A}Oe1 \cup \mathcal{A}Oe2 \cup_{\dots} \mathcal{A}O_{n-1} \cup \mathcal{A}O_n$ = SHDO template.

The resulting $\mathcal{A}O_e$s are subsets of SHDO template.

We represent this as:

< SHDO template, τ1, τ2...τn-1, τn>

where τ1, τ2...τn-1, τn are n different formation functions for agents Ag1, Ag2...Agn-1, Agn respectively.

3. Result sharing and selection. Now their individual tasks are specified, *Information* agents are able to search for and retrieve the requested information. In this stage, sub-problems identified during the problem decomposition phase are solved by Information agents. Usually, a task assigned to an Information agent is composed of more atomic actions regarding a specific problem. The Information agents perform atomic actions and migrate from one to another database in order to accomplish their overall task. Because different databases are assigned to different agents, sharing of the information between different Information agents within the system can be very useful. Agents are also cooperatively exchanging information covering different areas of the originally defined SHDO template and the solution is developed progressively. The final result progresses from the solution to small problems that are gradually refined into larger more abstract solutions.

The information that comes from different information resources via Information agents is sent to the 'Smart' agent. The 'Smart' agent compares and analyses the information coming from different information resources in order to select the most relevant information to be presented to the user.

As we see in Figure 6, information regarding 'DNA region of interest' is coming from three different Information agents. This information may be different or the same. If different information covers the same topic, the more precise information needs to be selected by 'Smart' agent and in the next stage, incorporated into SHDO template. In our case, 'DNA region of interest' contains information about regions in human DNA which may potentially contain a gene that may be responsible for the development of disease in question if mutation (abnormal change of gene structure) of this gene occurs. The information agents may provide for example, following information for the case of manic-depression [10]. A part of this information is presented in Table 1. The numbers represent chromosomes in human DNA that may contain the gene of interest (2, 10, 12, 17 and X chromosome) followed by the precise region of this chromosome where this gene is positioned (p13-16, q21- 24, q23-24, q11-12, q24-25 etc.).

Table 1: Information retrieved by different Information agents regarding DNA region of interest

Agent1	2, p13-19	**10, q21-24**	17, q11-14	**17, q11-12**	X, q24-27
Agent2	10, q21-26	10, q21-25	**12, q23-24**	17, q11-13	**X, q24-25**
Agent3	2, p13-17	**2, p13-16**	**12, q23-24**	12, q23-26	17, q11-13

'Smart' agent compares this information on two levels. Firstly, it assembles all information together such as information regarding chromosomes 2, 10, 12, 17 and X respectively. For each of the chromosomes, it compares information regarding the chromosome regions. In Table 1, for example, for chromosome 17 we have regions: q 11-13 (information provided by Agent 2), and q11-12 and q11-14 (information provided by Agent 1). In this context, a smaller DNA region of chromosome means being closer to the novel gene that, if mutated, causes disease in question. A researcher looking to exactly locate this gene is thus closer to his/her goal. That is reason why smaller regions of chromosomes are selected by 'Smart' agent to be incorporated into the SHDO template. In example of chromosome 17, region q11-12 would be selected.

4. Result assembly and presentation. Solutions to atomic problems are integrated into an overall solution by *'Smart'* agents. 'Smart' agent assembles the selected information into SHDO template. This step results in SHDO. In the example from Table 1, the following information would be selected and incorporated into SHDO template: chromosome 2, region p13-16; chromosome 10, region q21- 24; chromosome 12, region q23-24; chromosome 17, region q11-12 and chromosome X, region q24-25.

As in problem decomposition, this stage may be hierarchical with partial solutions assembled at different levels of abstraction. The use of ontology for representing the domain knowledge in a meaningful way is equally important in the solution synthesis stage. The retrieved information is compared, analyzed, assembled together and added on the SHDO template that was constructed at the beginning by the Interface agent. After the final reorganization of information within the resulting SHDO, the result is presented to the user as answer to his/her query.

7. State of art in biomedical domain

The multi-agent systems can be divided into two groups:

- Stationary multi-agent system, eg. Agent Cities [11] and AADCare [12]
- Mobile multi-agent systems, eg. BioAgent [13] and Holonic Medical Diagnostic System [14]

These systems have been designed for different communities. BioAgent is the only one discussed that is constructed for the purpose of being used by the bioscientific community as it is a mobile agent system designed to support bioscientists during the process of genome analysis and annotation. Other systems are designed to be used by the medical community mostly by physicians and patients.

The multi-agent systems differ from each other in the purpose they were designed for. Some of them are designed for a specific hospital, its physicians and patients and the available medical services. The information available to these systems is limited to a specific institution and these multi-agent systems rather help the management of the already available information. They do not have a purpose of gaining new knowledge regarding the disease in question. For example, *Agent Cities* is a multi-agent system composed of agents that provide medical services. The multi-agent system contains agents that allow the user to search for medical centres satisfying a given set of requirements, to access his/her medical record or to make a booking to be visited by a particular kind of doctor. *AADCare* agent architecture is a decision support system for physicians. It connects patient's records with the domain knowledge such as knowledge regarding a specific disease, a knowledge base of clinical management plans, a database of patient records etc. The other two multi-agent systems have specific tasks but the information that needs to be retrieved is not limited to a specific hospital or institution but is retrieved from the Internet. *BioAgent* is a mobile agent system where an agent is associated to the given task and it travels among multiple locations and at each location performs its mission. At the end of the trip, an information integration procedure takes place before the answer is deployed to the user. *Holonic Medical*

Diagnostic System architecture is a medical diagnostic system that combines the advantages of the holonic paradigm, multi-agent system technology and swarm intelligence in order to realize Internet-based diagnostic system for diseases. All necessary/available medical information about a patient is kept in exactly one comprehensive computer readable patient record called computer readable patient pattern (CRPP) and is processed by the agents of the holarchy.

The BioAgent system could be used by our system with some modifications. We can use the same principle of agent migration among multiple locations, information retrieval from each location and information integration at the end of the trip. Only the information we are interested in is not regarding the genome analysis and annotation but human diseases. There is need to design a multi-agent system for the purpose of dynamic information retrieval regarding common knowledge of human diseases as such a system does not exist yet. Holonic Medical Diagnostic System is Internet-based system but it operates on the basis of the information specified in the patient record. We aim to develop a system that has a broader scope encompassing information regarding disease types, symptoms, causes and treatments of a disease in question.

The ontologies are up to date the most expressive way to represent the knowledge. Ontologies were in the first place brought into the computer and information society for the purpose of being used by the agents. In Service Agent Layer of the BioAgent, Ontology Service is mentioned and is only used to provide semantics on data that is locally stored and to allow foreign agents to comprehend the meaning of local data. It is also mentioned that this service may not be present so that the agent would need to use its own ontology. However, we aim to develop an ontology that represents the general knowledge regarding human diseases that could possibly be used by various applications. Moreover, we aim to make the resulting multi-agent system in such way that the most of functions operating within the system would be based on the ontologies.

8. Conclusion

This methodology unifies the advantages of multi-agent system technology with those of an integrated ontology for the purpose of intelligent and dynamic information retrieval and representing the active knowledge about human diseases. Ontologies represent the domain knowledge and are used for the purpose of intelligent information retrieval. The multi-agent system provides the system dynamic, which requires distributed collaborative platform and easy access to resources. The information agents are specialized in retrieving information from different information resources, analyzing the obtained data, filtering redundant or irrelevant information, selecting the relevant information and presenting it to the user.

Ontologies are high expressive knowledge models and as such increase the system expressiveness and intelligence. The ontologies support the following processes involved in the information retrieval:

- constructing queries by users
- problem decomposition and task sharing
- result sharing and analysis
- information selection and integration
- structured presentation of assembled information to users

We introduce an ontology-based multi-agent model designed to aid medical researchers, physicians and patients in retrieving relevant information regarding human diseases. But we believe that the way we approach our problem is applicable to other knowledge domains as well.

The innovation in our work lies in the combination of ontology and multi-agent technology for managing unstructured biomedical research results into structured disease information for the end users. Involving intelligent agents to integrate dispersed knowledge sources into the ontology template results in a powerful mechanism for dynamic building of new knowledge. However, lots of work still remains, such as implementation of local agent interactions, security concerns and development of user view interfaces. The issue of relating the information agents to the legacy data sources at the receiving nodes is outside the scope of this paper.

References

[1] Baxevanis, A. D., Ouellette, F. B. F. 2001, BIOINFORMATICS: A Practical Guide to the Analysis of Genes and Proteins, Wiley-Interscience.

[2] Hadzic M., Chang E. 2005, 'Ontology-based Support for Human Disease Study', *Proceedings of the 38th Hawaii International Conference on System Sciences (HICSS-38)*.

[3] Noy, N.F., Musen, M.A. 2000, 'PROMPT: Algorithm and Tool for Automated Ontology Merging and Alignment', *Proceedings of the seventeenth national conference on Artificial Intelligence (AAAI-2000)*, pp. 450-455.

[4] Montyne, F. 2001, 'The importance of formal ontologies: a case study in occupational health', *Proceedings of the international workshop on Open Enterprise Solutions: Systems, Experiences, and Organizations (OES-SEO2001)*.

[5] Stevens R., Baker P., Bechhofer S., Ng G., Jacoby A., Paton N.W., Goble C.A., Brass A. 2002, 'TAMBIS: Transparent Access to Multiple Bioinformatics Information Sources', *Bioinformatics*, vol. 16, no. 2, pp.184-186.

[6] Bodenreider, O. 2004, 'The Unified Medical language System (UMLS): integrating biomedical terminology', *Nucleic Acids Res*, vol. 32, no. 1, pp. 267-270.

[7] Wooldridge, M. 2002, *An Introduction to Multiagent Systems*, John Wiley and Sons.

[8] Hadzic M., Ulieru M., Chang E., "Ontology-Based Holonic Diagnostic System for the Research and Control of New and Unknown Diseases", *Proceedings of the IASTED International Conference on Biomedical Engineering*.

[9] Spyns P., Meersman R. & Jarrar M. 2002, 'Data modelling versus Ontology engineering', *SIGMOD Record*, vol. 31, no. 4, pp. 7-12.

[10] Liu J, Juo SH, Dewan A, Grunn A, Tong X, Brito M, Park N, Loth JE, Kanyas K, Lerer B, Endicott J, Penchaszadeh G, Knowles JA, Ott J, Gilliam TC, Baron M (2003): Evidence for a putative bipolar disorder locus on 2p13-16 and other potential loci on 4q31, 7q34, 8q13, 9q31, 10q21-24, 13q32, 14q21 and 17q11-12. *Mol Psychiatry*, vol. 8, no. 3, pp. 333-342.

[11] Moreno, A., Isern, D. 2002, 'A first step towards providing health-care agent-based services to mobile users', *Proceedings of the first international joint conference on autonomous agents and multiagent systems (AAMAS'02)*, pp. 589-590.

[12] Huang, J., Jennings, N. R., Fox, J. 1995, 'An Agent-based Approach to Health Care Management', *International Journal of Applied Artificial Intelligence*, vol. 9, no. 4, pp. 401-420.

[13] Merelli, E., Culmone, R., Mariani, L. 2002, 'BioAgent: A Mobile Agent System for Bioscientists', *Proceedings of the Network Tools and Applications in Biology Workshop Agents in Bioinformatics (NETTAB02)*.

[14] Ulieru M. 2003, "Internet-Enabled Soft Computing Holarchies for e-Health Applications", *New Directions in Enhancing the Power of the Internet*, Springer-Verlag Heidelberg, pp. 131-166.

Self-Organization and Autonomic Informatics (I)
H. Czap et al. (Eds.)
IOS Press, 2005

Intelligent MAS for Electronic Customer Relationship Development

Javad SOROOR*
Electronic Commerce Division, Postgraduate Information Technology Engineering Group, Industrial Engineering Department, K. N. Toosi University of Technology, Tehran, Iran

Abstract. Customer relationship management encompasses all the aspects of the interaction with customer and it joins all the customer related elements within an organization together in an intelligent manner. Despite its numerous benefits for organizations, there are some serious concerns regarding the implementation of CRM. Customer relationship management projects are usually complicated, long-term and resource-consuming with outstanding results for far future. The new conceptual model of CRM introduced here is a Multi-Agent System (MAS) called 'agent-based CRD'. It is the process of developing electronic context and content within customer relationships on a collaborative basis using intelligent agents. Agent-based CRD comprises three essential building blocks: single view, intelligent electronic dialogue, and opportunity spotting. A general structure for agent-based CRD framework, which is easy to understand and implement is presented graphically, theoretically, and technically in full details.

Keywords. CRM; customer relationship development; intelligent agent technology; agent-based CRD; Multi-Agent System

1. Introduction

Customer relationship management (CRM) is the infrastructure that enables delineation of and increase in customer value and the correct means by which to increase value and motivate valuable customers remain loyal [14]. CRM technology is an enabling factor for processes that convert the strategies to commercial results [1]. This paper sets out to explain the fundamental meanings of agent-based CRD from both a business and a technology perspective.

1.1. Intelligent Agent Technology

A (software) agent is defined as an autonomous problem solving unit that may collaborate with other agents, and that tries to achieve optimized results in its problem area [5]. Agents differ from "traditional" software in that they are autonomous, pro-active, and adaptive. These qualities make agents particularly useful for the

* Javad Soroor was with Computer Science Department of Amirkabir University of Technology, Tehran, Iran. He is now with Postgraduate IT Engineering Group, K. N. Toosi University of Technology, Tehran, Iran, P.O.B.: 16315-989 (phone: +98.21.88465030,32, Fax: +98.21.88465031), J_Soroor@sina.kntu.ac.ir.

information-rich and process-rich environments [3]. Agent architectures are widely discussed since the late 1950s as a means to automate computer-supported tasks [4].

Agent technology is not a new, single technology but consists of an integrated (and rapidly evolving) use of multiple technologies: language and protocols for logic programming, content definition and agent communication, transport mechanisms, etc [8].

Agent technology may be classified as single-agent and multi-agent systems [10]. In single-agent systems, an agent performs task on behalf of a user or some process, while performing a task, an agent may communicate with a user and local or remote system resources. The mandatory attributes of single-agent systems are autonomy, decision making, temporal continuity, and goal-oriented; in addition, they may also posses intelligence attribute [7]. A multi-agent system is seen as a system that consists of a group of agents that can potentially interact with each other [13].

The agents in static agents based MAS are distributed in an environment to perform distributed problem solving, which coordinate with each other to solve a given task [11].

In order to design and construct agents it will be necessary to take the following points into account [9]: (i) Agents theory (formal specification that describes what properties the agents must fulfill), (ii) Agent languages (tools for the design and construction of agent-based systems, e.g.: Agents can be written in Java, TCL[1], Perl and XML languages.) and, (iii) Agent architecture (agents' internal structure which can be logic-based, reactive, Belief-Desire-Intention-based or layered).

1.2. Customer Relationship Development (CRD)

CRM technology is an enabling factor for the processes that convert the strategies into commercial results [1]. Customer relationship development can be defined as [12]: "The process of developing context and content within customer relationships on a collaborative basis, resulting in fast, reliable and foreseeable benefits to company and customer". CRD is an established and operational customer communication concept offering a fast-track approach to getting started is Customer relationship management. It aims at maximizing the return on the project investment through an efficient and effective development and implementation process. Table 1 [6] outlines some of the key features of a focused CRD approach.

By proposing an intelligent agent-based architecture for CRD, we intend to shed more light on this alternative approach.

Table 1. The key features of a focused CRD approach

Target group	Existing customers
Purpose	Develop relationships, retain customers
Mindset	Customer enabling
Scope	Achieving first base in CRM
Time perspective	Fast-track development and implementation— 3 to 6 months
Primary driver	Customized communication
Critical success factors	Single view, intelligent electronic dialogue, opportunity spotting
Evaluation	Return on investment, customer satisfaction and retention

[1] Tool Command Language

2. CRD Model

As organizations embrace the ideas of CRM, agent-based CRD offers a variable route forward, using a development process by which goals can be achieved:

- With a relatively small investment
- Within set budgets, and
- Within timescales of just a few months.

Agent-based CRD comprises three essential building blocks as shown in figure 1. They are discussed in more detail below [6].

◎ **Single view.** This is the capability to extract information on the whole online customer relationship and present it in a convenient way to both customers and software agents. It can be achieved by creating a single reference record of all the relevant customer, product and relationship information, including activities. It does not have to be a real-time view— often it is a snapshot view within predetermined parameters. The company then shares this single view with the customer to increase the understanding of both parties and identify the possibilities to further enhance the relationship.

◎ **Intelligent electronic dialogue.** This is the capability to initiate and maintain an electronic dialogue with the customer based on mutual knowledge and understanding of the online relationship. It ensures customer learning, thereby enhancing customer value. An intelligent electronic dialogue approach enables the agents to manage the exchange of information with the customer more effectively. Such an exchange takes place whether following up an identified opportunity, managing a service encounter, or aiding the customer in collecting and interpreting information. However, an intelligent electronic dialogue takes place only when information is put to work in an agent-based environment where customers are helped to achieve their objectives. An intelligent electronic dialogue delivers relevant content in a meaningful context on a continual basis.

◎ **Opportunity spotting.** This is the capability to identify and manage opportunities for the benefit of both parties. It can take place on either a proactive or reactive basis. It uses customer data to identify, prioritize and act on opportunities to develop the online relationship. Through opportunity spotting, the company is able to identify the potential in each individual customer relationship through applying strategic and tactical rules, and translate these rules into meaningful actions.

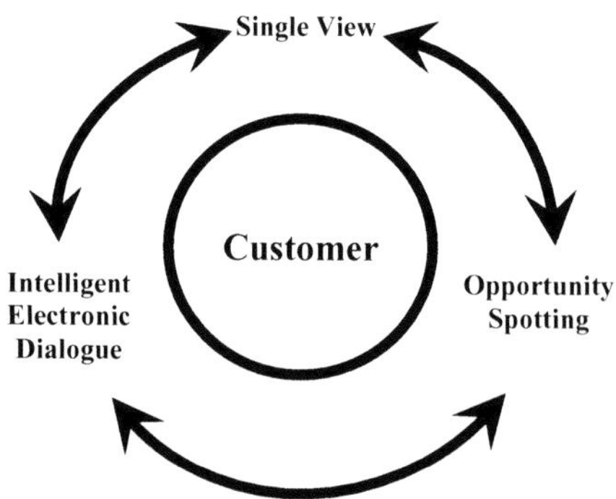

Figure1. The CRD model

3. The Agent-based CRD Framework

In the proposed structure in figure 2, four intelligent agents are used; three of them— known as 'main intelligent agents'— are responsible for the essential building blocks of the agent-based CRD: single view, intelligent electronic dialogue, and opportunity spotting. The other one is the 'data collector'. The main intelligent agents communicate with each other by means of 'ACL[1] messages'. These agents must register in the Agent Platform (AP), which specifies several types of agents that can facilitate the multi-agent communication [9]. The Agent Platform type necessary to get registered in is the Agent Management System (AMS), which provides platform-typical management functions (such as life cycle monitoring, checking of all the entities' content behavior, etc).

The data collector agent uses two types of references to gather data: online customer touch points and offline customer touch points. All the agents deal with the data base management system (DBMS) directly. Customer data warehouse which consists of three data bases is managed by DBMS. The web server acts as an intermediary between the customers' browser (company's website) and the three main intelligent agents.

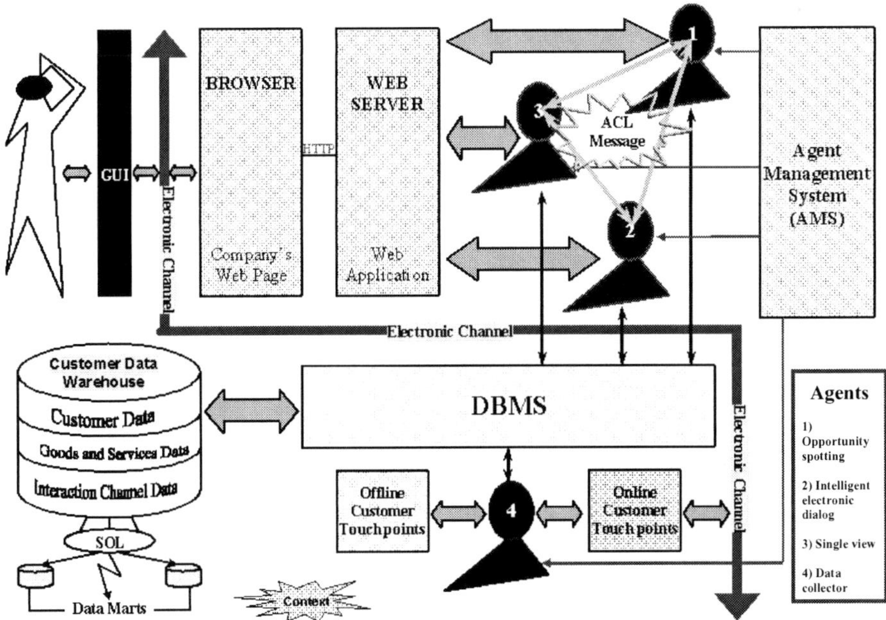

Figure 2. A general structure for agent-based CRD framework

[1] Agent Communication Language

4. Creating a Single View of the Online Customer Relationship

In agent-based CRD, the single view is first and foremost a powerful enabler of customized communications, albeit also a useful tool for customer service and relationship analysis. The customer, product and online relationship dimensions together contribute the information necessary for creating that agent-based context (figure 3).

Full-scale customer relationship management projects put enormous emphasis on creating a single view, and on its capacity to deliver real-time, all-products information to the workflow environment. But it has proved difficult for many firms, for both technical and organizational reasons.

Agent-based CRD offers an alternative route forward. The approach is to identify what key elements are needed to support the customer in decision-making, as well as the appropriate software agents in developing the online relationship. The single view has a number of key roles to play, both in establishing an intelligent electronic dialogue with the customer, and in spotting the opportunities within the online relationship. The aim, in the context of agent-based CRD, is to achieve an increased understanding of the online relationship through a consolidated record of information covering product holdings and customer activity, including a basic evaluation of the status of the online relationship. This information should be available prior to and during online discussions with the customer. The approach also aims to provide the customer with relevant, readily available information as a means to their greater self-management as the online relationship develops.

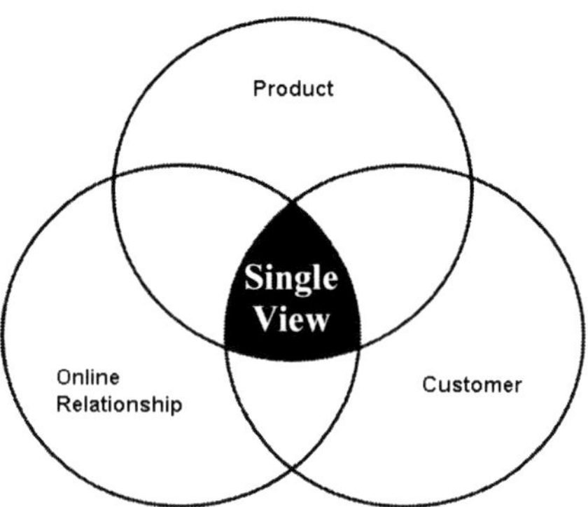

Figure 3. Key dimensions of a single view.

5. Intelligent electronic dialogue

In this section, we describe the structure, process and content issues associated with developing an intelligent electronic dialogue.

5.1. Characteristics of an intelligent electronic dialogue

Intelligent electronic dialogue is about creating communication process that are based on customers' understanding and their wish to buy— instead of what the company wants to tell them about, and sell them. This is achieved by balancing the content of the dialogue to each individual customer's needs and preferences.

Intelligent electronic dialogue should be [2]:

- Customer-driven— not initiated by product owners or functions;
- relevant— based on identified and articulated customer needs and preferences;
- online and ongoing— albeit varying in intensity, scope and frequency;
- adaptive— acknowledging that different customers have different needs for exchanging information;
- comprehensible— based on consistent use of vocabulary and expressions that are easy to understand.

In agent-based CRD, the learning dialogue builds on information about the present status of the online relationship, but also takes into account online and offline relationship history and potential identified through opportunity spotting agents. This requires a true customer focus.

In the case of an online bank, based on a customer-centric approach to intelligent electronic dialogue, the bank may create an 'electronic dialogue checklist'. Through it, the bank can ensure that electronic communications were driven by customer benefit and value— not products, functions or campaigns.

The electronic dialogue checklist supports the ongoing development of the intelligent electronic dialogue. It can be put on the bank's home page or send to the customers' electronic mailboxes so that they can fill out and submit the dialog checklist electronically.

Using this checklist as a decision support tool for the day-to-day management of the customer dialogue by means of agents, enables the bank to reduce the 'waste', and achieve a better co-ordination of electronic messages, and a more efficient use of channels.

5.2. Setting the dialogue boundaries

The level of customer interest in even the most important of product or service propositions can vary significantly, according to the type of proposition, the circumstances or life-stage of the customer, and many other factors [12]. From an agent-based CRD perspective, it is imperative that the agents manage the dialogue content based on the customer's interest level, in order to make the information accessible and relevant. By doing this, the customer will perceive a higher value from the dialogue. The model used to guide intelligent electronic dialogue, is predicted on four levels of customer interest in figure 4.

5.3. The Infrastructure for Intelligent Electronic Dialogue

An intelligent electronic dialogue has to evolve over the customer's life cycle to ensure that all potential profit opportunities are realized. Five key guidelines regarding the intelligent electronic dialogue are [2]:

- it would be challenging— and should not be attempted— for the company to try and control all the events throughout the customer life cycle that influence customer behavior;
- the company should, however, strive to create an online or offline agent-based environment where knowledge exchange with customers is both open and ongoing, so as to allow for rapid adaption to their changing circumstances;

- customer interest levels tend to rise significantly if and when critical events occur; be they initiated by the customer, the company, or external circumstances;
- the electronic dialogue is the company's responsibility, but should be conducted under conditions agreeable to the customer;
- through continually providing the customer with relevant, proactive advice, the 'surprise effects' of sudden online relationship changes can be reduced.

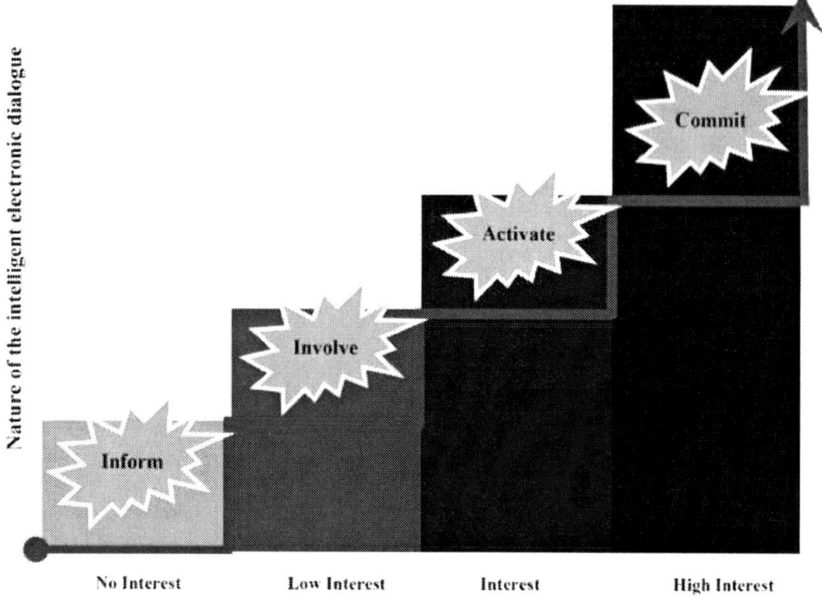

Figure 4. Four levels of customer interest and nature of the dialogue

6. Opportunity Spotting

This section outlines the framework and basic building blocks for opportunity spotting, in conjunction with a single view and the management of an intelligent electronic dialogue. It describes the components within the opportunity spotting framework, including online customer scoring and segmentation.

6.1. From sales-driven to customer-centric opportunity spotting

Successful agent-based CRD is a result of the company's ability to identify and act on opportunities that arise as the relationship develops and matures. In undertaking agent-based CRD projects, however, the company mindset should be to enable and empower the online customer. Indeed, the emphasis in the agent-based CRD approach is to encourage and facilitate a customer-driven approach to the development of the online relationship. This is achieved by a structured agent-based approach where the company guides and prompts customers through the online relationships— proactively advising them on opportunities. Agent-based CRD then gives customers both the knowledge to understand the opportunity, and the mechanism to take it.

6.2. Agent-based CRD Approach to Opportunity Spotting

There are a number of triggers that can be used to originate the electronic dialogue with the online customer.

- **Behavior.** The agent-based CRD methodology collects basic information on which channels a customer uses, and which offers he or she has responded to through which channel. This has two functions:
  - It guides the way in which the customer prefers to be contacted;
  - It indicates the general cost of servicing the customer— this is important as, when coupled with information on the customer's general economic circumstances, it can generate an online profile of the customer and determine which products he or she would use profitably.
- **Usage.** At its simplest, the single view should include information on when contracts must be renewed to generate an event-based electronic dialogue with the customer. It can also look for other key indicators, such as large transaction volumes. When the online customer calls up product information on the website, or looks at price calculator, it might well be an indication for intelligent agents that he or she is entering an event, or an online decision-making period. This may generate an event-based electronic dialogue, and knowledge of a customer's online behavior will indicate the channel through which agents should conduct this electronic dialogue.
- **Potential.** Guiding all online initiatives with customers in an understanding of their potential.

7. Discussion and conclusion

In this paper, the fundamental meanings of agent-based CRD were discussed from both technical and business points of view. The conceptual model was explained in detail thoroughly. We did our best to make the text clear and understandable for the reader with the least background on this subject matter. A number of useful tables, figures, and diagrams were provided to accomplish this goal.

By studying this paper, it is expected that one gains:

- basic knowledge of intelligent agents and their general characteristics;
- a fundamental understanding of CRD and the key features of a focused CRD approach;
- a clear perception of agent-based CRD model and its general structure;
- a realization of the benefits of the agent-based CRD approach;
- the ability to develop an agent-based CRD framework;
- the concept of managing context and content in an agent-based CRD setting;
- an awareness of online customer experience framework;
- the accurate understanding of three building blocks essential for agent-based CRD: single view, intelligent electronic dialogue, and opportunity spotting;
- skills to build an agent-based CRD capability, including the twelve-step process to successful agent-based CRD;
- necessary managerial and technical skills regarding this new CRM solution;

In short, this paper lays the foundation of the agent-based CRD and by proposing some instructive models and frameworks, tries to create the skeleton of this modern, fast-track CRM methodology. The last few sections were especially provided as a protective roof against failure over the agent-based CRD's foundation and skeleton.

Possible future studies on agent-based CRD models may be concentrated on one or some of the following issues:

- The key successful factors of agent-based CRD;
- Implementing some case of agent-based CRD;
- Agent Communication languages (ACL) for agent-based CRD;
- Second-level[1] agent-based CRD;
- Agent-based CRD and interactive marketing;
- Technological enablers of agent-based CRD;

These issues shall be reviewed in our next papers in full depth.

References

[1] P. Greenberg, *CRM at the speed of light*, McGraw-Hill/Osborne, 2004.
[2] J. Soroor, *Successful Implementation of Agent-based CRD*, The North American Technology and Business Conference, Montreal, Canada, 2005.
[3] A. Moukas, R. Guttman, P. Maes, *Agent-mediated electronic commerce*, MIT Media Laboratory, International Conference on Electronic Commerce, 1998.
[4] N. R. Jennings, M. J. Wooldridge, *Agent technology: foundations, applications, and markets*, Springer, Berlin, 1998.
[5] J. M. Bradshaw, *Software agents*, AAAI Press, Menlo Park, pp. 3-46, 1997.
[6] M. J. Tarokh and J. Soroor, *An Agent-based Electronic Customer Relationship Management Solution*, International Industrial Engineering Conference, Tehran, Iran, 2005.
[7] M. J. Tarokh, J. Soroor, *ie-SCM: intelligent electronic supply chain management*, IEEE SOLI Conference, Beijing, China, 2005.
[8] G. Tomas, M. Hult, *Global supply chain management: an integration of scholarly thoughts*, Industrial marketing management 33, pp. 3-5, 2004.
[9] F. Garcia-Sanchez et al., *An integrated approach for developing e-commerce applications,* Expert Systems with Applications 28(2), pp. 223-235, 2005.
[10] S. S. Manvi, P. Venkataram, *Applications of agent technology in communications: a review*, Computer communications 27, pp. 1493-1508, 2004.
[11] G. Weiss, *Multiagent systems: a modern approach to distributed artificial intelligence*, MIT Press, Boston, USA, 2001.
[12] R. Blomquist, J. Dahl, T. Haeger, *Customer relationship development*, Financial World Publishing, UK, 2002.
[13] N. Vlassis, *A concise introduction to multiagent systems and distributed AI*, 2003.
[14] J. Dyche, *The CRM handbook*, Addison-Wesley, 2002.

[1] The model discussed in this paper is considered to be a first-level agent-based CRD.

Self-Organization and Autonomic Informatics (I)
H. Czap et al. (Eds.)
IOS Press, 2005

Java-Based Mobile-Agent Systems in the Real World: Security Highlights

Martin ŠIMEK[1]
University of West Bohemia, Department of Computer Science and Engineering, Pilsen, Czech Republic

Abstract. This document presents a study for the securing of Java-based systems based in the Mobile Agent paradigm. Security is an important issue for the widespread development of applications based on software agent technology. It is generally agreed that without the proper countermeasures in place, use of agent-based applications will be severe impeded. This document gives an overview of the threats associated with software agent systems focused on the elements of our simplified model: Agents and Agent Platforms and also describe a general method for controlling the behavior of mobile agent-system entities through the allocation of privileges. This approach overcomes a number of problems in existing agent systems and provides a means for attaining improved interpretability of agent systems designed and implemented independently by different manufacturers.

Keywords. Mobile-agent system, Security management, Java language

1. Introduction

Over years computer systems have evolved from centralized monolithic computing devices supporting static applications, into networked environments that allow complex forms of distributed computing. A new phase of evolution is now been developed based on software agents.

The main advantages of the mobile agent paradigm lie in its ability to move client code and computation to remote server resources, and in permitting increased asynchrony in client-server interactions. Agents can be used for information searching, filtering and retrieval, or for electronic commerce on the Internet, thus acting as personal assistants for their owners. Agents can also be used in low-level network maintenance, testing, fault diagnosis, and for dynamically upgrading the capabilities of existing services.

Mobile Agents offer a modern paradigm for distributed computation, but their potential benefits must be weighed against the very real security threats they pose. It is usually assumed that a Mobile Agent System is subject to the fundamental threats of disclosure of information, denial of service, corruption of information, misuse, abuse and repudiation.

A secure mobile agent system must provide services to counter these threats. However, there is no countermeasure if the attacker exploits system flaws or security weak-

[1] University of West Bohemia, Univerzitní 8, 30614 Pilsen, Czech Republic; E-mail: simekm@kiv.zcu.cz

nesses such as bypassing controls, exploiting trapdoors, or introducing Trojan horses. Because of that, most agent systems rely on the assumption that the home platform and other equally trusted platforms are implemented securely, with no flaws or trapdoors that can be exploited, and behave non-maliciously. A secure architecture must therefore confine key security functionality to a trusted core that enforces the essential parts of the security policy.

1.1. Security Threats in Mobile Agent Systems

As the sophistication of mobile software increases, so do the associated security threats and vulnerabilities. As noted before, threats to the security of mobile agents generally fall into four comprehensive categories: disclosure of information, denial of service, corruption of information, and interference or nuisance.

Many of the threads that will be under study have always existed in some form in the past (e.g. executing code from an unknown source) and have counterparts in classical client-server systems [4, 5, 6]. However, mobile agents offer a greater opportunity for abuse and misuse, making the scale of threats broader. Contrary to the usual situation in computer security where the owner of the application and the operator of the computer system were the same, the agent's owner and system's operator are different.

2. Threat Categories

After studying the system, four threat categories have been identified: an agent attacking the agent platform where is been executed, the agent platform attacking the agent that is been executed, and agent attacking another agent on the agent platform and other entities attacking the agent system which covers also the cases of an agent attacking an agent on another agent platform and of an agent platform attacking another platform. The last category is based on attacks to the communication capability of the platform trying to exploit potential vulnerabilities.

2.1. Agent against Agent Platform

The mobile agent paradigm needs an agent platform to accept code developed elsewhere and execute it. An incoming agent has two main lines of attack that the security policy should avoid: to gain unauthorized access to information residing at the agent platform or to use its authorized access in an unexpected and disruptive way.

Unauthorized access can be avoid implementing strong access control mechanisms that are able to identify and authenticate the agents making it impossible for and agent to masquerade as one trusted by the platform. A bad use of the authorized access can cause deny of platform services to other agents so the resource assignment has to be tightly set and controlled.

2.2. Agent Platform against Agent

A receiving agent platform can easily isolate and capture an agent and may attack it by extracting information, corrupting or modifying its code or state, denying requested services, or simply reinitializing or terminating it completely. It must be remarked that an agent is very susceptible to the agent platform because the latter provides the entire computational environment where the agent will execute. Modification of the agent by the platform is a particularly insidious attack, since it can radically change the agent behavior or the accuracy of the computation Examples of these attacks are turning a trusted agent into a malicious one or changing the collected data to cause incorrect results.

Repudiation should be avoided so that every agent platform is responsible for the actions it made. The agent's code should be hidden to the agent platform to allow software protection and secrecy. Some data should be hidden to the agent platform and some should be impossible to modify.

2.3. Agent against Other Agents

An agent can attack another agent in the same agent platform in several ways: falsify transactions, eavesdrop upon conversations, or interfere with agent's activity.

Diverse are the possibilities of attack for an agent. If the agent platform is not correctly designed the agent could access and modify another agent's data or code, or interfere with it by invoking its public methods. It could also use platform services to eavesdrop on intra-platform messages. Even with reasonable control mechanism in place, an agent can attempt to send messages repeatedly o other agents in an attempt to deny the ability to communicate.

Repudiation should be avoided too in this case so that every agent is responsible for the actions it made. This would avoid, for example, an agent could deny that a legitimate transaction occurred.

The executing agents should be isolated from one another to avoid not allowed interactions between them to take place. Communications between agents should be ciphered and the agents involved in the communication properly authenticated.

2.4. Other Entities against Agent System

Even assuming that the locally active agents and the agent platform are well behaved, other entities both outside and inside the agent framework may attempt actions to disrupt, harm, or subvert the agent system. The methods involve attacking the inter-agent and inter-platform communications through masquerade (e.g., through forgery and replay) or intercept.

For example, an entity may eavesdrop at a level of protocol below the agent-to-agent or platform-to-platform protocol on messages in transit to and from a target agent or agent platform to gain information. An attacking entity may also intercept agents or messages in transit and modify their contents, substitute other contents, or simply replay the transmission dialogue at a later time in an attempt to disrupt the synchronization or integrity of the agent framework. Denial of service attacks through available network interfaces is another possibility.

Communications between agent platforms should go through a secure channel with both extremes of the communication authenticated and with the contents ciphered. It could be a good idea to allow agent-to-agent communication only within the same agent platform unless the communication channel is secure enough.

3. Design of a Security Policy

Once identified the security threats presented by a mobile agent system and decided to use Java language because of it's inherit advantages, a complete security policy must be built taking advantage of the tools provided by the language and developing other tools where required.

An agent trusts the home platform where it is instantiated and begins execution. The home platform and other equally trusted platforms are implemented securely, with no flaws or trapdoors that can be exploited, and behave non-maliciously.

Public key cryptography, primarily in the form of digital signature but also for encrypting in some situations, is utilized through certificates and revocation lists managed through a public key infrastructure (PKI).

3.1. Research Challenges in Security

There is no FIPA [2] specification to be followed for the design of a security policy for a system based in the mobile agent paradigm. There are a wide variety of solutions proposed, each of them focusing in a different point depending on the importance given by the designer to the different security threats. However, there is a lack of an overall framework that integrates these techniques into an effective security model. It is widely believed that this is because sometimes the countermeasures are not compatible with one another, due to their redundancy in purpose and overlap in functionality. It is important to differentiate the different techniques between those oriented toward detection and those oriented toward prevention. Detection implies that the technique is aimed at discovering unauthorized modification of code or state information. Prevention implies that the technique is aimed at keeping the code and state information from being affected in a meaningful but negative way, as opposed to random alterations. To be effective, detection techniques are more likely than prevention techniques to depend on a legal or other social framework to penalize misbehavior.

Besides that, the complexity and the overhead involved in these techniques make it difficult for them to be accepted and widely used. The designer of the security model must decide which countermeasures must be embodied with an agent system, and which ones can be applied independently within the context of a particular application.

A security model should be as much as possible FIPA compliant, that is, it should respect all the specifications released by now and must predict what will be proposed as a specification in the future.

3.2. Limitations in Java Cryptographic Mechanisms

Java provides several security APIs which implement a limited set of cryptographic primitives [8]. For example, one-way hashing using the Secure Hash Algorithm (SHA) is supported and so is public-key based digital signing using the Digital Signature Algorithm (DSA). If we refer to the techniques available to encrypt, we notice that, while symmetric encryption using the Digital Encryption Standard (DES) algorithm is implemented, there is no implementation of any asymmetric (public-key based) encryption algorithm. The RSA algorithm [3] is used for public-key based signatures and is not reversible. Therefore, it cannot be used for public-key encryption. However, we can have advantage of the existing implemented security primitives for designing the secure protocols necessary in a mobile agent system.

3.2.1. The BouncyCastle Crypto Package

The Bouncy Castle Crypto package [1] is a Java implementation of cryptographic algorithms. The package is organized so that it contains a lightweight API suitable for use in any environment with the additional infrastructure to conform the algorithms to the JCE (Java Cryptography Extension) framework.

This API has been specifically developed for those circumstances where the rich API and integration requirements of the JCE are not required. But the most important feature is that a RSAEngine has been implemented for asymmetric block ciphering (it takes as default key size 2048 bits). This allows asymmetric ciphering, not implemented in the default JCE cryptography provider. The Bouncy Castle provider is a JCE compliant provider that is a wrapper built on top of the lightweight API. The Bouncy Castle package has three implementation of a keystore.

The first "BKS" is a keystore that will work with the keytool in the same fashion as the Sun "JKS" keystore. The keystore is resistant to tampering but not inspection.

The second will only work with the keytool if the password is provided on the command line, as the entire keystore is encrypted with a PBE based on SHA1 and Twofish. This makes the entire keystore resistant to tampering and inspection, and forces verification. The Sun JDK provided keytool will attempt to load a keystore even if no password is given, this is impossible for this version.

In the first case, the keys are encrypted with TripleDES. The third is a PKCS12 compatible keystore [3]. PKCS12 provides a slightly different situation from the regular key store, the keystore password is currently the only password used for storing keys. Otherwise it supports all the functionality required for it to be used with the keytool. In some situations other libraries always expect to be dealing with Sun certificates, if this is the case PKCS12-DEF must be used, so that the certificates produced by the key store will be made using the default provider.

4. SMAS Mobile Agent System

The SMAS project was motivated mainly to investigate an agent programming system architecture to address the requirements stated above. In SMAS, the mobile agent implementation is based on the generic concept of a mobile object [11]. Agents are active mobile objects, which encapsulate code and execution context along with data. The SMAS is implemented using the Java language and its security mechanism. It also makes use of several other facilities provided by Java, such as object serialization, reaction, and remote method invocation, etc.

Security and robustness are among the most important requirements of an agent-programming environment. In order to build any agent-based application, a set of hosts have to run a facility that receives the visiting agents and provides to them an execution environment and access to its services. We refer to such entities "engines" in the consider architecture. Service providers willing to host agents originating from unknown sources require a high level of confidence in the protection mechanisms to enforce the desired security policies for preventing unauthorized or malicious agents from leaking, destroying or altering the server's resources or disrupt its normal functioning. Malicious agents can also cause inordinate consumption of host resources to mount "denial of service" attacks. Security mechanisms are thus necessary to safeguard hosts' resources.

4.1. Basic Elements of the SMAS System

A host in the Internet can provide services to mobile agents by running the engine. The engine creates a confined execution environment for visiting agents and allows the local resource owner to grant access of its resources to agents in a selective manner. It provides primitive operations such as those, which allow agents to migrate to other hosts, communicate with each other, query their environment, etc. A set of these engines collectively constitute the mobile agent programming environment, in conjunction with ancillary servers such as name resolves and public-key repositories. Following are the important elements of SMAS's computation model.

Client – the client is liable for creation and dispatch of an agent. The client can communicate with agent during its lifecycle. The agent can be retracting to the client after termination.

Code Base Server – other information included in the credential is the URL for agent's code base server, which is a server that provides the code for the classes required by the mobile agent during its lifecycle.

Name Server – the name server keeps database of engines and agents. For each of them server contains its current location.

Engine – the engine provides services for agents, such as those which allow migrating to other hosts, communicate with each other, etc. The engine contains one or more agent repositories. Inside these repositories are situated places.

Agent Repository – the most common view of the agent repository is that it is a context in which the agent can execute. You can regard it as an entry point for visiting agent that wishes to execute. One of repositories on the host is called default, which provides only limited set of services and basic security policy. Without specification of destination repository every arrival agents are stored in default repository. Individual repositories are differing in a set of offered services and security policy. Although set of services of the host is given, not all services are offered to all agents. The reason of this architecture is separate different application.

Place – the place is responsible for interaction between the engine and the agent. Towards agent the place represents services of the engine. The place provides a uniform set of services (individual services) that the agent can rely. Agent is closed in context. Whole interaction between the agent and neighborhood is provided by the proxy scheme.

5. SMAS Security Model

We have developed a security model that provides an overall framework for agent security. The model supports the flexible definition of various security policies and describes how and where a secure system enforces these policies. Security policies are defined in terms of a set of rules by one administrative authority. The policies specify

- the conditions under which agents may access objects,
- the authentication required of users and other principals, which actions an authenticated entity is allowed to perform, and whether entities can delegate their rights,
- the communications security required between agent and agent repository, including trust.

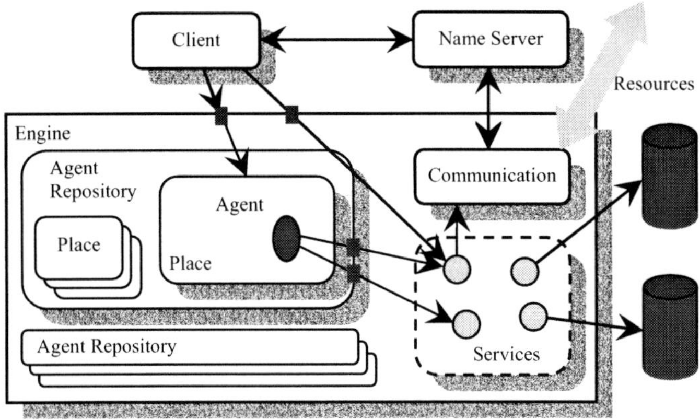

Figure 1. Structure of the SMAS mobile agent system

5.1. Identity of the Owner and the Creator

For authorization of agent actions is used entirely identity of the owner and the creator. These identities are stored in the agent metadata as classes that include two objects principals and credentials.

Principal – this is the object with the name of the entity. SMAS defines four types of principals. SimpleNamePrincipal includes name of the entity as the string. It is only secondary form which is useful for display the name of the entity. LoginNamePrincipal represents login name the entity. Verification of this name can be used e.g. Kerberos system. In the RealNamePrincipal is included real name of the user. X500Principal represents X.500 name, which can be authenticated by X.509 certificates.

Credential – for the authentication of the entity are used credentials. In the SMAS is defined only one type of credentials - Certificate. This certificate contains identity of the owner signed by the certification authority. Thus we can verify its authenticity.

5.2. Agent Authorization

Java language contains quite sophisticated security architecture based on the permissions [7, 11]. It is proposed only for the mobile code and therefore this authorization is usable only on programming code level. JAAS (Java Authentication and Authorization Service) extend this architecture by authorization based on the identity, but only on the thread-level. This approach is hard to use, therefore the SMAS system extend security policy by constraints.

Permissions – this system is useful for "static" authorization of objects. Configuration of permissions is included in the configuration of the engine.

Constraints – using permissions for control of the resource access is very difficult. Therefore, general mechanism is used. Constraints are assigned by the agent repository during agent inserting. Static constraints are derived from standard class Properties; dynamic constraints are created during agent execution. One constrain is closely defined – maxThreadCount.

5.3. Separately Stored Parts of the Agent

The agent must know and store some information about you e.g. self name or identity of the creator. These particulars could be clearly internal parts of the agent, but this is dangerous, because agent can modify your facilities. Upon these grounds SMAS store this information separately.

Metadata – include basic information about the agent – identity of the creator and the owner and location of the source code. The whole particulars are signed by the client.

Name and Address – these items are generated by the engine upon agent dispatch and therefore they cannot be as part of metadata.

Source Code – information about the source code is in metadata and is called codebase. Each agent have own class loader.

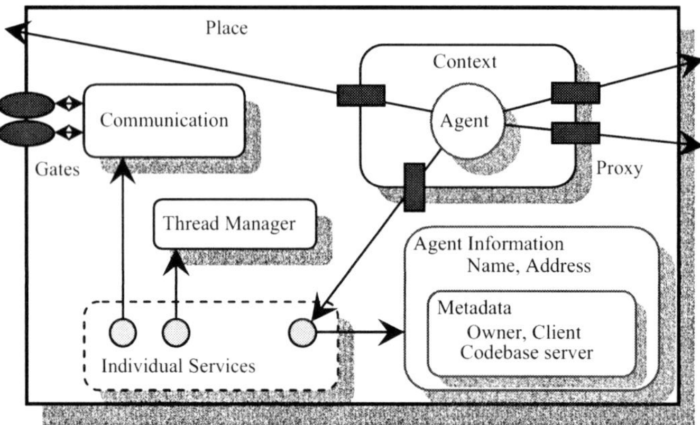

Figure 2. Inner structure of the place

5.4. Agent Communication

As mentioned earlier, Java's cryptography package does not provide any algorithms for public key based encryption and decryption. Since this was necessary for some of the security mechanisms we needed, we implemented the ElGamal cryptosystem [12]. This is an asymmetric cryptosystem whose security is predicated upon the difficulty of computing discrete logarithms in a finite field. It is reversible, i.e., it can be used for encryption as well as digital signature generation and verification. The Java security library provides a cryptographically secure pseudorandom number generator. These classes proved to be very useful for implementing cryptographic algorithms.

Before the communication, agents must authenticate of its identity. For the authentication is used challenge-response algorithm with automatic key (certificate) distribution from [12].

For the authentication is used algorithm from [11] (optimized version) with automatic public keys (certificates) distribution. The protocol for this authentication was developed using a challenge-response mechanism, with randomly generated nonces to prevent replay attacks [13]. It operates at the application level, i.e., it is not a network level protocol for creating authenticated network connections wherein the endpoints know each other's host names securely.

6. Conclusions

It is certainly true that the policy designed throughout this document is a powerful weapon against almost all the security threats in the systems designed based on the mobile agent paradigm, but then is also true that this countermeasures have to be carefully analyzed to study whether the performance lost makes it worth to use all of them of not. Not only are all these countermeasures not needed in most of the applications, but they can also imply that the important advantages of the mobile agent paradigm could remain hidden under the extensive use of resources and the excessive overhead that is introduced. Flexibility has to be offered to the developer so that the level of security can be customized for each particular design allowing balancing the security obtained with the performance lost.

Attention must be paid also in the future standardization of the security requirements. In the near future new specifications for the design of a security policy that allows the interconnection of different mobile agent systems will appear. Work is being made in this area. And certainly it will help to a widespread deployment of applications based in the mobile agent paradigm.

We also described the design and implementation of the Simple Mobile Agent System (SMAS). In particular, we presented the SMAS object and component models and described its components, such as resource naming and locating, communication and mobility.

Acknowledgement

This research and work was supported by the Grant Agency of Czech Republic (GAČR); project No. 201/03/P093 "Security policy in mobile agent systems".

References

[1] BouncyCastle Crypto Package. http://www.bouncycastle.org.
[2] L. Chiariglione, *FIPA Specification*, Foundation for Intelligent Physical Agents, 2000.
[3] RSA Laboratories, *Cryptographic Message Syntax Standard, Public Key Cryptography Standards (PKCS) #7*, November 1993.
[4] W. Jansen, *Countermeasures for Mobile Agent Security*, Computer Communications, 23 (2000), Elsevier, October 2000, pp. 1667-1676.
[5] W. Jansen, *Determining Privileges of Mobile Agents*, National Institute of Standards and Technology.
[6] W. Jansen, *Intrusion Detection with Mobile Agents*, National Institute of Standards and Technology.
[7] G. Karjoth, D. Lange, M. Oshima, *A Security Model for Aglets*, IEEE Internet Computing, pp. 68-77, July-August 1997.
[8] W. Stallings, *Cryptography and Network Security: Principle and Practice*, Prentice Hall, 1998.
[9] M. Šimek, *Security Threads in Mobile Agent Systems*, Proceedings of 5th Workshop on Agent Based Simulation ABS 2004, SCS, Lisbon, Portugal, 2004
[10] M. Šimek, *Extension of Security Policy in Java-based Mobile Agent Systems*, Proceedings of 7th World Multiconference on Systemics, Cybernetics and Informatics SCI 2003, IIIS, Orlando, Florida, USA, 2003
[11] N.M. Karnik, *Security in Mobile Agent Systems*, Department of Computer Science, University of Minnesota, 1998.
[12] T. ElGamal, *A Public Key Cryptosystem and a Signature Scheme based on Discrete Algorithms*, IEEE Transactions on Information Theory, IT 31(4), pp. 469{472, July 1985.
[13] M. Abadi, R. Needham, *Prudent Engineering Practice for Cryptographic Protocols*, IEEE Transactions on Software Engineering, 22(1), January 1996.

Self-Organization and Autonomic Informatics (I)
H. Czap et al. (Eds.)
IOS Press, 2005

Agent Based Computational Model of Trust

Alexander GOROBETS [a,1] and Bart NOOTEBOOM [b]
[a] *Sevastopol National Technical University, Ukraine*
[b] *Tilburg University, The Netherlands*

Abstract. This paper employs the methodology of Agent-Based Computational Economics (ACE) to investigate under what conditions trust can be viable in markets. The emergence and breakdown of trust is modeled in a context of multiple buyers and suppliers. Agents adapt their trust in a partner, the weight they attach to trust relative to profitability, and their own trustworthiness, modeled as a threshold of defection. Adaptation occurs on the basis of realized profit. Trust turns out to be viable under fairly general conditions.

Keywords. Agent-based computational economics, transaction costs, trust

1. Introduction

The viability of trust between firms in markets is a much-debated issue (for a survey, see [1]). Economics, in particular transaction cost economics, doubts the viability of trust. Thus, refuting skepticism from TCE would make the strongest case for trust, and that is the project of this article. In this article we employ TCE logic, but we also deviate from TCE in two fundamental respects.

First, while TCE assumes that optimal forms of organization will arise, yielding maximum efficiency, we consider that problematic. The making and breaking of relations between multiple agents with adaptive knowledge and preferences may yield complexities and path-dependencies that preclude the achievement of maximum efficiency. Second, while TCE assumes that reliable knowledge about loyalty or trustworthiness is impossible [2,3], so that opportunism must be assumed, we expect that under some conditions trust is feasible, by inference from observed behaviour, and that trustworthiness is viable, in yielding profit. To investigate this, the methodology of ACE enables us to take a process approach to trust [4,5,6], by modeling the adaptation of trust and trustworthiness in the light of experience in interaction.

The analysis is conducted in the context of transaction relations between multiple buyers and suppliers, where buyers have the option to make rather than buy, which is the classical setting for the analysis of transaction costs. We employ a model developed from an earlier model from Klos and Nooteboom [7]. In this model, agents make and break transactions relations on the basis of preferences, based on trust and potential profit.

---

[1] Corresponding Author: Sevastopol National Technical University, Management Department, Streletskaya Bay, Sevastopol 99053, Ukraine; E-mail: alex-gorobets@mail.ru.

The article proceeds as follows. First, further specification is given of technical details of the model. Next, we specify the experiments. The article closes with conclusions.

2. The Model

2.1. Preference and Matching

In the literature on trust distinctions are made between different kinds of trust, particularly between competence trust and intentional trust [1]. Intentional trust refers, in particular, to (presumed) absence of opportunism. That is the focus of TCE and also of the present article. We focus on the risk that a partner will defect and thereby cause switching costs. In our model trust is interpreted as a subjective probability that expectations will be fulfilled [8], which here entails realization of potential profit. Thus, expected profit (E) would be: E = profitability•trust. In the model, agents are assumed to have differential preferences for different potential partners [9], on the basis of a generalized preference score:

$$\text{score}_{ij} = \text{profitability}_{ij}^{\alpha_i} \cdot \text{trust}_{ij}^{1-\alpha_i}, \qquad (1)$$

where: score_{ij} is the score *i* assigns to *j*, $\text{profitability}_{ij}$ is the profit *i* can potentially make 'through' *j*, trust_{ij} is *i*'s trust in *j* and $\alpha_i \in [0, 1]$ is the weight *i* attaches to profitability relative to trust, i.e. the 'profit-elasticity' of the scores that *i* assigns; *i* may adapt the value of α_i from each timestep to the next.

At each time step, all buyers and suppliers establish a strict preference ranking over all their alternatives. Random draws are used to settle the ranking of alternatives with equal scores. The matching of partners is modeled as follows. On the basis of preferences buyers are assigned to suppliers or to themselves, respectively. When a buyer is assigned to himself this means that he makes rather than buys. In addition to a preference ranking, each agent has a 'minimum tolerance level' that determines which partners are acceptable. Each agent also has a quota for a maximum number of matches it can be involved in at any one time. A buyer's minimum acceptance level of suppliers is the score that the buyer would attach to himself. Since it is reasonable that he completely trusts himself, trust is set at its maximum of 1, and the role of trust in the score is ignored: $\alpha = 1$. The algorithm used for matching is a modification of Tesfatsion's (1997) deferred choice and refusal (DCR) algorithm [10] and it proceeds in a finite number of steps, as follows:

(1) Each buyer sends a maximum of o_i requests to its most preferred, acceptable suppliers. Because the buyers typically have different preference rankings, the various suppliers will receive different numbers of requests.

(2) Each supplier 'provisionally accepts' a maximum of a_j requests from its most preferred buyers and rejects the rest (if any).

(3) Each buyer that was rejected in any step fills its quota o_i in the next step by sending requests to next most preferred, acceptable suppliers that it has not yet sent a request to.

(4) Each supplier again provisionally accepts the requests from up to a maximum of a_j most preferred buyers from among newly received and previously provisionally accepted requests and rejects the rest. As long as one or more buyers have been rejected, the algorithm goes back to step 3.

The algorithm stops if no buyer sends a request that is rejected. All provisionally accepted requests are then definitely accepted.

2.2. Trust and Trustworthiness

Trust, taken as inferred absence of opportunism, is modelled on the basis of observed absence of defection. Following [4], we assume that trust increases with the duration of a relation. As a relation lasts longer, one starts to take the partner's behavior for granted, and to assume the same behavior (i.e. commitment, rather than breaking the relation) for the future. An agent *i*'s trust in another agent *j* depends on what that trust was at the start of their current relation and on the past duration of their current relation:

$$t_i^j = t_{init,i}^j + (1 - t_{init,i}^j)\left(1 - \frac{1}{fx + 1 - f}\right), \qquad (2)$$

where t_i^j = agent *i*'s trust in agent *j*,

$t_{init,i}^j$ = agent *i*'s initial trust in agent *j*,

x = the past duration of the current relation between agents *i* and *j, and*
f = trustFactor.

This function is taken simply because it yields a curve that increases with decreasing returns, as a function of duration x, with 100% trust as the limit, and the speed of increase determined by the parameter f.

In addition, there is a base level of trust, which reflects an institutional feature of a society. It may be associated with the expected proportion of non-opportunistic people, or as some standard of elementary loyalty that is assumed to prevail. If an agent *j*, involved in a relation with an agent *i*, breaks their relation, then this is interpreted as opportunistic behavior and *i*'s trust in *j* decreases; in effect, *i*'s trust drops by a percentage of the distance between the current level and the base level of trust; it stays there as *i*'s new initial trust in *j*, $t_{init,i}^j$ until the next time *i* and *j* are matched, after which it starts to increase again for as long as the relation lasts without interruption.

The other side of the coin is, of course, one's own trustworthiness. This is modelled as a threshold τ for defection. One defects only if the advantage over one's current partner exceeds that threshold. It reflects that trustworthiness has its limits, and that trust should recognize this and not become blind [1,11]. The threshold is adaptive, as a function of realized profit.

2.3. Costs and Profits

In sum, the way profits are made is that buyers may increase returns by selling more differentiated products, and suppliers may reduce costs by generating production efficiencies.

There are two sources of production efficiency: economy of scale from a supplier producing for multiple buyers, and learning by cooperation in ongoing production relations. Economy of scale can be reaped only in production with general-purpose assets, and learning by cooperation only in production that is specific for a given buyer, with buyer-specific assets. This yields a link with the fundamental concept, in TCE, of 'transaction specific investments'. We assume a connection between the differentiation of a buyer's product and the specificity of the assets required to produce it. In fact, we assume that the percentage of specific products is equal to the percentage of dedicated assets. This is expressed in a variable $d_i \in [0, 1]$. It determines both the profit the buyer will make when selling his products and the degree to which assets are specific, which determines opportunities for economy of scale and learning by cooperation. Economy of scale is achieved when a supplier produces for multiple buyers. To the extent that assets are specific, for differentiated products, they cannot be used for production for other buyers. To the extent that products are general purpose, i.e. production is not differentiated, assets can be switched to produce for other buyers. In sum, economy of scale, in production for multiple buyers, can only be achieved for the non-differentiated, non-specific part of production, and economy by learning by cooperation can only be achieved for the other, specific part.

Both the scale and learning effects are modelled as follows:

$$y = \max\left(0, 1 - \frac{1}{fx + 1 - f}\right), \tag{3}$$

where

for the scale effect, f=scaleFactor, x is general-purpose assets of supplier j summed over all his buyers and scale efficiency $y = e_{s,j}$;

for the learning effect, f=learnFactor; x is the number of consecutive matches between supplier j and buyer i and learning efficiency $y = e^{i}_{l,j}$.

Function (3) expresses decreasing returns for both scale and experience effects. For the scale effect, it shows positive values along the vertical axis only for more than 1 general-purpose asset. This specifies that a supplier can be more scale-efficient than a buyer producing for himself only if the scale at which he produces is larger than the maximum scale at which a buyer might produce for himself. For the learning effect, a supplier's buyer-specific efficiency is 0 in their first transaction, and only starts to increase if the number of transactions is larger than 1. If a relation breaks, the supplier's efficiency due to his experience with the buyer drops to zero.

All this results in the following specification of profit. The number of general-purpose assets that a supplier j needs in order to produce for a buyer i, is equal to $(1-d_i)(1-e_{s,j})$. The number of buyer-specific assets that a supplier j needs, to produce for a buyer i, is equal to $d_i(1-e^i_{l,j})$. Thus, the profit that can potentially be made in a transaction between a buyer i and a supplier j is:

$$p^j_i + p^i_j = (1+d_i) - (d_i(1-e^i_{l,j}) + (1-d_i)(1-e_{s,j})). \tag{4}$$

The first part of (4) specifies returns and the second part specifies costs. It is assumed that the agents involved share the profit equally. In other words, we allow for defection but not for the threat of defection with the purpose of increasing one's share in jointly produced added value.

2.4. Adaptation

An agent is adaptive if 'the actions of the agent in its environment can be assigned a value (performance, utility, or the like); and the agent behaves in such a way as to improve this value over time' [12]. In this model, agents adapt the values for $\alpha \in [0, 1]$ (weight attached to profit relative to trust) and $\tau\, [0, 0.5]$ (threshold of defection) from one time step to the next, which may lead to changes in the scores they assign to different agents. Here, adaptation takes place on the basis of past, realized profit. While τ could conceivably rise up to 1, a maximum of 0.5 was set because initial simulations showed that otherwise relations would get locked into initial situations, with little switching. Note that this biases the model in favour of opportunism.

At each time step, each agent assigns a 'strength' to each possible value of α and τ. This expresses the agent's confidence in the success of using that particular value. The various strengths always add up to constants C_α and C_τ, respectively. At the start of each timestep, the selection of values for α and τ is stochastic, with selection probabilities equal to relative strengths, i.e. strengths divided by C_α and C_τ, respectively. The strengths of the values that were chosen for α and τ at the start of a particular timestep are updated at the end of that timestep, on the basis of the agent's performance during that timestep, in terms of realized profit: the agent adds the profit obtained during the timestep to the strengths of the values that were used for α or τ. After this, all strengths are renormalized to sum to C_α and C_τ again [13]. The idea is that the strength of values that have led to high performance (profit) increases, yielding a higher probability that those values will be selected again. This is a simple model of 'reinforcement learning' [13,14,15,16].

2.5. The Algorithm

The algorithm of the simulation is presented by the flowchart in Figure 1. This figure shows how the main loop is executed in a sequence of discrete time steps, called a 'run'. Each simulation may be repeated several times as multiple runs, to even out the influence of random draws in the adaptation process. At the beginning of a simulation, starting values are set for certain model parameters. The user is prompted to supply the number of buyers and suppliers, as well as the number of runs, and the number of timesteps in each run. At the start of each run, all agents are initialized, e.g. with starting values for trust, and selection probabilities for α and τ. In each timestep, before the matching, each agent chooses values for α and τ, calculates scores and sets preferences. Then the matching algorithm is applied. In the matching, agents may start a relation, continue a relation and break a relation. A relation is broken if, during the matching, a buyer does not send any more requests to the supplier, or he does, but the supplier rejects them.

After the matching, suppliers that are matched to buyers produce and deliver for their buyers, while suppliers that are not matched do nothing. Buyers that are not matched to suppliers produce for themselves ('self-matched', in 'make' rather than 'buy'). Afterward, all buyers sell their products on the final-goods market. Profit is shared equally with their supplier, if they have one. Finally, all agents use that profit to update their preference rankings (via α and τ), used as input for the matching algorithm in the next timestep. Across timesteps realized profits are accumulated for all buyers and suppliers, and all the relevant parameters are tracked.

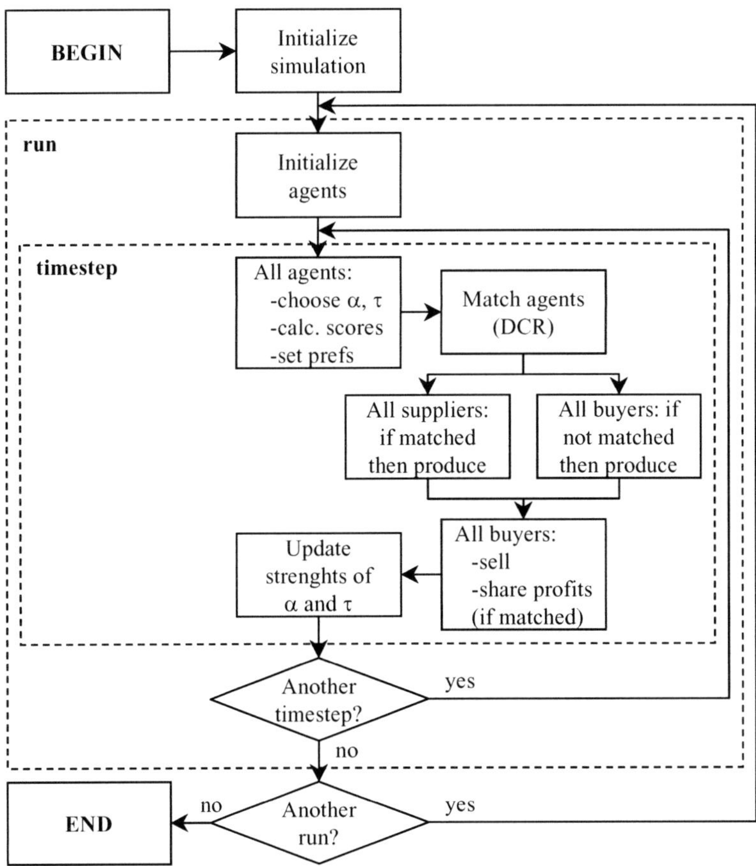

Figure 1. Flowchart of the simulation.

3. Experiments

3.1. Hypotheses

The goal of the experiments is to test the following hypotheses. While according to TCE maximum efficiency can be achieved, we expect:

Hypothesis 1: due to complexities of interaction maximum efficiency can rarely be attained.

According to TCE, high asset specificity leads to more make rather than buy. We expect the same result in our extended framework, according to the same argumentation.

Hypothesis 2: When trust is low, higher asset specificity/differentiated products yields less outsourcing.

In agreement with TCE we expect: *Hypothesis 3:* The more trust, the more collaboration in 'buy', rather than 'make'. More specifically:

Hypothesis 3a: The lower the weight attached to profit relative to trust (α), the more collaboration (buy rather than make), and the more loyalty (less switching).

Hypothesis 3b: The higher the threshold of defection (τ), the more collaboration (buy rather than make), and the more loyalty (less switching).

Counter to TCE we expect:

Hypothesis 4: Even in markets, where profit guides adaptation, high trust (low α; high τ) may be sustainable.

Recall that if during the matching between buyers and suppliers a buyer decides to 'buy' rather than 'make', he can follow two different strategies. One is an opportunistic scale strategy, where the buyer seeks a profit increase on the basis of economy of scale, by trying to find a supplier who serves more than one buyer. This entails much switching and less emphasis on loyalty and trust. The other strategy is the learning by cooperation strategy, seeking an increase of profit in ongoing relations. This entails less switching and more emphasis on loyalty and trust. Thus, in manipulating the strength of the scale effect relative to the effect of learning by cooperation, we can bias the model towards opportunism or loyalty. This interacts with the degree of asset specificity/specialization, since economy of scale applies only to general purpose assets, and learning by cooperation only to specific assets. Note that there is an overall bias towards the opportunistic scale strategy, in that economy of scale is immediate, thus yielding a more immediate return in profits, while learning by cooperation takes time to build up. Thus, we are stacking the odds in favour of the TCE theory that we criticize. However, this does seem to be a realistic feature, supporting the intuition that trust is more viable in a long-term perspective.

3.2. Model Parameters

Each simulation run involves 12 buyers and 12 suppliers and continues for 100 timesteps. In order to reduce the influence of random draws, each run is repeated 25 times and results are averaged across all runs. Initially, results are also averaged for the two classes of agents: buyers and suppliers, in order to explore systematic effects. Each buyer's offer quota was fixed at 1, and each supplier's acceptance quota was set to 3. In previous experiments with each supplier *j*'s acceptance quota set to the total number of buyers, the system quickly settled in a state where all buyers buy from a single supplier. For this reason, suppliers were only allowed to have a maximum of three buyers. This limits the extent of the scale economies that suppliers can reach. A maximum number of buyers may be associated with competition policy setting a maximum to any supplier's market share.

To test Hypothesis 1, we analyse outcomes in terms of cumulative profit, to see to what extent maximum attainable profits are in fact realized, and how this varies across runs of the model. To test Hypothesis 2, we consider different values for the percentage of specific assets/differentiated products: d = 25, 45, and 65%. In addition, to test Hypotheses 3 and 4, we vary initial trust in the range 10, 50 and 90%, initial threshold for defection (τ) from 0 to 0.5, initial weight attached to profit relative to trust (α) from 0.0 to 1.0.

We present and discuss averages, across runs as well as agents (all buyers, all suppliers), as an indication of overall results. We present the results in the order of different starting values of trust. This reflects different institutional settings, from high to low trust 'societies'. Here, we can see to what extent those are stable or shift. In particular, the question is whether high initial trust can be sustained (Hypothesis 3), and whether perhaps distrust can evolve into trust.

3.3. High Initial Trust

First, we consider an initial situation of high, 90% trust across all agents. This reflects a society with a general assumption of high trustworthiness. First, we take intermediate initial expected values for α (0.5) and τ (0.25). Next to the variation of degree of specificity (d = 0.25, 0.45, 0.65), we vary the strength of economy of scale (scale factor sf) and learning by cooperation (lf), as follows:

- both medium strength (lf = sf = 0.5);
- high learning (lf = 0.9), medium scale (sf = 0.5). This is expected to favour a learning by cooperation strategy, with high loyalty;
- medium learning by cooperation (lf = 0.5), high scale (sf = 0.9). This is expected to favour a scale strategy, with less loyalty.

The results are given in Table 1.

Table 1. Buyers' maximum normalized profits for different learn and scale factors

d	#S.per buyer	Buyers max. normal. Profit			
		l.f.=0.5; s.f.=0.5	l.f.=0.9; s.f.=0.5	l.f.=0.5; s.f.=0.9	l.f.=0.9; s.f.=0.9
0.65	1	0.98	0.994	0.978	0.99
0.45	1	0.91	0.92	0.89	0.90
0.25	1	0.80	0.81	0.77	0.78

High initial trust is sustained, and in fact increases from 0.9 to the maximum of 1.0

Table 1 supplies maximum normalized profit actually achieved in the course of time. It is obtained by dividing the buyers' profits by the maximum attainable (theoretical) profit they can potentially make in each experiment, which depends on differentiation, and on learn and scale factors. Maximum attainable profit is the profit a buyer makes when he has an infinite relation with a supplier who produces to an infinite number of buyers. The latter factor is limited because suppliers' acceptance quota is set to 3. Usually maximum actual profit is achieved at the last steps of simulation because of adaptation processes in relations between buyers and suppliers. At the start point the normalized profit is about 52% for high d and 61% for low d.

Table 1 shows that buyers perform better for high d then for low d. This is partly built-in: differentiated products are assumed to give higher profit margins. The outcome also results from the effect of economy of scale for general purpose (non-specific assets) and learning by cooperation for specific assets. Maximum scale effect is achieved when d is low. Here, the maximum arises in a situation where which 12 buyers together buy from only 4 suppliers (each, i.e. one third of all suppliers producing for the maximum of three buyers). Because the optimal network configuration, where suppliers produce for 3 buyers, emerges rarely, buyers organize closer to the optimum when d is higher. Maximum profit is approached more closely when products are more differentiated, because then buyers are less sensitive to the optimal configuration of network between agents. Then, a buyer has less scope for increased efficiency by getting into an arrangement of one supplier producing for him as well as two other buyers.

Now we turn to the hypotheses. Maximum actual profit never achieves maximum attainable profit, which confirms Hypothesis 1. The high levels of initial trust are sustained, and in fact increase, on average, from 0.9 to the maximum of 1.0, which confirms Hypothesis 3. Other results, not specified in the table, show that here there is maximum outsourcing: each buyer has a supplier, even at high levels of asset specificity (d=0.65). This is in agreement with Hypothesis 2. At high levels of trust, outsourcing takes place even at high levels of asset specificity. For all levels of asset specificity (d), in each run at least one supplier produced for the maximum of 3 buyers, on average across runs 10% of suppliers did this, 15% of suppliers produced for 2 buyers, 40% for 1 buyer, and 35% for 0 buyers. The results indicate that in this high-trust society buyers follow the strategy of learning by cooperation and loyalty for all d, without switching between suppliers, even for the low value d=0.25, where only 25% of assets are subject to learning by cooperation.

So far, we assumed intermediate levels for the initial weight attached to profit (α) and for the threshold of defection (τ). Now we analyze the effects of varying those values: $\alpha = 0.0$ and 1.0; $\tau = 0.0$ and 0.5. Learn and scale factors are fixed at the average level, i.e. 0.5. The purpose of this experiment is to explore the effects on outsourcing and profit of initial values of focus on profit and threshold of defection. According to hypotheses 3a and 3b, a low value for α and a high value for τ would favour a trust strategy of durable collaboration and loyalty, but the net effects are difficult to judge a priori. A high threshold of defection would tend to favour outsourcing and stable relations, particularly when initial trust is also high, provided such relations are profitable. However, a high weight attached to profit may prevent outsourcing, whereby effects of loyalty do not get a chance to arise. Here, we use the power of the ACE methodology to explore outcomes of processes that are too complex to compute analytically. The results are given in Table 2. Here, we also supply the average number of suppliers per buyer, as an indicator of the extent of outsourcing.

Table 2. Buyers' maximum normalized profits for different α and τ

d	Buyers max. norm. Profit & #Suppl. per buyer							
	α=0.0; τ=0.0		α=0.0; τ=0.5		α=1.0; τ=0.0		α=1.0; τ=0.5	
0.65	0.96	1	0.99	1	0.96	0	0.99	0.3
0.45	0.91	1	0.91	1	0.85	0	0.92	0.4
0.25	0.80	1	0.80	1	0.82	0.5	0.84	0.6

When α=0, agents put their emphasis on trust and follow the strategy of learning by cooperation for all d. The distribution of suppliers between buyers in this case is the same as before (Table 1). Each buyer has ongoing transactions with the same supplier but when loyalty is equal to zero (τ=0) buyers sometimes break relations with suppliers for high d because then profit doesn't exceed the level of when they make. These buyers try to switch to other suppliers but they don't succeed because all agents are concentrated on trust built up in the past of their current relation. Opportunistic buyers then return to their initial partners and as a result they lose in profit slightly, for high d, because of switching costs. If loyalty is high (τ=0.5) there is no switching for any level of d, and agents try to generate as much profit as possible in stable relations by using the advantage of loyalty and trust, in learning by cooperation.

When α=1, agents focus on profitability rather than on trust, and buyers follow two strategies simultaneously: some of them buy from suppliers and others make themselves. If τ=0.0 approximately half of buyers have suppliers for d=0.25 and these buyers follow the scale strategy, seeking a supplier who already serves two buyers, and trying to match with him. As a result, in this case 17% of all suppliers produce for three buyers. For d=0.45 and d=0.65 buyers prefer to make themselves, mostly because outsourcing is only preferred as relations with suppliers last longer and generate economies of learning, but this is unlikely to happen at zero loyalty. However, because of high initial trust buyers try to reach suppliers sometimes and then lose profit a little because of switching costs. If τ=0.5 the proportion of buyers who have suppliers increases for all d: 60% of buyers have suppliers for d=0.25, 40% for d=0.45 and 30% for d=0.65. However, the distribution of suppliers over buyers is different for all d. When d=0.25 approximately 20% of suppliers produce for three buyers and therefore profit is higher than in the case with τ=0.0. When d=0.45 about 12% of suppliers produce for three buyers and 5% of suppliers produce for one buyer and when d=0.65 suppliers produce only for one buyer and it is about 30% of them. Therefore for low and average d more buyers follow the scale strategy because high loyalty allows them to keep stable relations with matched suppliers and generate higher profit than in the case with zero loyalty. For high d one part of buyers (70%) produce themselves and other part (30%) follow the strategy learning by cooperation because economies of learning are more important than scale effect.

In sum, overall the results confirm hypotheses 3a and 3b, but there are complicated interaction effects that can be calculated only by simulation. Counter to hypothesis 3a, a high weight attached to profitability relative to trust (α) does not always favour opportunism. Once a buyer accumulates efficiency by learning by cooperation, an emphasis on profit also favours loyalty, not to lose the benefit of learning by cooperation.

3.4. Average and Low Initial Trust

Now we turn to 'societies' with a lower, average level of initial trust: X = 0.5. Learn and scale factors are again fixed at the average level, i.e. 0.5. The main outcome here is that buyers make for high and average levels of specific assets (d), and buy only for low levels. This confirms Hypothesis 2. The results are specified in Table 3.

Table 3. Buyers' maximum normalized profits for average initial trust

d	#Suppl. per buyer	Buyers max. normal. Profit
0.65	0	0.99 trust remains at 0.5
0.45	0	0.87 trust remains at 0.5
0.25	1	0.80 trust increases to 1.0

At first sight, it may seem counter-intuitive that trust increases from an average to the highest level under low specific assets (d = 0.25), since then the effect of learning by cooperation is lowest, so that the rewards of a trust strategy seem lowest. The explanation is as follows. Under average trust, suppliers are more attractive than buyers consider themselves only for low d, because potential losses in a case of switching are smaller for low d than for high d. For high levels of specificity, buyers never enter into relations with outside suppliers, and thus never profit from collaboration and forego opportunities for the build-up of trust.

Compared with the corresponding case in the high trust world (first column, Table 1), normalized profits are the same for high and low values of d, but lower for intermediate values. The network configuration of suppliers and buyers for low d is the same as in the case of high initial trust: 10% of suppliers produce for 3 buyers, 15% of suppliers produce for 2 buyers, 40 % for 1 buyer, 35% for 0 buyers. Buyers follow the learning by cooperation strategy in ongoing relations without switching.

In the case of low initial trust, i.e. X=0.1, buyers produce themselves (have no suppliers) even for a low level of specific assets. This again confirms hypothesis 2. The results are specified in Table 4. The result is a drop of normalized profits for low d, compared to the medium and high trust cases. All opportunities for learning by cooperation in collaboration are foregone.

Table 4. Buyers' maximum normalized profits for low initial trust

d	#Suppl. per buyer	Buyers max. normal. Profit
0.65	0	0.99 trust remains at 0.1
0.45	0	0.87 trust remains at 0.1
0.25	0	0.63 trust remains at 0.1

Overall, the results show that under high trust (Table 1, column 1) maximum realized normalized profit is higher than under low trust (Tables 3 and 4) for intermediate and low levels of asset specificity (0.91 and 0.80 vs. 0.87 and 0.63), and marginally lower for high asset specificity (0.98 vs. 0.99). Overall, this confirms the central hypothesis 4 that trust can well be viable in markets.

4. Conclusions

The overall outcome is that in interactions between agents maximum efficiency is seldom achieved and that both trust and opportunism can be profitable, but they go for different strategies. This suggests that there may be different societies, going for different strategies, of switching or of loyalty, which settle down in their own self-sustaining systems.

High initial trust dictates buy relative to make for all levels of specific investments. For high specific investments, buyers' maximum profit is almost the same as in the cases of average or low initial trust. Low initial trust imposes make relative to buy, but buyers' maximum profits for low specific investments are smaller than in the case of high initial trust. Overall, across all parameter settings, profit tends to be higher under high than under low trust.

In addition to the expected results, incorporated in the hypotheses, the model yields a few unanticipated results. One is that buyers organize closer to maximum possible efficiency for high levels of specific investments. The reason is that for low levels of specific investments there is more scope for scale effects, but this is difficult to attain by having suppliers supply to the maximum number of buyers. A strong effect of learning by cooperation, a high weight attached to trust, and high loyalty favour the learning by cooperation strategy for high levels of specific investments, while a high weight attached to profit and high loyalty favour the scale strategy for low and average levels of specific investments. Finally, it is not always the case that a high weight attached to profitability relative to trust does not always favour opportunism. Once a buyer begins to profit from learning by cooperation, an emphasis on profit may also lead to loyalty, in an ongoing relationship.

References

[1] Nooteboom, B. (2002), *Trust: forms, foundations, functions, failures and figures*, Edward Elgar: Cheltenham UK.

[2] Williamson, O.E. (1975), *Markets and Hierarchies: Analysis and Antitrust Implications*, The Free Press: New York.
[3] Williamson, O.E. (1993), Calculativeness, trust, and economic organization, *Journal of Law and Economics*, 36/1, 453-486.
[4] Gulati, R. (1995), Does familiarity breed trust? The implications of repeated ties for contractual choice in alliances, *Academy of Management Journal*, 38/1, 85-112.
[5] Zucker, L.G. (1986), Production of trust: Institutional sources of economic structure, 1840-1920, in: B.A. Staw and L.L. Cummings (eds.), *Research in Organizational Behavior*, Vol. 8, JAI Press: Greenwich, Conn, 53-111.
[6] Zand, D.E. (1972), Trust and managerial problem solving, *Administrative Science Quarterly*, 17/2, 227-239.
[7] Klos, T.B. and B. Nooteboom, (2001), Agent based computational transaction cost economics, *Journal of Economic Dynamics and Control*, 25, 503-526.
[8] Gambetta, D. (ed.) (1988), *Trust: The Making and Breaking of Cooperative Relations*, Basil Blackwell: Oxford.
[9] Weisbuch, G., A.P. Kirman and D.K. Herreiner (2000), Market organisation and trading relationships, *Economic Journal*.
[10] Tesfatsion, L.S. (1997), A trade network game with endogenous partner selection, in: H.M. Amman, B. Rustem and A.B. Whinston (eds.), *Computational Approaches to Economic Problems, Advances in Computational Economics*, Vol. 6, Kluwer: Dordrecht, 249-269.
[11] Pettit, Ph. (1995), The virtual reality of homo economicus, *The Monist*, 78/3, 308-329.
[12] Holland, J.H. and J.H. Miller (1991), Artificial adaptive agents in economic theory, *American Economic Review*, 81/2, 365-370.
[13] Arthur, W. Brian (1993), On designing economic agents that behave like human agents, *Journal of Evolutionary Economics*, 3/1, 1-22.
[14] Arthur, W. Brian (1991), Designing economic agents that act like human agents: A behavioral approach to bounded rationality, *American Economic Review*, 81/2, 353-359.
[15] Kirman, A.P. and N.J. Vriend (2000), Evolving market structure: An ACE model of price dispersion and loyalty, *Journal of Economic Dynamics and Control*, ...
[16] Lane, D.A. (1993), Artificial worlds and economics, part II, *Journal of Evolutionary Economics*, 3/3, 177-197.

Part III

Self-Organization/Adaptation for Grid Computing

Self-Organization and Autonomic Informatics (I)
H. Czap et al. (Eds.)
IOS Press, 2005

Autonomic Pervasive Grids: A Session Manager Service for Handling Mobile Users

A. Coronato[a] and G. De Pietro[b]
[a]*DRR-CNR, Via Castellino 111, 80131 Napoli, Italy*
{coronato.a@na.drr.cnr.it}
[b]*ICAR-CNR, Via Castellino 11,, 80131 Napoli, Italy*
{giuseppe.depietro@na.icar.cnr.it}

Abstract. Grid computing environments are being extended in order to present some features that are typically found in pervasive computing environments. In particular, Grid environments have to allow mobile users to access to their services and resources, and self-adapt based on mobile user location and context. Moreover, mobile users have to have the possibility of changing their location without wondering about their pending computations. This requires that the environment be able to self-manage mobile users implicit disconnections, who may also reappear later and be willing of resuming their computations.

In this paper, we present a session manager service for distributed federations of pervasive grids. The service makes mobile users able of leaving a pervasive grid and resuming later, and even in another grid of the federation, their computations. The service relies on the mobile agents technology. In particular, a personal agent is associated to every mobile user active in the environment. The personal agent offers the list of application services available at the location of the user and handles the list of activated services. If the user moves in another grid, the personal agent migrates in the new grid, handles the list of pending services, and updates the list of available services for the new location.

Keywords. Pervasive Grid Computing, Middleware Services, Autonomic Computing.

1. Introduction

In the late 1990s, the Grid computing model has emerged and rapidly affirmed as the new computing paradigm for large-scale resource sharing and high-performance distributed applications.

The term "The Grid" was primarily introduced by Foster and Kesselman to indicate a distributed computing infrastructure for advanced science and engineering [1]. Successively, it has been extended to denote the virtualization of distributed computing and data resources such as processing, network bandwidth and storage capacity to create a single system image, granting users and applications seamless access to vast IT capabilities.

As a result, Grids are geographically distributed environments, equipped with shared heterogeneous services and resources accessible by users and applications to solve complex computational problems and to access to big storage spaces.

Recently, Grid computing environments are being extended in order to present some characteristics that are typically found in pervasive computing environments.

The goal for Pervasive Computing is the development of environments where highly heterogeneous hardware and software components can seamlessly and spontaneously interoperate, in order to provide a variety of services to users independently of the specific characteristics of the environment and of the client devices [2]. Therefore, mobile devices should come into the environment in a natural way, as their owner moves, and transparently, that is owner will not have to carry out manual configuration operations for being able to approach the services and the resources, and the environment has to be able to self-adapt and self-configure in order to host incoming mobile devices.

The conjunction of the two paradigms is leading towards Pervasive Grid environments [3]. A key feature for Pervasive Grids is the possibility of making mobile users able to get access to the Grid and to move both among different areas within the same Grid environment and among different Grid environments. For this kind of environments, the list of available services may depend on the physical location the user is in (as an example, a video conference can be produced only in sites equipped with video cameras and streaming systems). In addition, users could leave the virtual environment while having pending computations. After that, the environment has to decide whether to let computations go on, or to suspend them and free their allocated resources. For an example, if the user disappears from the environment after having started a real-time video streaming, likely the user is no more interested in the service, then the environment has to free all allocated resources. Differently, if he is no longer active in the environment after having launched a computational process for solving a complex differential equations system, probably he will reappear later to pick up results. In this case, allocated resources must be kept up working even though the user is not active in the environment.

These characteristics call for advanced sessions manager services, which must confer to environment autonomic behaviours as self-management, self-configuration, and self-adaptation. As a matter of fact, when a user is no longer active in a site, it is possible that he has definitively left the environment, or his device has failed, or he has left the physical site in order to reach another site, or more simply he has turned off his device to save battery or to have a break. Actually, since active devices affect the environment in terms of in use/available resources and services, we can conclude that the environment must reliably self-manage resources and services when users move around. It must be able to predict whether the user will require to resume its computation later, or not. Then, active resources must be kept allocated, or freed, depending on what user is supposed to do.

This paper describes strategies for conferring autonomic characteristics to a federation of pervasive grids and services that make the environment able to handle sessions for mobile users. In particular, we propose an agent-based session manager service architecture that can apply to federation of grids and confer pervasive and autonomic characteristics to the environment.

The rest of the paper is organized as follows. Section 2 presents the architecture of a federation of pervasive grids. Section 3 describes the strategies for handling sessions in the environment and the high-level service architecture. In section 4 the implementation details are outlined. Finally, section 5 concludes the paper.

2. A Federation of Pervasive Grids

Grid applications can be physically placed in different sites, which are topologically organized to compose intra-Grids, extra-Grids or inter-Grids [4,5]. Intra-Grids pool resources offered by different divisions existing within a single organization, which defines policies to share, access and use such resources. Computers of a single organization use a common security domain and share data internally on a private/LAN network. Extra-Grids bring together more intra-Grids using a remote/WAN connectivity. They are based on multiple organizations, characterized by more than one security domain, which cooperate among them through external partnerships in order to address about business, academic and other topics. For this reason, these partners have to agree common policies, in addition to or extending the existing single organization ones, to enable the access to services and resources of the extra-Grid. Finally, inter-Grids are based on many organizations distributed in the world, which pool their own resources using an Internet/WAN connectivity in order to sell, for example, computational or storage capacity. Because of the great number of organizations involved in this type of Grid, a partner agreement is very difficult to realize and so a more complex management about the offered services and their integration and collaboration in a secure context is required.

The environment consists of an extra-Grid, which is a federation of two intra-Grids. These latter have been developed over the Globus Toolkit 4.0 [6] platform and deployed in different physical sites (namely, Site1 and Site2). Both physical environments are equipped with wireless access points, and interconnected via the internet.

The intra-Grid 1 is equipped with a cluster of Linux PCs, printers, a multimedia room, and other resources. In the intra-Grid 2, several computational resources, a multimedia laboratory, a multimedia room, and an e-testing room are available. In detail, intra-Grid 2 is made up by the following computational resources:

- *Rendering Station* – This is a workstation for rendering row motion data in 3D graphic applications; it is a 12 processors Silicon Graphics workstation supported by the IRIX operating system;
- *Projector* - This is a projector, driven by a pc, to project multimedia presentations;
- *E-Testing Server* - This is a 2 processors server that hosts an E-Testing application;
- *Streaming Server* - This server hosts a video streaming application;

The application services available in the environment are the following:

- *RenderingService* - This service enables users to submit row motion data and to build 3D graphic applications. This is a computational service that executes processes on the Rendering Station. The service can be accessed from both Intra_grid 1 and intra-Grid 2;
- *PresentationService* - This service enables a user to project its presentation and to control the presentation flow in a multimedia room. For this reason, this service must be available only in the multimedia room of the intra-Grid 2;

- *VideoConferenceService* - This service enables attendees to follow a presentation on their personal mobile device. A video server streams the presentation captured by the PresentationService over the network;
- *E-TestingService* - This service performs on-line evaluation tests for courseware activities. When a session test starts, students must be in a multimedia room of the intra-Grid 2. Evaluation tests are synchronized and students have a predefined period of time for completing each test section. Students can interrupt their test by explicitly closing the service or by leaving the multimedia room. This service must be available only in the testing room of the intra-Grid 2.

All application services have been deployed as a Grid Service over the Globus platform [7].

Figure 1 shows services and resources available in the environment.

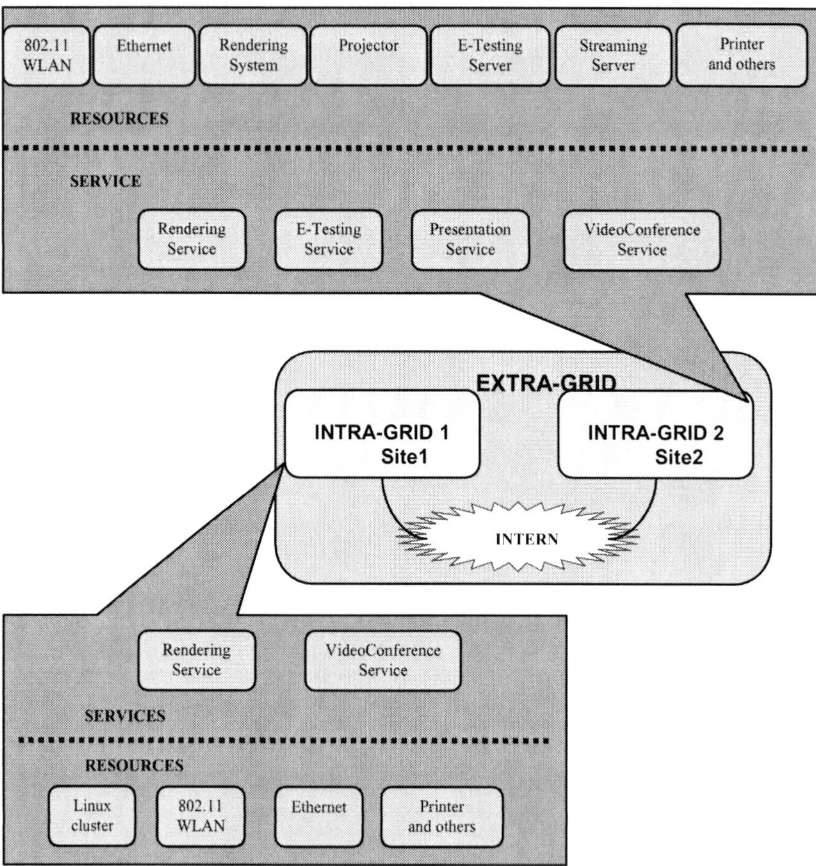

Figure 1 – Architecture

In the environment, mobile users can dynamically enter, leave, or move around. Every time a mobile user enters in a site, the environment locates and tracks him. Location and tracking services have primarily been developed for pervasive environments [8] and successively adapted to grid environments. In addition to this, the environment has to provide the mobile user with the list of services available at his location and has to reliably handle user movements within/outside the environment.

To clarify the desired behavior, let consider a possible scenario. A mobile user, say for an example John Smith, enters Site2 with his laptop. He is supposed to present a new research project in the multimedia room. The presentation will be streamed by the *VideoConferenceService* for the people located in the multimedia room of Site1. As he comes in the environment, he gets wireless access by an access service that transparently recognize and authenticate registered mobile users. Before starting his presentation, John launches the *RenderingService*, which is a long computational service, for an elaboration related to another research project. After having concluded his presentation, John simply turns off his device, he doesn't wait for rendering results since that elaboration process could require a few hours. Differently, he wants come back in the environment the day after to pick up results.

In this scenario, there are several relevant events/moments, which require specific actions from the environment. Such events are:

(1) John enters in Site2 – The environment has to locate and provide him with the list of available services;
(2) John launches the *RenderingService* – The environment has to update the list of activated services;
(3) John moves in the multimedia room – The environment has to update the list of available services;
(4) John launches the *PresentationService* – The environment has to update the list of activated services;
(5) John concludes the presentation and turns off his device – The environment has to detect the implicit disconnection, then it has to update the list of active services. In particular, the environment has to free the resources allocated for the *PresentationService*, but it has to let the *RenderingService* continue to elaborate;
(6) John returns the day after in the environment in any Site1 or Site2 – The environment has to locate him and to provide him with the list of available and active services, as well as with the elaboration results.

The desired behavior for the environment is granted by the collaboration of basis service like the location service, the access service, and the session manager service. In the next session, we briefly describe services functionalities and collaborations. After that, we focuses on the session manager service architecture.

3. The Session Manager Service

The session manager service is in charge of handling available and active services for every user in the environment. In order to accomplish its task, it strictly interacts with other environmental services. In particular, it interacts with the *AccessService* and the *LocationService*.

The *AccessService* is able to recognize incoming mobile devices. As a matter of fact, it includes a DHCP component that assign IP addresses to registered mobile devices. When an incoming mobile device obtains an IP address, the *AccessService* communicates such an event to the environment.

The *LocationService* is in charge of locating mobile devices within the environment and of detecting implicit disconnections that occur when a localized mobile device becomes unavailable in the environment. Any location change, as well as any disconnection, is communicated to the environment by the *LocationService*.

The *SessionManagerService* has two major functions. It handles the list of services available at every location of the environment, and it handles the list of services activated by every user. In addition, it decides which services have to continue when the mobile user is no longer active in the environment, and which services have to be terminated. Up to date, a simple policy is adopted. Services are classified as interactive (real-time and highly-interactive services) or computational (lowly-interactive services). Interactive services are freed, whereas computational services are kept up working.

Before introducing the architecture of the session manager service, it is important to clarify the difference between location and site. As a matter of fact, in Site2 three different locations are outlined. In every location, the *RenderingService* and the *VideoConferenceService* are available, whereas the *PresentationService* is available only in the multimedia room (Location1) and the *E-TestingService* is available only in the testing-room (Location2). The rest of the physical area of Site2 is aggregated in the Location3.

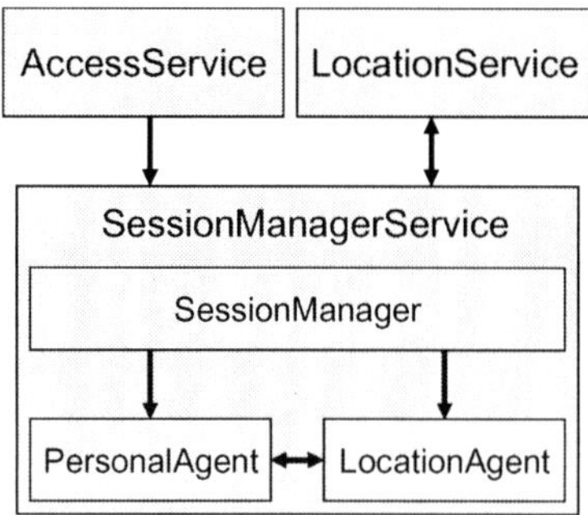

Figure 2 – Session Manager Service Architecture

The service architecture is illustrated in figure 2. The service has four main components:

- *PersonalAgent* – This agent is created when a new user appears in the environment. There's a *PersonalAgent* per every user. It handles the list of services activated by the user. Moreover, it migrates accordingly with user movements;
- *LocationAgent* – This agent is deployed at a specific location. There's a *LocationAgent* per every location. It handles the list of application services available at its location. It interacts with the *PersonalAgent*;

- *SessionManager* – This component is an agent container. There's a *SessionManager* per every site. It creates both Personal and Location Agents, as well as it requires that they move;
- *SessionManagerService* – This component is a centralized unit that coordinates all *SessionManagers*. It interacts with the *LocationService* and the *AccessService*. Consequently, it communicates with the proper *SessionManager*.

4. Implementation Details

All components for the *SessionManagerService* have been fully developed in Java choosing the JADE [9] framework for mobile agents. The JADE framework has been developed by TILAB and supports the implementation of multi-agent systems through a middleware that is fully compliant with the FIPA specifications.

In the following, all interactions that occur among software components, with respect to the scenario described in section 2, are presented. Let's resume and explain them.

John enters in Site2 – John's device requires an IP address to the *AccessService*. When the *AccessService* accords an IP address to John's device, it communicates such an event to the *SessionManagerService*. After that, the *SessionManagerService* interrogates the *LocationService* in order to get John's device location. Then, the proper *SessionManager* component (the one active in Site2) creates a new *PersonalAgent*, which gets the list of available application services for that location from the *LocationAgent*;

John launches the *RenderingService* – The *PersonalAgent* updates the list of activated services;

John moves in the multimedia room – The *LocationService* communicates to the *SessionManagerService* that John's device has been detected in a new location. Therefore, the *SessionManager* component forces the *PersonalAgent* to move in the new location within the same site;

John launches the *PresentationService* – The *PersonalAgent* updates the list of activated services;

John concludes the presentation and turns off his device – The *LocationService* detects the implicit disconnection and communicates such an event to the *SessionManagerService*. The *SessionManager* of Site2 requires that the *PersonalAgent* update its list of activated application services. In particular, the *PersonalAgent* shuts down the *PresentationService*, which is an interactive service, and lets the *RenderingService* working. It's worth noting that the *PersonalAgent* remains active because an application service has to continue its elaboration and the mobile user is supposed to return back in the environment to pick up results;

John returns the day after in the environment in any Site1 or Site2 – The *AccessService* communicates such an event to the *SessionManagerService*. In the case John returns in Site2, the *SessionManager* of Site2 simply move the *PersonalAgent* to the correct location. Differently, if John's device appears in Site1, the *SessionManager* of Site2 creates a clone of the *PersonalAgent* and forces it to move in Site1. From now on, both personal agents cooperates to serve john until each one has active application services to handle.

5. Conclusions and Directions for Future Works

In this paper we presented a session manager service that introduces autonomic and pervasive features in grid environments. This facility provides the grid with support for granting access to mobile users as well as for customizing services depending on the user location.

Future work will aim to improve the session manager service making it able to customize services depending on the user's device characteristics. In addition, security aspects, which have not been considered yet, will be explored.

References

1. Foster, C. Kesselman, "The Grid: Blueprint for a New Computing Infrastructure", Morgan Kaufmann, 1999.
2. D. Saha and A. Murkrjee, "Pervasive Computing: A Paradigm for the 21st Century", IEEE Computer, March 2003.
3. V. Hingne, A. Joshi, T. Finin, H. Kargupta, E. Houstis, "Towards a Pervasive Grid", International Parallel and Distributed Processing Symposium, IPDPS 2003.
4. L. Ferreira, V. Berstis, J. Armstrong, M. Kendzierski, A. Neukoetter, M. Takagi, R. Bing-Wo, A. Amir, R. Murakawa, O. Hernandez, J. Magowan, N. Bieberstein, "Introduction to Grid Computing with Globus", IBM RedBooks, September, 2003.
5. J. Joseph, M. Ernest, C. Fellenstein, "Evolution of grid computing architecture and grid adoption models", IBM Systems Journal, December, 2004.
6. I. Foster, "Globus Toolkit Version 4: Software for Service-Oriented Systems", IFIP International Conference on Network and Parallel Computing, Springer-Verlag LNCS 3779, pp 2-13, 2005, also available on-line at: www.globus.org.
7. Foster, C. Kesselman, J. Nick, S. Tuecke, "The Physiology of the Grid: An Open Grid Services Architecture for Distributed Systems Integration", Open Grid Service Infrastructure WG, Global Grid Forum, June 22, 2002.
8. Coronato and G. De Pietro, "Location and Tracking Services for a Meta-UbiComp Environment", in the proc. of the 3rd International Metainformatics Symposium (MIS 2004) as Lecture Notes in Computer Science, LNCS 3511, Springer Verlag, 2005.
9. F. bellifemmine, A. Poggi and G. Rimassa, "Jade Programmers Guide", http://sharon.cselt.it/projects/jade

Self-Organization and Autonomic Informatics (I)
H. Czap et al. (Eds.)
IOS Press, 2005

Avalanche Dynamics in Grids: Indications of SOC or HOT?

A Vijay Srinivas [a,1], D Janakiram [a] and M Venkateswar Reddy [a]
[a] *Distributed & Object Systems Lab, Dept. of CS & E, Indian Institute of Technology Madras*

Abstract. Complex systems such as those in evolution, growth and depinning models do not evolve slowly and gradually, but exhibit avalanche dynamics or punctuated equilibria. Self-Organized Criticality (SOC) and Highly Optimized Tolerance (HOT) are two theoretical models that explain such avalanche dynamics. We have studied avalanche dynamics in two vastly different grid computing systems: Optimal Grid and Vishva. Failures in optimal grid cause an avalanche effect with respect to the overall computation. Vishva does not exhibit failure avalanches. Interestingly, Vishva exhibits load avalanche effects at critical load density, wherein a small load disturbance in one node can cause load disturbances in several other nodes. The avalanche dynamics of grid computing systems implies that grids can be viewed as SOC systems or as HOT systems. An SOC perspective suggests that grids may be sub-optimal in performance, but may be robust to unanticipated uncertainties. A HOT perspective suggests that grids can be made optimal in performance, but would then be sensitive to unanticipated perturbations. An ideal approach for grid systems research is to explore a combination of SOC and HOT as a basis for design, resulting in robust yet optimal systems.

Keywords. Distributed Systems, Grid Computing, Self-Organized Criticality (SOC), Highly Optimized Tolerance (HOT), Complex Systems

1. Introduction

Evolution in biological systems, earthquake study in geological systems, stock markets of economical systems and engineered systems such as power grids are examples of complex systems [1]. A key property of such systems is the existence of punctuated equilibria or so called avalanche dynamics [2]. Small perturbations lead to only small localized effects in normal equilibrium systems such as a flat beach or the atoms in a gas. The equilibria or steady state in complex systems is punctuated with avalanches, when small perturbations lead to global changes. The frequency of such avalanches follows a power law behaviour, implying that the large events are equally probable as small events.

Self-Organized Criticality (SOC) has been proposed as a theory to explain the power laws in complex systems [3]. The basis for SOC is phase transitions from statistical physics, in which the system undergoes a transition into the critical region and exhibits

[1]Correspondence to: A Vijay Srinivas, Distributed & Object Systems Lab, Department of CS & E, Indian Institute of Technology Madras, Chennai - 600036, India; Email: avs@cs.iitm.ernet.in; Website: http://dos.iitm.ac.in;

interesting properties. SOC systems tend to be robust to unanticipated perturbations but are not amenable to optimizing any design objective. An alternative theory, namely Highly Optimized Tolerance (HOT) has been proposed [4] to explain complex behaviour in designed systems. A critical feature of HOT is that systems are robust to *designed for* (predictable) uncertainties and optimal, but are vulnerable to design flaws and unanticipated perturbations.

Large scale distributed systems can also be perceived as complex systems. A clustered distributed data storage mechanism has been viewed as a complex system in [5]. The authors explain why existing design techniques based on designing for predictable operational conditions (known as precognition) may fail in the face of unexpected perturbations. This may lead to butterfly effects, where a small perturbation could result in global changes.

Grid computing systems are also large scale distributed systems, albeit with a few special characteristics. These include the presence of virtual organizations, which are autonomous entities that aggregate their resources to enable wide-area resource sharing [6]. Scalability and robustness become key issues and both have not yet been addressed extensively by the grid community [7].

Electrical grids, being similar to computing grids, have been perceived as complex systems. The theoretical basis for electrical grids have been explored in [8,9]. The power law behaviour of electric grids (the frequency of blackouts plotted against time in US electric grid) have been shown as evidence for SOC in [8]. It has been argued that as electric grids are designed systems, the complex behaviour arises due to design optimization [9]. However, to the best of our knowledge, theoretical basis for grid computing systems have not been explored.

We focus this paper on the theoretical foundation for grid computing systems, especially focusing on robustness and scalability issues. First, we study avalanche dynamics in two real grid computing systems: Optimal grid [10] and Vishva, our own P2P grid system [11]. The avalanche dynamics are explained theoretically by both SOC and HOT theories of complex systems. We also explore the consequences of applying SOC and HOT, resulting in an important implication for grid computing system designers: A combination of both SOC and HOT theories is the ideal basis for system design and needs to be explored.

2. Complex Systems Theory: SOC and HOT in Brief

2.1. SOC: A Theory of Complex Systems in Nature

A simple sand pile model has been studied in [3]. Assume that sand is dropped on to a table (at any random position in the table) grain by grain. Initially, the sand will be stationary. This is the flat state or equilibrium state. As the amount of sand dropped increases, over a long period of time, the sand forms several bunches into a big pile. As sand continues to be dropped, small (big) piles could break down, depending on the slope, causing sand to fall off. Sometimes, this could result in sand being added to an adjacent pile. If the slope of this pile also exceeds a certain value, it also breaks down. Thus, addition of even a single grain of sand could result in avalanche effect. The system is said to have attained criticality. At this stage, addition of even one grain of sand could result in an avalanche (large events) or it could only result in local effects. The frequency of occurrence of large events in the sand pile obeys a power law. Thus, this has been touted as the perfect example of a so called self-organized critical system.

Complexity originates from the ability of these dynamic systems to self-organize into a critical state (edge of chaos) where minor perturbations may lead to large events or avalanches. One example of such large events is the extinction of dinosaurs in biology. A parameter called density leads the system from normal state to Edge of Chaos (EOC)[1] and beyond criticality to chaos. An important characteristic of SOC is the universality principle, which says general features like avalanches or critical components do not depend on model details. Further, external forces serve only to initiate events. Self similarity is also another key property of SOC based systems, with fractals being perfect examples.

Systems that are *not* complex include systems in equilibrium such as gases or economics (Nash equilibrium theory), where small perturbations lead to only small localized effects. Chaos systems are also not complex systems, in the sense that the complexity is exhibited only at the critical density. Beyond the critical density, the system goes into a chaotic state, where small disturbances can lead to unpredictable or chaotic behaviour. Chaos systems do not produce fractal structures, another indication of complexity [3]. Examples of systems that are complex include the game of life, a cellular automata that produces complex behaviour with simple rules.

SOC theory does not have a firm mathematical foundation [3]. The approach to criticality has been modelled mathematically using the Gap equation [2]. This is explained below using the Bak-Sneppen evolution model [13]. Representatives for different species are arranged in a circle. These have fitness levels represented by f, which represent the fitness of the species on the whole. The initial fitness levels for the species is a uniform random distribution. The rules for updating the fitness levels are 1. select the least fit species. 2. mutate it by replacing its fitness with a new random fitness. 3. mutate two random neighbours. These rules are applied repeatedly. Mutation of neighbours represent the interaction of species with others, for instance through a food chain. The fitness function follows the Gap equation [2] given by $f(t) = f_c - A * (t/N)^{\frac{-1}{\gamma-1}}$, where f_c represents the critical fitness value, t is time or the number of update steps, N is the number of species and γ is a critical exponent. Equations to determine f_c and the critical exponent for a particular system are given in [2].

2.2. HOT: Complexity in Designed Systems

Physicists have proposed HOT as an alternative theory, especially in designed systems. The canonical example used to describe HOT is the so-called forest fire model. Consider a (virtual) forest managed by a designer. Trees can be grown by the designer anywhere. The aim of the designer is to maximize yield in the presence of fire risk. Lightening may strike any part of the forest anytime and destroy the set of trees that are clustered together. The designer can construct fire breaks between clusters to prevent fire from spreading too far. It is assumed that the probability distribution of lightening strike is known. The basic idea is that the resources in the system (fire breaks) tend to get concentrated in regions that are more susceptible to fires. Regions that are less susceptible tend to get larger

[1] SOC theory is similar to the Edge of Chaos theory [12] but with some key differences, including the fact that complexity occurs only at EOC whereas in SOC systems, complexity is more robust.

cluster of trees. The distribution of fire sizes follows a power law behaviour for various distributions of sparks. This shows that power laws (and hence complex behaviour) arise due to an external agent optimizing a design objective (yield) [14].

The mathematical perspective of the HOT theory has been dealt with in respect of the forest fire model [4]. The main goal is to achieve optimal resource distribution (one that minimizes the cost of avalanches), given the constraints on resources, resource breaks and an expected spark distribution. The mathematics behind HOT can be explained using the following parameters: X, an abstract d-dimensional space, considered as an event map of interconnected events in the system; p(x) is the probability distribution of initiating events $\forall x \varepsilon X$; A(x) is the region affected by an event at x. $E(A^{\alpha})$ is the expected cost of an avalanche initiated; R(x) is the resource that restricts the size of events; There may be constraints on the resources, say $\int_X R(x)dx =$ a constant. The overall objective is to minimize the expected cost of an avalanche $E(A^{\alpha})$ given by $\int_X p(x)A^{\alpha}(x)dx$. A few assumptions such as that $A(x) = R^{-\beta}(x)$ have been made. Using variational methods, it has been shown that $P_{cum}(A)$, which is the sum of $P(A)$ for regions of size greater than or equal to A, exhibits power law behaviour.

The key difference between SOC and HOT is that in SOC complexity emerges and is independent of model details, whereas in HOT systems, complexity arises due to optimization of design objective. HOT systems tend to be very complicated (in terms of interactions), giving rise to complex behaviour. Another example of a HOT system is Internet routing. Internet routing has been shown to exhibit power law behaviour [15]. Initially, this was attributed to SOC as the complex behaviour can be seen to naturally emerge in a self-organized way, as new machines can be plugged into the system and flow control is performed transparent of applications. However, it can also be observed that the TCP/IP protocol, the basis for Internet routing, is very complicated and has emerged out of decades of research. Further, the power law behaviour of Internet routing is independent of density (congestion level), implying that there is no critical density and hence SOC is not applicable. It has been argued that power laws arise due to optimizing throughput and does not emerge as in SOC [4].

3. Avalanche Dynamics in Grid Computing Systems

This section gives a brief view of two key grid computing systems, Optimal grid and Vishva. It then explains how they exhibit avalanche dynamics.

3.1. Optimal Grid: Failure Avalanches

Optimal grid [10] is an effort at building an autonomic middleware infrastructure for grid applications. It targets applications with connected problems, meaning that they require inter-task communication at certain pre-specified intervals. The Autonomic Program Manager (APM) is the problem coordinator for that application instance. It partitions the problem into Original Problem Cell (OPC) collections. A set of OPC collections represents chunks (called Variable Problem Partitions or VPPs) of the n-dimensional problem space. The APM defines dependencies between the VPPs. It uses T-Spaces, a distributed shared object space based on Java, for the actual inter-task communication. The architecture of optimal grid is depicted in figure 1.

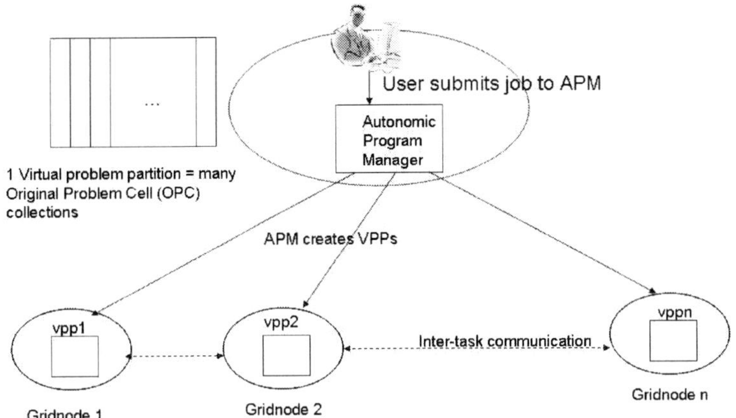

Figure 1. Optimal Grid: A Pictorial Representation

Autonomic features of optimal grid include self-configuring, self-optimizing and self-healing. Self configuration refers to the automatic initial problem space partition based on available nodes and their capabilities in terms of computing power and memory. A chunk goes to each compute agent (CA), with possibly multiple CAs per node. Self-optimizing feature of optimal grid refers to the APM reassigning at each round, the chunks or the number of OPC collections for each CA based on load or compute performance. Self-healing in optimal grid has been conceptualized as the ability to deal with failure of CAs by storing intermediate results in the tuple spaces.

The current implementation available from IBM (http://alphaworks.ibm.com) is not exactly self-healing as it cannot handle failures of both CAs and the APM. The current version of optimal grid exhibits failure avalanches, as a single CA failure (or APM failure) results in stalling the entire computation[2]. This is due to the inherent coupling in the application (such as Finite Element Model (FEM)), which stipulate that the value at a particular point in space-time depends on values of neighbours in the previous time step. Thus, if one node fails, the CAs on neighbouring nodes will not receive data and will start busy waiting. Their neighbours in turn will also start busy waiting. Thus, the entire computation stalls. This is what we refer to as avalanche effects, as a small perturbation (failure of a single node or a CA) results in global effects (stalling the entire computation). This can be seen in table 1, where we have shown the time taken for computation in the case of the eden problem for a 256*256 map, sequence interval of 100 seconds and project duration of 600 iterations.

The table and other studies in the paper have been made on a Intranet Institute wide testbed consisting of about 35 nodes running Linux/Solaris operating systems. The nodes are dispersed in four different clusters, with nodes within the same cluster connected by a 10/100 Mbps ethernet switch, with 10Mbps link across clusters. The nodes are heterogeneous in architecture (some are Intel, some are Sparc), CPU speeds (varies from 133Mhz to 3.2 Ghz) and memory capacity (varies from 256 MB to 1GB).

[2]The effect is observed in spite of the *edgeLogging* parameter set to true in grid.cfg. This is the mechanism

Table 1. Optimal Grid: Computation Time on Eight Nodes

Problem Size (Iterations=600) (Seq Interval=100)	Average Time Taken (seconds)	
	Without Failures	With one CA Failure
256*256 Map	870	∞
512*512 Map	2160	∞

3.2. Vishva Peer-to-Peer Grid: Load Avalanches

Programs which are tied up with specific machines will not be resilient to the dynamic conditions of a grid. Hence, we have designed Vishva, a P2P grid model that explores transparent programmability of parallel tasks in a grid. It enables location independent intertask communication among processes on machines spread over a wide area distributed system. By using P2P techniques, Vishva overcomes certain limitations of existing grids, such as the assumption of a few "almost failure free" dedicated components. To the best of our knowledge, Vishva is one of the few attempts at using P2P techniques in grids, other than organic grid [17] and OurGrid [18]. It must be noted that Entropia [19] is a desktop grid system with a layered sandboxing architecture, but does not use P2P concepts. Further, organic grid does not support inter-task communication. Vishwa uses both structured and unstructured P2P concepts that makes its architecture different from organic grid or OurGrid. The two layered architecture of Vishwa is depicted in figure 2.

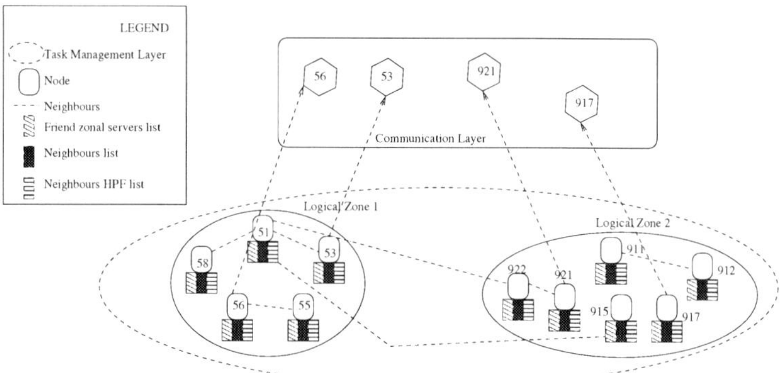

Figure 2. Two Layered Architecture of Vishwa

The bottom layer is a task management layer. It is responsible for initial task deployment on the best available under-utilized nodes as well as the runtime migration of tasks to handle grid dynamics. It can be viewed as an unstructured P2P overlay that groups the participating nodes into proximity based logical zones [20]. Horse Power Factor (a measure of load and capability of a node) advertisement is handled in this layer. Each node maintains HPF information of neighbours. This information is updated by the HPF

to handle CA failures in [16].

advertisement protocol of the unstructured P2P layer. When a task arrives, if there are sufficient number of neighbour nodes, subtasks are spawned on these nodes. If the task requires more nodes, a P2P search is conducted over the unstructured layer within k-hops. Thus, this layer is used for task mobility based on HPF factors. This layer suffices for purely data parallel applications that do not require inter-task communication.

The second layer, the communication layer, enables inter-task communication among the user tasks by using the concept of Distributed Pipes (DP) [21] extended with structured P2P concepts. It also enables Vishwa to handle both coordinator and grid node/network failures. DP provides the abstraction of UNIX pipes across nodes of a distributed system for inter-task communication. Pipe information is maintained among peers over a structured P2P layer (an abstraction of Distributed Hash Tables [22]). This layer also maintains task coordinator information such as donor list, subtask list, associated timeouts etc. It helps to mask coordinator failures and enables the model to scale. The structured P2P layer is capable of automatically configuring itself in the face of failures, leading to the self-configuration property, essential for autonomic grid computing systems.

Vishva does not exhibit failure avalanche effects due to the redundancy introduced by the P2P layers. The structured P2P layer maintains coordinator and donor information. Thus, if a donor fails, the status of its computation is recovered from the structured P2P layer. The task is continued on a different node, thus handling donor failures. Even coordinator information is maintained in $k - 1$ other nodes in the P2P layer and hence, coordinator failures are masked from the donors and applications.

Vishva interestingly exhibits load avalanches at critical load density. Consider a computation in which all nodes in the system have load nearing the critical load (we use CPU utilization of 90% as a threshold for task migration), nearing say $60 - 70\%$ utilization. If load on any one machine increases beyond the threshold (this is the small perturbation), the node tries to migrate the task to its neighbour. The neighbour's node also goes beyond the threshold due to the new migrated task. Hence, it also tries to migrate the task to its neighbour and so on. Thus, a small perturbation leads to avalanche effects. We have experimentally verified this effect by instrumenting the load using a cpuhog[3] process on all nodes.

We have considered not just load, but a factor known as Horse Power Factor (HPF) which is calculated as a summation of processor speed multiplied with percentage of idle time and percentage of free memory [23]. The HPF is a very useful metric for task distribution in clusters or even grids. We have found that the critical HPF density is around 30 units. If all nodes in the system have HPF of 30-40 units, the avalanche effect was observed. A small disturbance (load variation in one node) causes task migration to another node. As a result of this new task, the new node's HPF also comes down, resulting in further task transfer. This is shown in figure 3. It must be noted that higher the HPF, the more tasks a node can take. As the average HPF reduces (due to load increases), small events cause small effects. Beyond a critical HPF, it is visible as an avalnche effect.

[3]A process which deliberately hogs the CPU and generates configurable loads. Check http://www.xmailserver.org/linux-patches/lnxsched.html for more details.

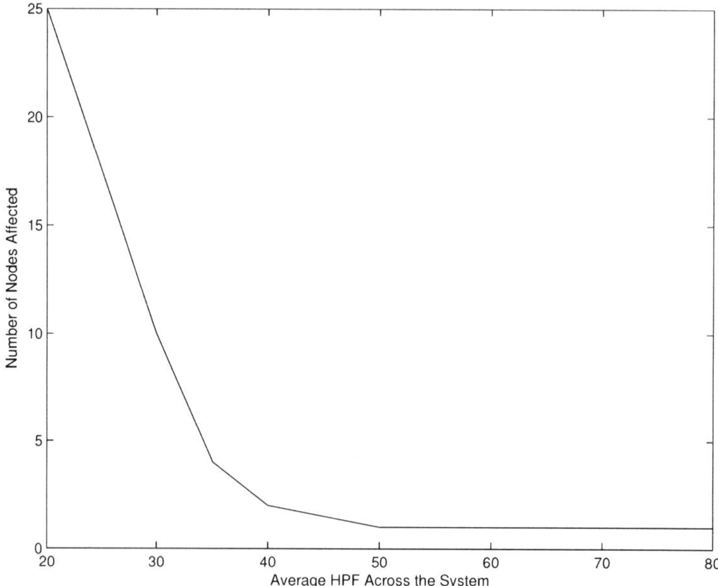

Figure 3. Vishwa: Load Avalanche Effect

4. Are Grids SOC or HOT Systems?

In this section, we argumentatively show that grids may be considered as both SOC and HOT systems. It is difficult, if not impossible to prove the same formally and is left for future research.

The avalanche dynamics exhibited by both the practical grid systems Vishva and optimal grid is an indication that grids may exhibit power law behaviours. In complex system theory, power law behaviour can be attributed conceptually to both SOC and HOT. In the case of SOC, an essential requirement is the separation of time scales, between evolution and actual avalanches [24]. This means, if the system evolves over a long period of time, it may go into criticality and produce avalanches that lasts for much shorter times. Thus, for systems that have been running over long periods, such as the electric grids, it is easier to show SOC behaviour [8]. We can only speculate that grid computer systems may eventually attain criticality.

Computational grid systems can become SOC systems over a period of time. Consider long running computations that may run for days on the grid. During the course of the computation, the load variations in the system may follow the gap equation. This implies that the load density will keep increasing and reach a critical value, beyond which load avalanches may be exhibited. If users start regularly submitting jobs, there may be several such jobs running parallely. As user demand increases, load increases and system capacity needs to increase consequently, either by addition of new nodes/networks or by upgrading existing nodes/networks. Thus, the two forces - driving force of user demand and the relaxing force of capacity addition results in the system state (in terms of loading pattern) operating near criticality. This is similar to electric grids, where the driving force of customer load and relaxing force of response to blackouts through capacity additions drives the system to criticality [8]. In addition, many grids are incorporating self-organizing properties in order to make them autonomic, for instance organic grid [17]. This makes them more susceptible to attain criticality. Thus, it can be argued that grid computing systems may become SOC systems.

A main consequence of an SOC perspective is that the system is likely to be sub-optimal with respect to yield (in terms of throughput, response time or scalability). However, a good aspect of SOC system is that they are insensitive to $p(x)$, the initial probability distribution of events. Thus, if an SOC system is subjected to a completely different distribution, the yield would still be almost the same. The difficulty is that the yield in both cases is lower, when compared to a HOT system that is subject to the same probability distribution for which it was optimized. SOC systems give lower yields mainly at densities higher than the critical density. An important implication of the SOC perspective is that grids must be operated below criticality, in order not to exhibit avalanche effects. A rough indication of the criticality point was obtained for the Institute cluster and that was around 30 HPF units (as explained in section 3.2).

Grid computing systems can also be perceived as HOT systems. The abstract space X in HOT theory can be mapped to the abstract space of load events in the grid. The overall idea is that using an expected job distribution and constraints on resources, an optimal resource distribution can be arrived at. Load events include parallel job submission, sequential load addition and load becoming greater a pre-defined threshold. p(x) is the probability distribution of initiating events $\forall x \varepsilon X$. A(x) is the number of nodes affected by x and measures the impact of load events at x. R(x) is the Horse Power Factor (HPF) of x, that includes its computing power, memory and load [23]. R(x) can be considered as the resource that restricts the size of load events. Constraints can be imposed on R(x), like $\int_X R(x)dx$ = a constant, implying that the sum total of all HPFs in the system is a constant. We can minimize the expected cost of load avalanches $E(A^\alpha) = \int_X p(x)A^\alpha(x)dx$, giving $p(x)R^{-\alpha\beta-1}(x)$ = constant. For batched jobs such as those driven by the physics experiments [25], HOT based design achieves optimality. However, for non-batched jobs, if the actual job distribution is different from the expected one, HOT may be sub-optimal.

A key assumption made in HOT is the optimal allocation of resources to spaces, concentrating more resources where load events are more likely. This implies that the HOT system is optimized for a given probability distribution of events. If a HOT system is subjected to a distribution of events same as the one it assumed initially, it gives optimized yield event at high density. However, if it is subjected to a distribution of events completely different from the one assumed and optimized for initially[4], it results in sub-optimal yields. The yield may be much lower than an equivalent SOC system that is subjected to unexpected probability distributions [4]. A HOT system may also be vulnerable to large unexpected events at places where events are unlikely. An unexpected event could be a tree falling across clusters in the forest fire model or a simple JVM garbage collection problem in distributed systems [5] or even software failures.

[4]For instance, a HOT system might have been optimized for a Gaussian probability distribution of events and subjected to a uniform event distribution.

5. Conclusions

We have presented a different perspective of large scale distributed systems (especially grid computing systems) based on theory of complex systems from statistical physics. Two theories, SOC and HOT are applicable. An SOC perspective of grid computer systems indicates that the system may be sub-optimal in performance, but resilient to unexpected event distributions. An HOT perspective suggests that grids can be optimized for performance and scalability but may be sensitive to unanticipated uncertainties such as design faults or software failures. This has an important implication for grid designers: A combination of the two theories SOC and HOT should form the basis for grid design. It must have the best elements of both: robustness to unanticipated event distributions from SOC; yield optimization and robustness to known uncertainties from HOT.

We are exploring ways to introduce elements of randomness into the HOT optimization process, so that the resultant system is nearly optimal and is also robust to uncertainties. The unstructured P2P layer of Vishva gives that randomness. So, a HOT based design realized over this unstructured layer could be one way of building robust and scalable grid computing systems.

Efforts such as [17] are underway to build biologically inspired grid computing systems based on ant colony behaviour. We plan to use the randomness and other self-organizing properties from such systems and optimize yield in Vishva based on HOT. We will also try to estimate the robustness and scalability of the new design. The hybrid theory also opens up research in large scale distributed systems, as fundamental trade-offs and limits of scalability and robustness can be formulated.

References

[1] Manfred Schroeder. *Fractals, Chaos, Power Laws: Minutes from an Infinite Paradise.* W H Freeman and Company, 1991.

[2] Maya Paczuski, Sergei Maslov, and Per Bak. Avalanche Dynamics in Evolution, Growth, and Depinning Models. *Physical Review E*, 53(1):414–443, 1996.

[3] Per Bak. *How Nature Works: The Science of Self-Organized Criticality.* Springer Verlag, 1996.

[4] J M Carlson and John Doyle. Highly Optimized Tolerance: A Mechanism for Power Laws in Designed Systems. *Physical Review E*, 60(2):1412–1427, 1999.

[5] Steven D Gribble. Robustness in Complex Systems. In *Proceedings of the eighth Workshop on Hot Topics in Operating Systems (HotOS-VIII)*. Schloss Elmau, Germany, May 2001.

[6] Ian Foster, Carl Kesselman, and Steven Tuecke. The Anatomy of the Grid: Enabling Scalable Virtual Organizations. *International Journal on Supercomputer Applications*, 15(3), 2001.

[7] Ian Foster and Adriana Iamnitchi. On Death, Taxes, and the Convergence of Peer-to-Peer and Grid Computing. In *Proceedings of the 2nd International Workshop on Peer-to-Peer Systems (IPTPS'03)*, Berkeley, CA, February 2003.

[8] A Carreras, David E Neuman, and Ian Dobson. Evidence for Self-Organized Criticality in a Time Series of Electric Power System Blackouts. *IEEE Transactions on Circuits and Systems*, 51(9):1733–1740, 2004.

[9] M Stubna and J Fowler. An Application of the Highly Optimized Tolerance Model to Electrical Blackouts. *International Journal of Bifurcation and Chaos*, 3(1):237–242, 2003.

[10] Tobin J Lehman and James H Kaufman. Optimal Grid: Middleware for Automatic Deployment of Distributed FEM Problems on an Internet-Based Computing Grid. In *Proceedings of*

the IEEE International Conference on Cluster Computing (CLUSTER'03), pages 164– 171, Kowloon, Hong Kong, December 2003.

[11] M Venkateswara Reddy, M A Maluk Mohamed, Tarun Gopinath, A Vijay Srinivas, and D. Janakiram. Vishwa: A Paradigm for Anonymous Remote Computation and Communication for Peer-to-Peer Grid Computing. Technical Report IITM-CSE-DOS-2005-12, Distributed & Object Systems Lab, Department of Computer Science & Engineering, Indian Institute of Technology Madras, 2005.

[12] Stuart Kauffman. *At Home in the Universe: The Search for Laws of Self-Organization and Complexity*. Oxford University Press, 1995.

[13] Per Bak and K Sneppen. Punctuated Equilibrium and Criticality in a Simple Model of Evolution. *Physics Review Letters*, 74:4083–4086, 1993.

[14] Mark Newman. The Power of Design. *Nature*, 405(5):412–413, 2000.

[15] Walter Willinger, Ramesh Govindan, Sugih Jamin, Vern Paxson, and Scott Shenker . Scaling Phenomena in the Internet: Critically Examining Criticality. *Proceedings of National Academy of Sciences*, 99(1):2573–2580, Februrary 1999.

[16] Glenn Deen, James Kaufman, Tobin Lehman, and John Thomas. A Practical Introduction to Optimal Grid. Tutorial available at http://www.alphaworks.ibm.com/tech/optimalgrid.

[17] Arjav J Chakravarti, Gerald Baumgartner, and Mario Lauria. The Organic Grid: Self-Organizing Computation on a Peer-to-Peer Network. *IEEE Transactions on Systems, Man and Cybernetics, Part A*, 35(3):373–384, May 2005.

[18] Nazareno Andrade, Francisco Brasileiro, Walfredo Cirne, and Miranda Mowbray. Discouraging Free Riding in a Peer-to-Peer CPU-Sharing Grid. In *International Symposium on High Performance Distributed Computing (HPDC-13)*, pages 129–137, 2004.

[19] Andrew Chien, Brad Calder, Stephen Elbert, and Karan Bhatia. Entropia: Architecture and Performance of an Enterprise Desktop Grid System . *Journal of Parallel and Distributed Computing*, 63(5):597–610, 2003.

[20] Pradnya Karbhari, Mostafa Ammar, Amogh Dhamdhere, Himanshu Raj, George Riley, and Ellen Zegura. Bootstrapping in Gnutella: A Measurement Study. In *Proceedings of Passive and Active Measurement Workshop PAM 2004*, Antibes Juan-les-Pins, France, April 2004. Springer in the Lecture Notes in Computer Science (LNCS) series.

[21] Binu K J, R Karthikeyan, and D Janaki Ram. DP: A Paradigm for Anonymous Remote Computation and Communication for Cluster Computing. *IEEE Transactions on Parallel and Distributed Systems*, 12(10):1052–1065, October 2001.

[22] Antony Rowstron and Peter Druschel. Pastry: Scalable, Distributed Object Location and Routing for Large-Scale Peer-to-Peer Systems. In *Proceedings of the 18th IFIP/ACM International Conference on Distributed Systems Platforms (Midleware 2001)*, pages 329–350, Heidelberg, Germany, November 2001.

[23] Rushikesh K Joshi and D Janaki Ram. Anonymous Remote Computing: A Paradigm for Parallel Programming on Interconnected Workstations. *IEEE Transactions on Software Engineering*, 25(1):75–90, January/February 1999.

[24] Henrik J Jensen. *Self-Organized Criticality: Emergent Complex Behaviour in Physical and Biological Systems*. Cambridge Lecture Notes in Physics. Cambridge University Press, 1998.

[25] F Carminati, P Cerello, C Grandi, E Van Herwijnen, O Smirnova, and J Templeton. HEP Common Application Layer. HEPCAL RTAG Report, November 2004.

Self-Organization and Autonomic Informatics (I)
H. Czap et al. (Eds.)
IOS Press, 2005

Combining Virtual Organization and Local Policies for Automated Configuration of Grid Services

Shishir BHARATHI, Beom Kun KIM, Ann CHERVENAK and Robert SCHULER
USC Information Sciences Institute

Abstract. We investigate approaches for the automated configuration of distributed Grid services. In particular, we implement several approaches for combining configuration information specified by a group of collaborating institutions (a *Virtual Organization or VO*) with local configuration parameters. We describe our implementation of merging strategies for configuring the Globus Replica Location Service. Based on our initial work, we describe outstanding issues for merging local and VO configuration policies and for resolving conflicting policies.

Keywords. Distributed systems, Grid services, automatic configuration, policy management

1. Introduction

Over time, Grids are increasing in size with respect to the number of resources and service instances being deployed, and these Grid resources and services are cooperating to perform increasingly complex tasks. Grid services provide well-defined functionality to clients, and their implementation may be either centralized or distributed. Distributed services provide better load balancing and higher tolerance to failure than centralized services.

An example of such a distributed Grid service is the Globus Replica Location Service (RLS) [1, 2], a distributed registry used for management and discovery of replicated data items. RLS is used by an increasing number of data-intensive Grid applications for management of replicated data. For example, the Laser Interferometer Gravitational Wave Observatory (LIGO) project [3, 4] deploys a distributed RLS at ten locations that contains registered mappings for over 40 million files. The Earth System Grid project [5] deploys RLS servers at four locations and stores mappings for over 400,000 files.

For distributed Grid resources and services to be effectively utilized, they need to be suitably configured and managed. This has proven to be challenging considering the number of services deployed, the needs of different applications, quality of service requirements, and the dynamic nature of grids. In addition, since all resources in a Grid do not fall under a single administrative domain, the restrictions imposed by each of the local system administrations must be respected in addition to any policies that are specified for the collaborating institutions.

Typically, the configuration of a Grid service or a resource deployed at a site has been the burden of the local system administrator. But, with larger deployments of distributed Grid services and the need for coordination among multiple institutions, there is an increasing need for tools that provide flexible, scalable, automatic configuration of Grid services.

Earlier work has been done on automating configuration tasks. At one extreme is the scenario where all services fall under a single administrative domain (unlike a typical Grid), where automatic configuration is usually done with a centralized configuration manager that has global knowledge [6]. There might also be degrees of decentralization [7], but with the assumption that individual services or sites are not selfish and are interested mainly in the well being of the distributed system as a whole. At the other end of the spectrum, peer-to-peer file sharing systems [8-10] are composed of nodes that are individually administered. Each node is typically concerned with only its own well being without regard to the state of other nodes in the system. Grid services fall somewhere between these extremes.

In this paper, we first discuss the motivation for automated configuration of distributed Grid services. Next, we briefly describe the distributed Globus Replica Location Service and its configuration. Then we describe our initial approach to providing automated configuration of RLS servers and discuss alternatives for merging policy information specified by each local site and the collaborating institutions. Based on this initial work, we identify outstanding issues in merging conflicting configuration policies. We conclude the paper with a discussion of related work and our future plans.

2. Configuration of Distributed Grid Services

A Grid is built when several collaborating institutions contribute resources to solve common problems [11]. Such a collaboration of personnel and resources is known as a Virtual Organization (VO). Each institution in the VO has its own set of policies that it would like to enforce on its resources and services, such as the access permissions granted to each user. At the same time, since it is part of the VO, the institution is also interested in cooperating with other sites and individuals in the VO. Further, the VO might have policies of its own that it would like to enforce on resources at the collaborating institutions.

Each of the policies that the VO or an individual institution wants to enforce produces a set of configuration settings for the services and resources in the VO. To determine how all these policies affect each other, resolve possible conflicts in these policies and produce the optimal configuration quickly is a complicated problem. Our research is concerned with identifying issues that arise when dealing with different sets of policies for configuring a distributed Grid service and providing automated mechanisms for determining optimal configurations.

In particular, we are interested in providing the capability for a Virtual Organization to specify most configuration policies for the services deployed at its collaborating institutions, but also allowing local administrators to override these policies as required. The advantage of allowing the VO to provide most of the configuration settings to each institution is that each local site configuration need not be modified every time the VO configuration changes. For example, consider the case when a new set of users is added to the VO. Without automatic configuration at the VO level, each configuration file at each site must be modified by local administrators to add access permissions for the new users. With automated configuration, a change at the VO level can be propagated and applied at each local site automatically.

One important consideration for automated configuration at the VO level is that local system administrators often require the ability to impose their own policies that may conflict with VO policies. For example, if a local site does not trust a particular user, it will deny that user access to its resources, even if that user is trusted by the VO and granted those access permissions at the VO level. The conflict between the desire to allow the VO to provide convenient, automated configuration at multiple institutions but still maintain local autonomy raises several challenging questions that are addressed later in the paper.

3. Background: The Replica Location Service

In this paper, we use the example of the Globus Replica Location Service (RLS) to highlight the problems of merging VO-specified and institution-specified configuration settings. Next, we briefly describe the RLS design and implementation, with particular emphasis on configuration of the distributed service.

The Replica Location Service is one component of a data management system for Grid environments. It is a registry that keeps track of where replicas exist on physical storage systems. Users or services register files in the RLS when the files are created. Later, users query RLS servers to find these replicas.

The RLS is a *distributed* registry, meaning that it may consist of multiple servers at different sites. By distributing the RLS registry, we are able to increase the overall scale of the system and store more mappings than would be possible in a single, centralized catalog. We also avoid creating a single point of failure in the Grid data management system. If desired, the RLS can also be deployed as a single, centralized server.

A *logical file name* is a unique identifier for the contents of a file. A *physical file name* is the location of a copy of the file on a storage system. The job of the RLS is to maintain associations, or *mappings,* between logical file names and one or more physical file names of replicas. The user can query an RLS server to find the logical file names, physical file locations, or attributes associated with them.

The RLS design consists of two types of servers: the Local Replica Catalog and the Replica Location Index. The *Local Replica Catalog (LRC)* stores mappings between logical names for data items and the physical locations of replicas of those items and answers queries about registered mappings. For a distributed RLS deployment, we also provide a higher-level *Replica Location Index (RLI)* server. Each RLI server collects information about the logical name mappings stored in one or more LRCs and answers queries about those mappings. The RLI index can be deployed in a hierarchical fashion to provide aggregation and redundancy of index information, as illustrated in Figure 1.

3.1 Configuring the RLS

The configuration settings for each RLS server are specified in a configuration file that resides on the same host as the service. These configuration parameters include access control lists that specify access permissions for users, the username and password used to access the back end database, timeout values for requests, and parameters that specify whether an RLS server acts as a local catalog, a distributed RLI index or both. In the remainder of this paper, we discuss different approaches for constructing these local RLS server configuration files by combining an overall VO configuration file with local configuration parameters.

RLI RLI

RLI RLI RLI

LRC LRC LRC LRC

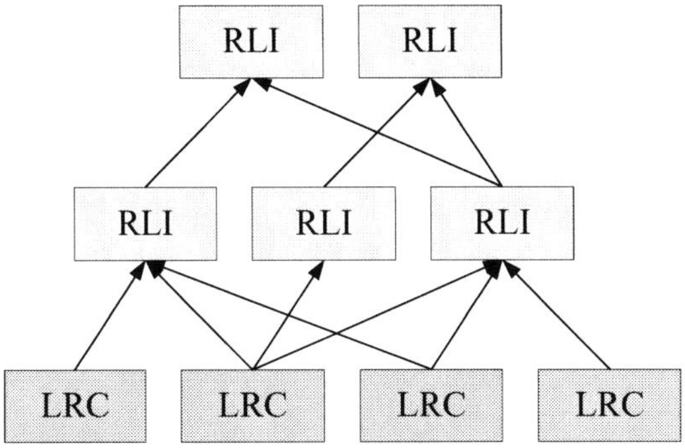

Figure 1: Example of a hierarchical, distributed RLS configuration

4. Implementation and Policy Alternatives

We have implemented a prototype for performing automated configuration of a distributed RLS service. First, we set up an HTTP server for the Virtual Organization that contains the current configuration file to apply to all RLS servers in the VO. Then, we deployed a script that runs at each RLS site and performs automated configuration for the local RLS server. The script periodically polls the VO HTTP server to determine whether there have been any updates to the VO configuration file. If so, the script downloads the VO configuration policy file, merges that with a local configuration policy file using one of the schemes described below, and then applies the merged configuration to the RLS server at the site. Thus, there may be some delay between the time that changes are made on the VO server and when those updates are applied to all the RLS servers in the distributed deployment.

Some RLS configuration parameters may be specified only in the local policy configuration, others only in the VO configuration, and some parameters may appear in both files before they are merged. For any parameter that appears in only one policy configuration file (local or VO), we apply the specified value to the final RLS configuration.

For some configuration parameters that appear in both local and VO configuration files, (e.g., the number of concurrent connections allowed, the number of threads servicing the requests, etc.), we let local policy override the VO policy. Some examples are shown in Table 1. The reason for this preference for local configuration parameters is that the local administrator is more likely to be aware of the limitations of the resources at that site and typically wants control over how local resources are used.

There are alternatives for handling these straightforward service configuration policies. For example, we could treat the policy value specified in the VO configuration file as the minimum value for that parameter, and value specified in the local configuration file as the maximum value; the final configuration value applied to the RLS service could then be avalue chosen from that range.

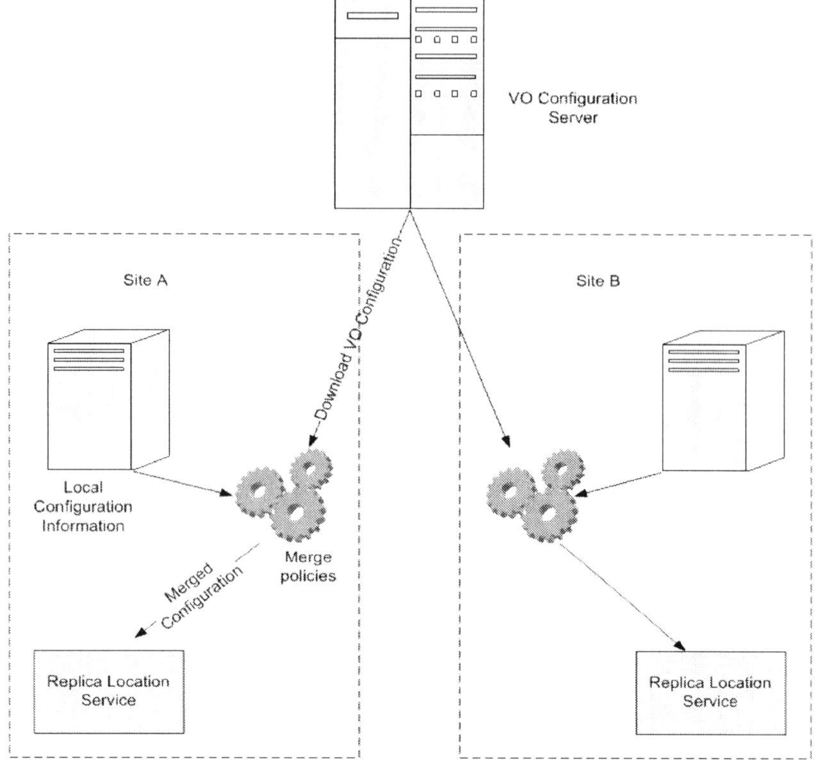

Figure 2 Merging of VO and local policies

Table 1: Merging of policies

VO Policy	*Local Policy*	*Final Policy*
max_connections: 20	max_connections: 50	max_connections: 50
max_threads: 30		max_threads: 30
	database_user: dbuser	database_user: dbuser
		timeout: <default>

For other parameters, particularly those specifying access privileges for users, we need more sophisticated mechanisms for determining the final configuration values. Table 2 shows one scheme that we implemented for merging VO and local policies. Here the final set of privileges for the user is the *union* of the sets of privileges listed in the VO policy file and the local policy file. If a privilege appears in either the VO or the local configuration file, then it is added to the final configuration for the RLS server. A consequence of this union operation is that after the merge, the user might end up with more privileges than either the VO or the local site desired. In this example, user Alice ends up with lrc_read privileges that the local site didn't want her to have and lrc_write privileges that the VO didn't want her to have. Thus, a simple policy of merging configuration files with a union operation is unlikely to satisfy either the VO or the local site administrators.

Table 2. Union of user privileges

VO Policy	*Local Policy*	*Final Policy*
Alice: lrc_read	Alice: lrc_write	Alice:lrc_read,lrc_write
Bob: lrc_read	Bob: lrc_read	Bob: lrc_read
	Carol: rli_read	Carol: rli_read
Dave:lrc_read,rli_read	Dave: lrc_read	Dave: lrc_read, rli_read

Table 3 shows another scheme that we implemented for merging policy files. Here, the final set of privileges for each user is the result of *intersecting* the privileges granted by the VO and the local site. This simple policy has the advantage that no access is granted to a user if it is denied by either the VO or the local site. In practice, however, this policy may be overly strict. For example, in Table 3 user Alice ends up with no privileges even though both the VO and local site wanted her to have certain privileges. The intersection merging scheme effectively requires that the local site must be involved in every authorization decision and must explicitly agree with any access rights granted by the VO before they can be applied at the RLS server. This eliminates one of the main advantages of doing automatic configuration, which is to allow a distributed service to be configured primarily based on VO policies without requiring the local site to be involved in every policy decision. For example, when a new user joins the VO and is added to the VO configuration file, the user will not be granted any access to local resources unless local configuration files are also updated to explicitly allow access for this user.

Table 3. Intersection of user privileges

VO Policy	*Local Policy*	*Final Policy*
Alice: lrc_read	Alice: lrc_write	Alice:
Bob: lrc_read	Bob: lrc_read	Bob: lrc_read
	Carol: rli_read	Carol:
Dave:lrc_read, rli_read	Dave: lrc_read	Dave: lrc_read

Our implementation and evaluation of these two schemes for merging local and VO configuration policies suggests that the union scheme is too permissive, while the intersection scheme is too restrictive. Next, we discuss outstanding issues related to merging these configuration files that will drive our future research.

Although we have used the examples of configuring Access Control Lists (ACLs) in this section, the issues involved in the merging of VO and local policies are just as relevant to other access control mechanisms. Role based access control, for example, introduces a level of indirection by mapping a user to a role and then assigning privileges to that role. Conflicts such as the ones described above arise when the VO and the local policies try to assign different roles to the same user or different privileges to the same role.

5. Outstanding Issues

We have noted the competing goals of providing the convenience of allowing most configuration management to be done automatically at the Virtual Organization level, while local sites still want to maintain autonomy and have the option of overriding VO policies. The desire for local site autonomy raises several interesting questions:

How should VO and local policies be merged? We have discussed several ways to merge VO and local policies. VO policies may take precedence over local policies or vice-versa, or we may choose one of the merging options (union or intersection or some other mechanism). Additionally, we might want some local policies to be merged with the VO policies, while other local policies should completely override the corresponding VO policies.

To what extent should local policies dominate? Allowing local policies to dominate might be desirable for some policy parameters. But should local access control policies be allowed to completely override the VO's policies? Some virtual organizations have expressed a desire for this functionality. However, considering that data stored in the local databases might belong to the VO and not to the individual institutions, it might not be advisable to allow local policies to override the access preferences of the VO. For example, in the case of the distributed Replica Location Service, if local policies allow access to the RLI index service to users that are not allowed those privileges by the VO, then the users may gain information about the location of data at other sites. This might not be what the VO or other sites want to allow.

There are additional security implications when a user is allowed to run a job at a site based on local access permissions despite a VO policy that tries to deny that access. This user job might need to access resources at other sites in the VO. The user's credentials have to be properly delegated to these other sites or else this will result in unauthorized operations.

What are the implications for accounting systems of allowing local polices to override VO policies? We must also consider other system management aspects such as accounting. If a local policy allows user Alice access to the VO resources even though the VO intended to deny that access, who should be charged for Alice's access? We will likely need to design accounting mechanisms that charge the VO when the policies are merged but charge the local site when local policies override VO policies.

6. Related Work

There has been considerable interest in the issues of dynamic resource configuration, authorization and access control and policy management. [12-16] We now briefly discuss some of the work in these areas that is related to the issues we have described.

Simple Network Management Protocol (SNMP) [6] based management systems are frequently used to manage routers, workstations and other kinds of network resources. The later versions of the protocol have addressed the shortcomings of the initial version by introducing decentralization, authentication etc. The Astrolabe management system [7] builds a hierarchy of resources and organizes them into zones. Agents running on each host communicate using Gossip protocols. Usually in such systems, all of the resources that are managed fall under the same administration and do not have to deal with the issues of conflicting policies. Yemini et al. [17] propose "management by delegation" which transfers the management responsibilities to the managed devices themselves in order to reduce the load on the manager. This approach requires that each resource be capable of self-management and reduces the burden on centralized managers. But, this in turn delegates the responsibility of policy conflict resolution to the individual resources.

Lupu and Sloman [18] describe the use of meta-policies to detect conflicts between policies that arise when using both negative and positive policies. They suggest that the conflict be resolved by giving a higher priority to negative policies. We could have potentially split up RLS ACLs into positive and negative ACLs and used a similar resolution. But, this would then prevent the local policy from allowing the user a certain privilege that the VO explicitly denies. Lupu and Sloman also briefly mention the conflicts that arise when multiple managers control the same resource. One open research question is whether detecting policy conflicts and passing them to a human administrator for resolution is sufficient or whether we can develop policies for resolving such conflicts automatically.

7. Future Plans

Many Grid services are now being designed according to the Web Service Resource Framework (WSRF) specifications [19, 20]. The WS-Resource Framework allows resources to export standard interfaces to query and modify resource properties. We expect to develop standardized configuration and management interfaces based on these interfaces. The framework also allows services and resources to *subscribe* to changes in the properties of other resources. The changes are propagated back to the subscriber as *notifications*. The subscriber can then react appropriately depending on the notification received. We expect this subscription/notification mechanism to be important in dealing with automatic configuration of Grid services.

The periodic polling operations performed by our current implementation could be replaced by a subscription/notification mechanism in WS-RF-based automated configuration management. Changes in the VO configuration would be propagated to the local sites faster and with less overhead. This approach would also make it easier to change resource configuration in response to dynamic events in the Grid. Standardized management interfaces based on WS-Resource Properties will allow for easier integration of new services into the VO management system.

While we expect the WS-Resource Framework to simplify the task of implementing automated management systems, the issues outlined in the previous sections still remain. We plan to continue discussions with the LIGO team as well as to discuss requirements with the Earth System Grid and TeraGrid projects and other virtual organizations. The large number of resources involved in the TeraGrid project and the sharing of resources by multiple VOs is expected to raise additional interesting issues for automated configuration.

We plan to explore the possibility of providing policy conflict information as feedback to other services in the grid. This information would be used to divert the flow of sensitive data away from the sites that have policy conflicts to those whose policies are a match with that of the VO's. In the case of projects like the Teragrid project, we expect policy conflicts to arise between different VOs sharing the same resources. We plan to investigate dynamic provisioning techniques to deploy and identify resources that match each VO's requirements.

References

1. Chervenak, A., et al. *Giggle: A Framework for Constructing Sclable Replica Location Services*. in *SC2002 Conference*. 2002. Baltimore, MD.
2. Chervenak, A.L., et al. *Performance and Scalability of a Replica Location Service*. in *Thirteenth IEEE Int'l Symposium High Performance Distributed Computing (HPDC-13)*. 2004. Honolulu, HI.
3. Abramovici, A., W. Althouse, et al., *LIGO: The Laser Interferometer Gravitational-Wave Observatory*. Science, 1992. **256**: p. 325-333.
4. LIGO Project, *Lightweight Data Replicator, http://www.lsc-group.phys.uwm.edu/LDR/*. 2004.
5. Bernholdt, D., et al., *The Earth System Grid: Supporting the Next Generation of Climate Modeling Research*. IEEE, 2005. **93**(3): p. 485-495.
6. Case, J., Fedor, M., Schoffstall, M., and J. Davin, *A Simple Network Management Protocol (SNMP)*. Internet Request for Comments, 1990.
7. Renesse, R.v., K.P. Birman, and W. Vogels, *Astrolabe: A Robust and Scalable Technology for Distributed System Monitoring, Management, and Data Mining*. ACM Transactions on Computer Systems, 2003. **21**(2): p. 164-206.
8. Stoica, I., et al. *Chord: A Scalable Peer-To-Peer Lookup Service for Internet Applications*. in *ACM SIGCOMM*. 2001.
9. Ratnasamy, S., P. Francis, M. Handley, R. Karp, and S. Shenker. *A Scalable Content-Addressable Network*. in *ACM SIGCOMM*. 2001.
10. Chawathe, Y., et al. *Making Gnutella-like P2P Systems Scalable*. in *ACM SIGCOMM 2003*. 2003. Karlsruhe, Germany.
11. Foster, I., C. Kesselman, and S. Tuecke, *The Anatomy of the Grid: Enabling Scalable Virtual Organizations*. International Journal of High Performance Computing Applications, 2001. **15**(3): p. 200-222.
12. Poladian, V., Sousa, J.P., Garlan, D., Shaw, M. *Dynamic Configuration of Resource-Aware Services*. in *ICSE*. 2004.
13. Kon, F., et al. *Dynamic Resource Management and Automatic Configuration of Distributed Component Systems*. in *COOTS*. 2001.
14. Marshall, A.D., et al., *A Policy-Driven Approach to Availability and Performance Management in Distributed Systems*. 1997.
15. Lesser, V.R., *Cooperative Multiagent Systems: A Personal View of the State of the Art*. IEEE Transactions on Knowledge and Data Engineering, 1999. **11**(1): p. 133-142.
16. Oppenheimer, D.L., A. Ganapathi, and D.A. Patterson. *Why Do Internet Services Fail, and What Can Be Done About It?* in *USENIX Symposium on Internet Technologies and Systems*. 2003.
17. Goldszmidt, G., Yemini, S., Yemini, Y. *Network Management by Delegation - the MAD Approach*. in *CAS Conference*. 1991.
18. Lupu, E.C. and M. Sloman, *Conflicts in Policy-based Distributed Systems Management*. IEEE Transactions on Software Engineering, 1999. **25**(6): p. 852-869.
19. Czajkowski, K., et al., *The WS-Resource Framework Version 1.0*. 2004.
20. Foster, I., et al., *Modeling Stateful Resources with Web Services version 1.0*. 2004.

Self-Organization and Autonomic Informatics (I)
H. Czap et al. (Eds.)
IOS Press, 2005

EXEHDA: adaptive middleware for building a pervasive grid environment[1]

Adenauer Yamin [a,2], Iara Augustin [b], Luciano Cavalheiro da Silva [c],
Rodrigo Araujo Real [a], Alberto E. Schaeffer Filho [c] and Claudio F. Resin Geyer [c]

[a] *Catholic University of Pelotas, RS, Brazil*
[b] *Federal University of Santa Maria, RS, Brazil*
[c] *Federal University of Rio Grande do Sul, RS, Brazil*

Abstract. We consider that the concept derived of the integration of pervasive, context-aware and grid computing is suitable for building the next generation of the grid applications which assist to nomadic and mobile users. To demonstrate our ideas we have implemented a pilot application, called GeneAl, using the API provided by EXEHDA middleware. EXEHDA is adaptive, service oriented and was conceived to support the execution of pervasive grid applications. The main concept embedded in middleware and application design is the context-awareness expressed by their adaptive behavior. This is also a key to provide functionality adapted to the constraints and unpredictability of the large-scale mobile environment. To achieve this objective, EXEHDA employs a lot of strategies in its services to allow the adaptation to the current state of the execution context, such as on-demand adaptive service loading and dynamic discovery and configuration. The middleware manages and implements the *follow-me semantics* for pervasive grid applications. In that sense, it provides services for distributed adaptive execution, context recognition, pervasive storage and access, anonymous and asynchronous communications.

Keywords. Pervasive Grid Environment, Large-scale Mobility, Adaptive Middleware, Context-aware Adaptation.

1. Introduction

The general view of a Grid Computing System is that of a three-layered system made up of computation/data, information and knowledge layers. Even though the computation/data layer of the Grid is the layer which is perhaps the most mature in terms of the time, experience and where most software is available and directly useable, it still lacks a lot of essential aspects that will allow the provision of seamless, pervasive and secure use of system resources.

---

[1]Sponsored by Rede Nacional de Ensino e Pesquisa (RNP), FINEP and CNPq Foundations.

[2]Correspondence to: Adenauer Yamin, Rua Felix da Cunha, 412, Centro, Pelotas, Brasil, CEP 96010-1000 Tel.: +55 53 9112 3478; Fax: +55 53 3225 8271; E-mail: adenauer@ucpel.tche.br.

Another side, the field of Pervasive Computing addresses issues regarding the provision of the next generation computing environments with information and communication technology everywhere, for everyone, all the time [5,4]. This way, human beings will be citizens of both the physical and multiple virtual worlds.

As a partial result of our research, we consider that the link between pervasive computing and grid computing brings up the possibility of a large-scale mobility for the user as an advantage of pervasive system field, and a wider use of the available computational resources, as an advantage for grid systems field. For us, the glue between these two fields is another research field: context-aware computing [2,8].

We have begun our research in ISAM project (http://www.inf.ufrgs.br/ isam) aiming to study alternatives to build pervasive applications with large-scale mobility [2,1]. As a partial result we found that the designed middleware to provide the large-scale pervasive environment is also suitable to program, execute and manage grid applications which are modelled by context-aware behavior. This way, we let grid applications adjust themselves to the current resources instead of look for resources that satisfy their needs.

In this paper, we show that the concept derived from the integration of pervasive, context-aware and grid computing is suitable for building the next generation of grid applications. The rest of the paper is structured as follows. Section 2 outlines the target scenario and the grid application model. Section 3 drafts the main features of the EXEHDA middleware to assist the need for pervasive grid applications and summarizes its services structured in subsystems. The dynamics of operation of the middleware services in supporting the execution of a pilot application is the subject of section 4. Finally, related works and concluding remarks are presented in sections 5 and 6, respectively.

2. Target Scenario: features and assumptions

The source of our proposal is the ISAM Software Architecture, which views mobile device as windows to access the computational environment. The large-scale computing environment formed by cells, starting from the union of the several mobile and stationary physical resources, is shown in the Figure 1. The computing environment components (data, code, devices, services and resources) are pervasive and managed by a middleware, which provides the pervasive access to them.

In this computing scenario, the mobile devices are more like interface devices. They do not store neither code nor data persistently (except for some caching strategy), but operate as portals that receive the code to be executed and can transfer the execution to other devices using proximity or resource availability as a selection criteria. Besides, each user has a **virtual environment** that can be accessed at any location, and with the available device. Moreover, the user's location in the pervasive environment has a deep effect in the way pervasive applications are executed. As the user physically moves (i.e., by carrying his current device - user mobility - or changing the device being used - terminal mobility), his currently running applications, in addition to the user's virtual environment, need to be continuously available for him, following the user's movements in the pervasive space. Such behavior defines what we call *follow-me semantics* of pervasive applications.

The main properties of ISAM applications are: distributed, mobile, adaptive and reactive to the context, pro-active (managed application) and express the follow-me se-

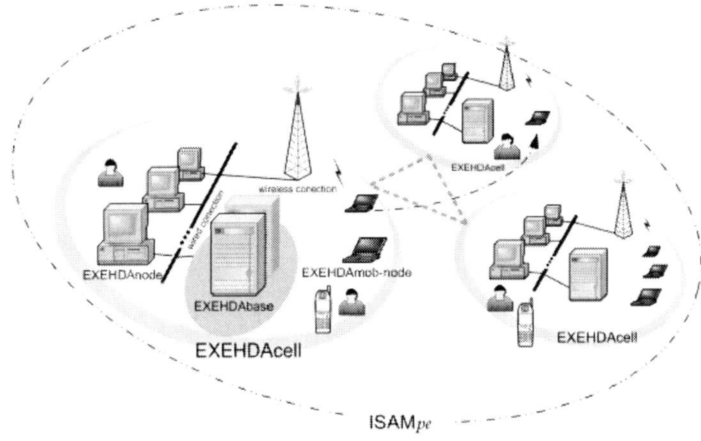

Figure 1. Pervasive Grid Environment (ISAMpe)

mantics - the application follows the user in his moving at large-scale space. The application's code is installed on-demand on the devices used and this installation is adaptive to context of each device.

2.1. ISAM Software Architecure

The ISAM software architecture, shown in figure 2, aims to provide an integrated solution, from language to execution environment, to build and execute large-scale pervasive applications. The upper layer corresponds to the abstractions ISAM provides to the application designer to ease the development of pervasive application, which includes a programming language, ISAMadapt[2], better tuned to the development of context-aware adaptive applications. The execution of such applications is supported by EXEHDA middleware.

The EXEHDA middleware was originally designed to provide an infrastructure for studying alternatives to support pervasive applications with large-scale scope. In this solution, context is a first order concept and it means *"all the relevant information to the application that can be obtained from the support system"*. The application explicitly identifies and defines the entities that characterize a situation and integrate its context. State alterations in these entities trigger an adaptation process in the application components. So, the context definition can then be refined to *"every entity's attribute for which a state alteration triggers an adaptation process in the application"*.

Another behavior present in pervasive executions is the notion of planned disconnections. At the moment, disconnections are considered as elective nature because they can be treated as foresee and prepared fails. They are planned. A mobile device, for example, would rather operate disconnected to reduce battery consumption and, at specific moments, reconnect to update the state of the global execution. Such disconnection/reconnection procedures should be, whenever possible, transparent to the applications.

This software architecture has also been shifted to deal with grid applications, which are designed according to the requirements explained next.

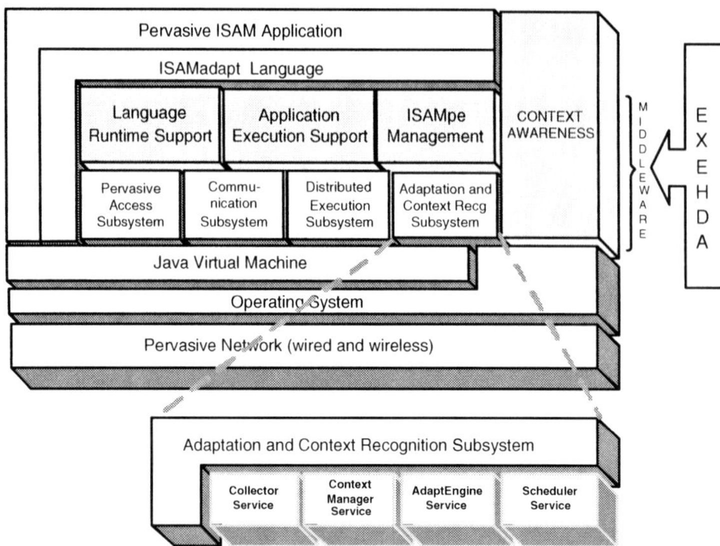

Figure 2. ISAM Software Architecture

2.2. *The concept of Pervasive Grid Applications*

Applications executing in the pervasive environment should not make undue assumptions neither on the devices which it will run on, nor the environment services it will use. So, environmental changes, due to dynamic intrinsic to the environment or user's mobility, must be reflected in the application behavior. To reduce the impact of these changes, the application must have a **context-aware adaptive behavior**. Context-aware adaptation is a fundamental concept for this application model.

Figure 3 illustrates our approach to the context-awareness of the pervasive grid computing. Both applications and the middleware were designed considering this adaptation model. The applications are managed by the middleware which is oriented by policies defined at the application development time [6]. We emphasize the collaboration between application and support layers to achieve the dynamic adaptation process.

In terms of conception and organization of the middleware, the following requirements were identified as necessary to assist the application demands:

- Adaptation to the current execution context: it takes place primarily at services loading-time. Such adaptation is guided by the device where the service will be loaded and aims to attend the situation of high heterogeneity of the resources observed in the pervasive environment;
- Reduced resource consumption: the number of services the middleware makes available may be large, due to the incremental incorporation of new features or even updates of the existing ones. Moreover, the capabilities of the resources in the pervasive environment may vary a lot from device to device. The perspective here is to install just the minimum necessary on each device, so that all nodes, even the ones with very limited resources, may stay operational;

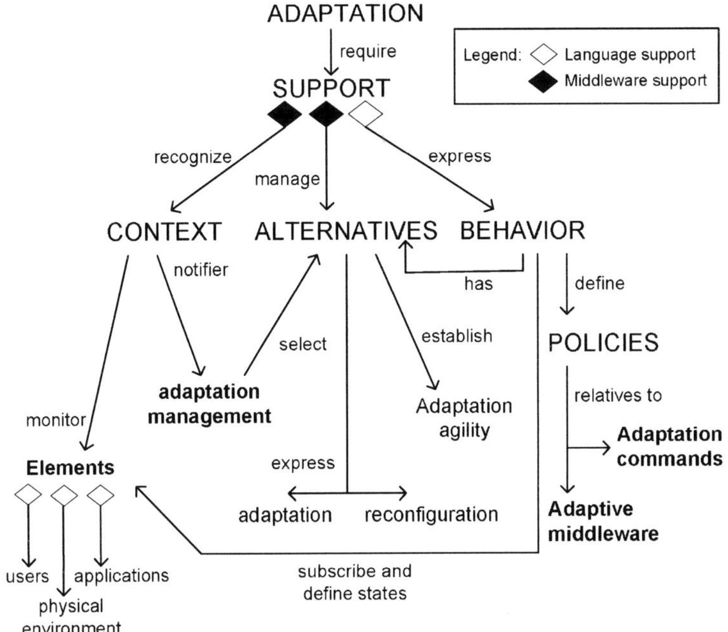

Figure 3. Adaptation Model

- Disconnected operation: while the device is in the planned disconnection mode, the application should keep its internal consistency and remain operational.

We argue that existing abstractions and services of traditional grid system are neither sufficient nor necessarily appropriate for our target applications. In our view, the key characteristic of pervasive grid systems is the need for context-aware adaptation to effectively support **follow-me applications**, this is in order to cope with varying resources and usage situations. To address this feature we define the EXEHDA middleware.

3. EXEHDA Middleware Features: general view

EXEHDA is based on services which assist in three perspectives: (i) ISAMpe management, by providing services to control the physical medium where the processing will take place; (ii) support for execution of application, by providing the services and abstractions necessary for implementing the follow-me semantics of pervasive grid applications; and (iii) support to application programming providing an API (Application Programming Interface).

Considering the defined application profile and the target environment, a middleware for supporting its execution must fill the following requirements: (a) support distributed execution; (b) anonymous and asynchronous communication; (c) pervasive access to data and code; (d) support large-scale logical mobility; (e) support large-scale physical mobility; (f) provide information about the execution context; (g) support dynamic adaptation of functional aspects; (h) support dynamic adaptation of non-functional aspects; (i) cooperative policy with the application in the control of adaptation.

For that, EXEHDA provides support to the adaptation at application level, and is itself adaptive. The adaptation makes the system adjust itself to changes, and the available resources can carry on working with as high as possible quality, at each moment. It is desirable that, when the system state changes, the middleware dynamically reallocates, reschedules and restructures the resources available for the application. The need for reallocating and rescheduling the system resources is determined by the execution context state [7].

3.1. Minimum Core

The requirement of operation in a high heterogeneous environment, where not just the hardware shows varying capacity of processing and memory, but also the software available on each device may vary, have motivated the approach by which a middleware **minimum core** has its functionality extended by **pluggable services**. Service loading is done on-demand and, moreover, is context adaptive. This way, we are able to use implementations of a service that are better tuned to each device and we reduce resource consumption by loading only services that are effectively used. Such scheme is possible because services are defined by their semantics and interface and not by a specific implementation.

The functionality provided is customizable in a per-node basis, being determined by the set of active services and controlled through **execution profiles**, and it's specified through an XML document.

An execution profile defines the set of services that should be activated in a node, assigning to each service an implementation, selected among the alternatives available for each service, and providing the parameters that will guide its execution.

Additionally, the execution profile of the middleware in a node also controls the loading policy that will be applied for each of the services defined in that profile. In this sense, the **loading policy** specifies one of two currently available modes: (i) bootstrap, meaning that the service should be loaded since the node startup; or (ii) demand, meaning the service will have been loaded by the time of its first use.

This loading policy for services is provided by the minimum core, which is installed on all the nodes that build up ISAMpe. The minimum core is formed by two components:

- **Profile Manager**, which is responsible for interpreting the information contained in the execution profiles, making them available at runtime for the other middleware services.
- **Service Manager**, which proceeds services activation in the node, based on the information provided by the Profile Manager. Service code is loaded on demand from a service repository, which may be local or remote depending on the device storage capacity and the nature of the service being loaded.

3.2. Tunning EXEHDA to the planned disconnection

The requirement for remaining operational during the periods when the planned disconnection is active have been motivated, additionally the conception of communication primitives tuned to this connectivity mode, the splitting of services into node and cellular instances. The former is local to each device, the latter executes on the base node. Hence, the local device will remain operational during the planned disconnection, once, for the satisfaction of a given request, the **node instance** of that service would renounce to access resources that are in the network. On the other hand, the **cellular instance** of the service, in execution on the base-node of the cell, acts as a reference point for services that require distributed, inter-node or inter-cell, coordination procedures.

3.3. Services Structured in Subsystems

The middleware services are conceptually organized in subsystems: data and code pervasive access, uncoupled spatial and temporal communication, large-scale distribution, context recognition and adaptation. The subsystem integration is shown in figure 4.

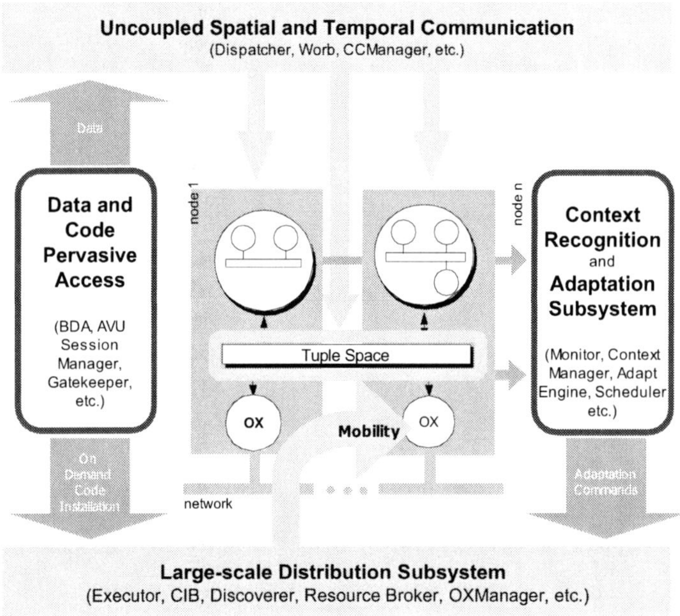

Figure 4. Middleware's Subsystems

EXEHDA is composed by several integrated services. Due to the limited space to describe them, the summary of main EXEHDA's services is related, in next section, based on a pilot application named GeneAl.

4. The Pilot Application

Simultaneously to middleware model and implementation, we are working towards implementing test pervasive grid applications that make use of the middleware services. In that sense, we highlight one of our in development applications: GeneAl, which is targeted to the genetic sequence alignment research scenario. This application explores the context-aware adaptation and follow-me semantics support provided by EXEHDA. As we defend the model of collaborative adaptation between application code and the execution system [6], the application was modelled considering this principle. It's introduced in two moments: programming and execution.

4.1. Programming the GeneAl Application

Genetic sequence alignment refers to the operation of nucleotide comparison that tries to find local similarities using biosequence databases. By a scoring scheme over compared sequences of nucleotides it is possible to identify which are the most similar. We implement a variant [Meidanis and Set ubal, 1994] of the traditional, and still broadly used, Smith-Waterman algorithm for global genetic sequence alignment [Smith and Waterman, 1981]. The idea of the algorithm is to find out the best alignment between two sequences. The sequences to be compared need not have the same length, but in this case it is necessary to insert gaps (-) in arbitrary places along the sequences. As the algorithm is only able to calculate the score between two sequences, to find the N best alignments between the given sequence and those contained in a database will require that it must be applied upon all peers <entry sequence, database sequence[i]>, where database sequence[i] is the ith biosequence in the database. As biosequence databases may contain thousands of stored sequences, this task can be pretty hard and time consuming, and a distributed infrastructure to help solving this problem is strongly recommended.

Using the API provided by EXEHDA middleware, we have developed the GeneAl application in Java. In order to allow new resources to be dynamically added, the problem was modelled according to a master-worker approach [7], associating a master object along with each biosequence database. A master is responsible for managing a number of workers that will perform the search for the best aligned sequences over a database. At the top of the individual databases, there is a management layer that gathers the partial results computed in each master to produce the final result. Figure 5 depicts such this organization. A master is responsible for splitting a biosequence database into a number of jobs, for distributing and keeping control of them and for generating the final result for a given execution cell. The workers get jobs from the master and process them. After the processing of a job has been finished, the worker will be responsible for sending the result back to the master and getting another job.

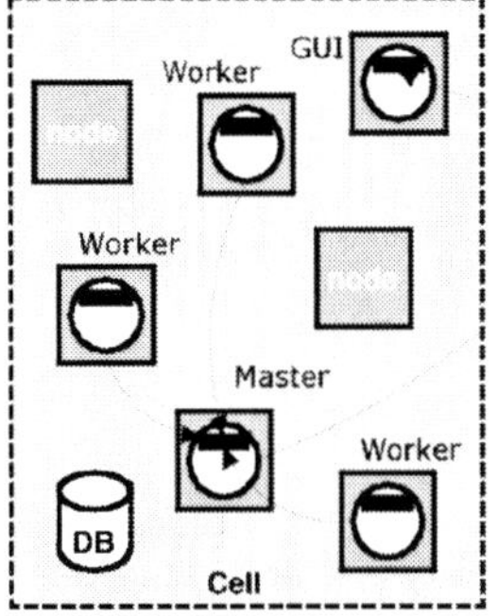

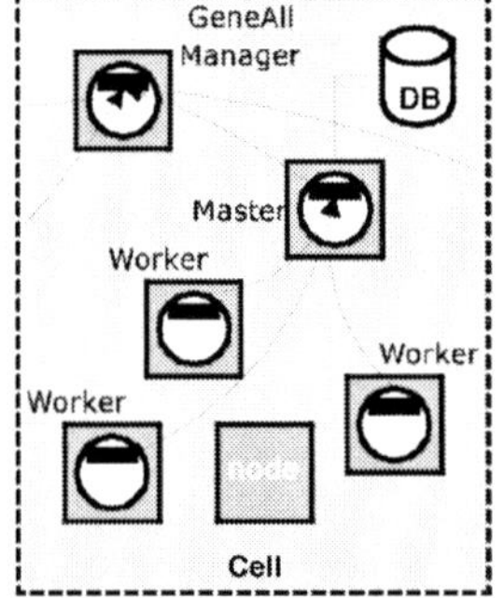

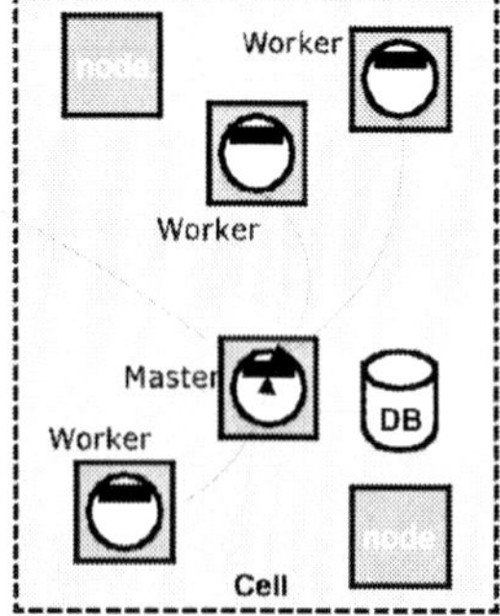

Figure 5. A View of Distribution of GeneAl Components

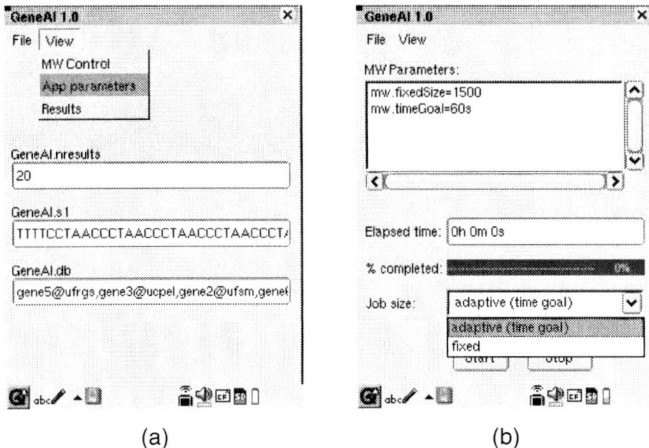

Figure 6. User Interface running in a Zaurus PDA

As the application can be launched from a regular desktop or a Zaurus PDA 5600, the interface adaptation was programmed in Java using Swing J2SE package and AWT J2ME package respectively. GeneAl does not require the databases to be located at a single site. This is the second degree of adaptation provided: the Discovery service allows the user to find the available databases at run-time for a given subject. Someone may then select among them where the processing should take place. The third degree of adaptation provided comes from its Scheduler service. The Scheduler is responsible for selecting the best nodes to host the workers. Finally, the fourth degree of adaptation comes from the data partitioning strategy: the size of a job can be defined statically before the beginning of the execution, or it can be adjusted dynamically along the execution according to the capacity/load of the host where the work is being done. This last approach is named *time goal*, and the jobs are resized according to the accomplishment of a given goal in a determined period of time. The size of the first job is an estimated value, but the size of the next job is decided by the performance of the previous job, using the relation about the expected time and the real elapsed time of the previous execution (reducing or increasing the number of biosequences per job). Figure 6 shows the Zaurus interface to (a) define the parameters and (b) choose the load partition strategy.

EXEHDA allows the worker throughput constraints to be modelled as context elements. This abstraction is supported by the context recognition subsystem of the middleware, which notifies the application components whenever a behavior change is observed and allows them to adapt by acting upon the size of the jobs.

4.2. *Starting the GeneAl Application*

Prior to actually being able to launch an application, the user needs to log on the ISAMpe. Application launching is done through the *isam-run* command line tool, which takes the original descriptor and the command line parameters and generates the extended descriptor communicated to the *Gatekeeper service*.

Typically, the first application effectively "pervasive" (it has the context adaptive and follow-me semantics properties) managed by middleware is the ISAMdesktop, a graphical user interface adapted to the context element: device type (PC or Zaurus 5600 PDA). The ISAMdesktop allows the user to manipulate its virtual environment, configuring user preferences or even launching applications installed there. The abstraction of a pervasive virtual environment to the users is supported by the *AVU service*.

Applications are not installed on the user's virtual environment in the traditional way, meaning that the application executable code is neither stored in nor managed by the AVU service. In fact, application installation consists only of copying the application launching descriptor to the users virtual environment. The executable code for the application is still provided on-demand by the *BDA service* by the time the application starts to execute on a given device.

Further, ISAMdesktop provides commands for the user to manipulate its current session. In this sense, a session is compose by the state of all the user's currently running applications. Session control commands include the traditional login/logout, save/restore session, as well as disconnect/reconnect operations which control middleware planned disconnection. Session control operations are managed by *Session Manager Service*.

4.3. Executing the GeneAl Application

The execution life-cycle of the application begins at the time the EXEHDA *Gatekeeper service* receives the application launching request, which consists of the extended application launching descriptor. The applications are characterized by an XML document, the *application launching descriptor*, which is generated at development-time and groups several meta-data about the application. The launching descriptor, maintains, among other data (e.g., functionality description, vendor, fixed execution arguments), a location independent reference to the application code, allowing the application to be launched from any EXEHDA node in the ISAMpe. Another service, the *BDA service* (Applications Data Base) is responsible for resolving, at run-time, the location independent references, in order to allow the code to be fetched and installed on the nodes used by the application. Therefore, the *BDA service* is responsible for implementing the pervasive access from the pointofview of the application executable code.

After the launching request is validated, the *Gatekeeper Service* delegates to another service, the *Executor service*, the effective launching of the application. At this time, the *Executor Service* instantiates and configures a class loader object for the application and begins the processing using the class defined in the <main> element of the launching descriptor. The installed class loader accesses the *BDA Service* in order to carry out the on-demand installation of application components (classes).

Once it has started, the application assigns an identifier (*ApplicationId*) to it, which is unique in the scope of the ISAMpe. Moreover, the application's extended launching descriptor is stored into the *CIB Service* (Cell Information Base). It maintains the attributes related to the management of the ISAMPe, describing the resources that constitute the cell and the neighborhood of the cells, as long as attributes that describe the running applications and the resources allocated to them. Additionally, the *CIB Service* keeps track of information about the users registered in the cell. Hence, whenever a new node is incorporated to the application distributed execution, the local instances of the EXEHDA services running on that node could recover from the *CIB Service* all the application execution attributes necessary for their operation.

Figure 7 shows (a) the execution accomplishment using the Zaurus PDA and (b) the results. Analysis of results about performance was published in [10].

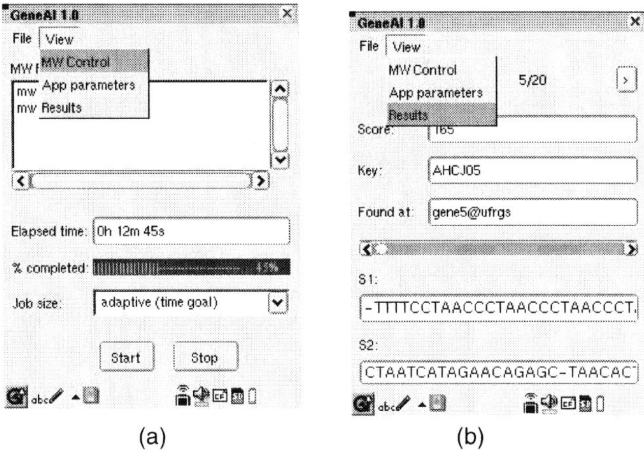

Figure 7. Accessing results in a Zaurus PDA

4.3.1. Context-aware Behavior

When a new application component, such as a worker in our application, is instantiated through the use of the *Executor Service*, other middleware services are activated. Even though the component being instantiated may not be adaptive itself, at least non-functional adaptations would take place. Thus, information about the current context state becomes necessary.

The assembling of the context state information, which guides many of the middleware operations and also the application adaptive behavior, is accomplished by the EXEHDA *context recognition subsystem*, through the cooperative operation of the *Monitor*, *Collector* and *Context Manager* services. The context recognition subsystem is presented in section 4.5. The produced context state information feeds both, functional (that modifies the code being executed) and non-functional (related to scheduling and resource allocation) adaptation processes, which are managed by the *AdaptEngine Service* and *Scheduler Services* respectively.

As previously stated in this paper, the adaptation model is collaborative. Such a collaborative adaptation process occurs in two forms: (i) adaptation commands, by explicit calls to some of the middleware services, and (ii) adaptation policies, which implicitly guide middleware operations. Adaptation policies are in the form of XML documents, deployed together with the application code when it is installed in the BDA pervasive repository. Typically, adaptation policies are defined at development-time by the application designer [6].

Specifically, the *Scheduler Service* is fed by an application provided *scheduling policy*. The scheduling policy defines not only the requirements for the resources to be allocated for the application, but also affinity criteria among the application components and between the application components and the external resources used by the application. The Scheduler service combines these abstract definitions with run-time gathered information, obtained from the context recognition subsystem, when deciding about component placement. We are using a probabilistic strategy based on bayesian networks to better estimate the system state, this strategy is implemented by TiPS [11], which integrates with the Scheduler. The Scheduler also negotiates resource allocation with the *Resource Broker Service*. Resource discovery and allocation are discussed in section 4.4.1.

Whenever a component migrates, the Scheduler service re-evaluates the affinity criteria defined for that component, triggering the migration of other components when necessary, thus contributing for the maintenance of the application cohesion and implementation of the follow-me semantics.

Besides, the functional adaptations, managed by the *AdaptEngine Service*, also receive meta-data generated at development-time in the form of *functional adaptation policies*. A functional adaptation policy, as shown in figure 8, binds component implementation to specific context element states. The *AdaptEngine Service* is activated when an adaptive component is instantiated or restored after a migration, in order to select the proper implementation for the current execution context state. Additionally, the *AdaptEngine Service* is also responsible for managing run-time functional adaptations by notifying the adaptive components when a context element for which they registered interest has changed its state. The information about the current state of the execution context used by the *AdaptEngine Service* is provided by the context recognition subsystem.

4.3.2. Multi-Level Communication Services

With respect to communications, EXEHDA currently provides, through the *Dispatcher*, *WORB*, and *CCManager* services, three flavors of communication primitives, each one addressing a distinct abstraction level.

The *Dispatcher Service* corresponds to the lowest abstraction level, providing message-based communications. Message delivering is done through per-application channels, which may be configured to ensure several levels of protection for the data being transmitted. Protection levels range from data integrity, using digital signatures, to privacy through encryption. Additionally, the *Dispatcher Service* uses a checkpointing/recovery mechanism for the channels, which is activated when a planned disconnection is in course. This feature may or may not be activated by the upper communication layers depending on its particular demands.

In order to make the development of distributed services easier, EXEHDA also provides an intermediary solution for communications, based on remote method invocations, through the *WORB Service*. The programming model supported is similar to Java RMI, though it is tuned to the pervasive environment while RMI is not. Specifically, WORB remote method invocations, differently from Java RMI, do not require the device to keep connected during the entire execution of the method on the remote node. Instead, it is build on the functionality provided by the Dispatcher service, including a per-invocation ID. The invocation ID remains valid during the disconnection, allowing the WORB to re-sync with the remote node after reconnection and recover the result of the invocation.

At a higher level, the *CCManager Service* provides tuple-space based communications. It builds on the *WORB Service*, which already handles planned disconnections, providing to applications an anonymous and asynchronous communication support. The asynchronous and anonymous communication model provided by the *CCManager Service* is better suited to the scenario where application components may migrate between nodes, since it does not require both sides to exist at the same time for the communication to take place.

```
<ADAPTERS>
 <CONTEXT n="Device">
  <STATES>
   <STATE n="console"/>
   <STATE n="pda"/>
   <STATE n="desktop"/>
  </STATES>
  <FILTER syntax="xml:StringMapping">
   <INDEX><SENSOR n="runtime.gui.awt.mode"/></INDEX>
   <MAPPINGS>
    <MATCH op="equals" v="none" ignoreCase="true" state="console"/>
    <MATCH op="equals" v="awt-low res" ignoreCase="true" state="pda"/>
    <MATCH op="prefix" v="swing" ignoreCase="true" state="desktop"/>
    <DEFAULT state="console"/>
   </MAPPINGS>
  </FILTER>
 </CONTEXT>
 <CLASS name="GeneAlGui">
  <ADAPTER contextElement="Device" default="desktop">
   <CONTEXT state="pda" impl="GeneAlGuiPda"/>
   <CONTEXT state="desktop" impl="GeneAlGuiDesktop"/>
  </ADAPTER>
 </CLASS>
 ...
</ADAPTERS>
```

Figure 8. Sample functional adaptation policy

When the application finishes its processing, the middleware uses the information stored into the *CIB Service* to execute a clean-up step, reclaiming resources previously allocated to that application.

4.4. Management of Physical Resources

From point of view of the middleware, ISAMpe resources fit in one of two categories: processing node or specialized resources. The former corresponds to the nodes, which effectively execute and whose access is managed by the middleware. The latter corresponds to unmanaged specialized devices, e.g. printers, scanners, etc., whose access is not done through one of the middleware services, but by using some specific libraries. Although unmanaged by EXEHDA, the specialized devices are also catalogued in the *CIB Service* in order to allow applications to locate and use such devices.

4.4.1. Resource discovery and allocation in the ISAMpe

The *Discoverer Service* is responsible for finding specialized resources in the ISAMpe given an abstract definition of the resource. Typically, this service interacts with the *CIB Service* from its own cell, aiming to satisfy the resource discovery request in the scope of the local cell. When the local resources fail in fulfilling the request, the *Discoverer Service* interacts with the *Resource Broker service* from the neighbor cells. The strategy adopted in the extra-cell search is characteristic of the particular *Discoverer Service* instance in use though the language used to describe resources and the service interfaces are standardized [9]. Since specialized resources are not managed by the middleware, the resource discovery provided by the *Discoverer Service* does not imply resource allocation or even resource reservation.

The *Resource Broker Service* implements the access control for resources in the perspective of discovery and allocation requests originated from other cells in the ISAMpe, controlling the resource visibility to the extra-cell universe, for both nodes and specialized resources.

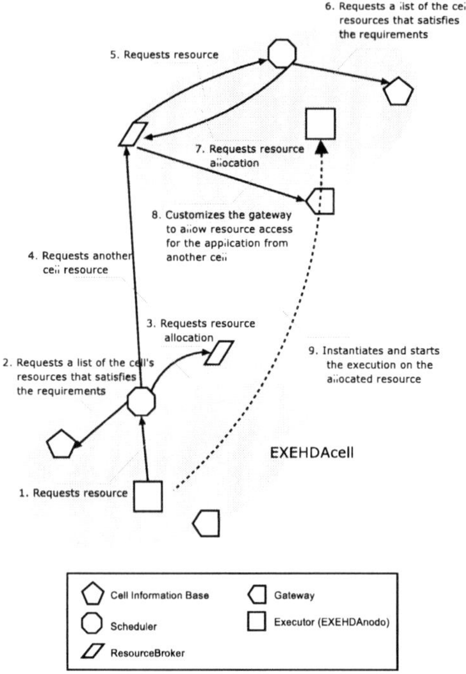

Figure 9. Resource Allocation Dynamics

4.4.2. Dynamic node configuration

DHC Service provides automatic configuration of the middleware execution profile in a node. The functionality is constructed from the cooperation between the local and cellular instances of the service, running on the nodes and base-node respectively. At node middleware bootstrap, a local instance of DHC is started and the node has no information about the cell configuration, neither about the location of the base-node. Therefore, DHC uses a base-node detection protocol, which is built on the top of IP multicast features. In case that a given number of tries is made and the base-node still could not be detected, the manager of the node is requested to enter such as information manually. Though, such procedure occurs only once in a given network segment, because after the local DHC instance knows about the location of the base-node, it is also able to answer peer requests, avoiding the need to enter the same information manually for the other nodes in that network segment.

4.5. Services to Build the Context Information

The *Monitor Service* implements a monitoring scheme based on sensors, that extracts a single index which describes a specific aspect. These sensors can be customized through parameters. The whole set of sensors installed on a node is part of the node description information registered into the *CIB Service*. The data generated by each of the sensors is gathered by the *Monitor Services*, which typically runs on the same node the sensors are installed. The gathered data are published by the *Monitor service* to a *Collector Service*, which typically runs on the base-node.

Both the extraction of the monitoring information by the sensors and the publication to the *Collector Service* by the Monitor occur in discrete multiples of a per-node configured *quantum*. The quantum parameter allows the resource owner to control, externally to the middleware, the degree of intrusion of the monitoring mechanism in the host. After a *quantum* of time, the *Monitor service* executes a pooling operation over the active sensors in the node. Then, it applies the publishing criteria specified for the sensor to the data, determining, or not, the generation of a publishing event for that sensor. Thus, the events generated inside a quantum are grouped into a single message, reducing the amount of dada the Monitor has to transmit to the Collector.

The *Collector Service* gathers monitoring information the several monitors in the cell generated and forwards them to the registered consumers. Among such consumers are other middleware services like the *Context Manager* and the *Scheduler*.

Aiming scalability of monitoring, when the amount of monitored data starts to become huge the task of multi-casting the gathered information may be delegated by the *Collector Service* to an auxiliary stand-alone instance of *Deflector Service*. In this configuration, the *Deflector Service* is registered as the unique consumer of the *Collector Service*, being responsible for multi-casting the monitoring data to the real consumers. Since *Collector Service* and *Deflector Service* are separated instances, they may run in distinct equipments in the base-node.

At the highest level of the context recognition subsystem is the *Context Manager Service*, which is responsible for transforming the raw data gathered by sensors in abstract (mnemonic) context information. Such context recognition chains are composed by three elements: (i) aggregator, (ii) translator and (iii) notifier. The aggregator element is responsible for composing data from one or more sensors in order to produce an aggregated value. The aggregated value feeds the translator element, which transforms it in an abstract (mnemonic) value, using the ranges defined by the application designer, whereas the notifier element detects changes in the values of the context element, notifying the *Adapt Engine Service* of the new value of the context element. Optionally, a predictor element may be included in the context recognition chain aiming to anticipate future context modifications and trigger adaptations in advance. Figure 10 shows the context recognition architecture organization supported by context recognition subsystem.

5. Implementation Status and Open Issues

Our current prototype has been developed considering two flavors of the Java Platform: J2SE, which is used in the desktop computers and the CDC profile of the J2ME Specification for Sharp Zaurus PDA's. Additionally, a few components are being developed in C/C++, notably the ones that extract monitoring information through direct interaction with the native operating system of the device. The EXEHDA main services are operational, and the first version of the middleware will be available to download by October 2005 at *Rede Nacional de Ensino e Pesquisa* Foundation website (http://www.rnp.br).

Besides from our implementation efforts, we envisage several aspects that need to be deepened and improved in future works. These involve the monitoring and translation process of context recognition service, security, pervasive service and resource discov-

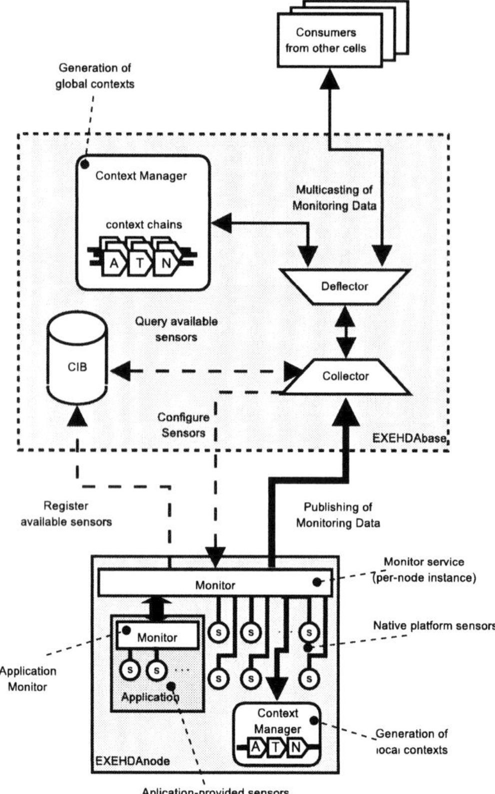

Figure 10. Context Recognition Architecture

ery, as well as the development of other applications and the including fault-tolerance features. Another field to explore is the ongoing development concept of Pervasive Object (PvO) to application design and implementation. Also, we intend to investigate the impact of ad-hoc network topology in the designed middleware services.

6. Related Works

Grid research and projects are currently focused on wired computers such as desktop PCs, supercomputers and clusters. So, mobile grid research is still weak. Some very recent initiatives aim to join mobility and grid computing, such as Akogrimo and Ubigrid. Akogrimo (http://www.mobilegrids.org) adds the mobility dimension to the European Grid. In this vision, grid services, pervasively available, are eventually meeting the *"everywhere at any time in any context"* paradigm. The services, comprising personalized knowledge and semantics, will allow ad-hoc, dynamic and possibly federated formation of problem solving scenarios, in business and science, to fixed, nomadic and mobile citizens. Ubigrid (http://ubigrid.lancs.ac.uk/) addresses the use of Grid technologies to support experiments in, and development and deployment of, ubiquitous computing systems.

Since Akogrimo and Ubigrid are recent projects, a comparison between EXEHDA and these projects is under development and the publication of the first results is expected by the end of 2005.

7. Conclusion

In this paper we presented a general view of EXEHDA middleware. It was designed to provide and manage the pervasive grid environment and control and execute the applications inside. The first results of our project indicate that is the to use of grid and pervasive technologies is possible, with advantage of both computing fields. The new generation of grid applications could extend their traditional application scenario to handle mobile users in their activities of working, learning, decision making, business and others.

References

[1] AUGUSTIN, I.; YAMIN, A.; BARBOSA, J.; GEYER, C. ISAM - a Software Architecture for Adaptive and Distributed Mobile Applications. In: IEEE SYMPOSIUM ON COMPUTERS AND COMMUNICATIONS, ISCC, 7, 2002. Taormina, Italy. Proceedings New York:IEEE Computer Society, 2002.

[2] AUGUSTIN, I.; YAMIN, A.; BARBOSA, J.; da SILVA, L.; REAL, R.; GEYER, C. ISAM, Joing Context-awareness and Mobility to Building Pervasive Applications. In: MAHGOUB, I.; ILYAS, M. (Ed). Mobile Computing Handbook. New York: CRC Press, 2004.

[3] GAMA, E. et al. Design Patterns: elements of reusable object oriented software. 10th Ed. Reading, MA: Addison-Wesley, 1997.

[4] SAHA, D.; MUKHERJEE, A. Pervasive Computing: a Paradigm for the 21st Century. IEEE Computer, New York:IEEE Computer Society, v.36, n.3, p.25-31, mar. 2003.

[5] SATYANARAYANAN, M. Pervasive Computing: Vision and Challenges. IEEE Personal Communications, New York, 2001.

[6] YAMIN, A.; AUGUSTIN, I.; BARBOSA, J.; da SILVA, L.; CAVALHEIRO, G.; GEYER, C. Collaborative Multi-level Adaptation in Distributed Mobile Applications. In: International Conference of the Chilean Computer Science Society (SCCC 2002), 12., 2002, Atacama, Chile. Proceedings. New York: IEEE Press, November 2002.

[7] YAMIN, A.; AUGUSTIN, I.; BARBOSA, J.; da SILVA, L.; REAL, R.; CAVALHEIRO, G.; GEYER, C. A Framework for Exploiting Adaptation in High Heterogeneous Distributed Processing. In: SYMPOSIUM COMPUTER ARCHITECTURE AND HIGH PERFORMANCE COMPUTING, 14, 2002, Vitória, ES, Brazil. Proceedings. New York: IEEE Press, 2002.

[8] YAMIN, A.; AUGUSTIN, I.; BARBOSA, J.; da SILVA, L. ; REAL, R.; CAVALHEIRO, G.; GEYER, C. Towards Merging Context-aware, Mobile and Grid Computing. *Journal of High Performance Computing Applications*. London: Sage Publications, 30f. 2003a.

[9] SCHAEFFER FILHO, A. E.; da SILVA, L. C.; YAMIN, A.; AUGUSTIN, I.; de MORAIS, L. L.; REAL, R.; GEYER, C. PerDiS, a Scalable Resource Discovery Service for the ISAM Pervasive Environment. In: International Workshop on Hot Topics in Peer-to-Peer Systems (HOT-P2P 2004), Volendam, The Nederlands, Oct., 2004.

[10] SCHAEFFER FILHO, A. E.; de MORAIS, L. L.; REAL, R.; da SILVA, L. C.; YAMIN, A.; AUGUSTIN, I.; GEYER, C. F. R. A Practical Grid Experience Using the ISAM Architecture for Genetic Sequence Alignment. 2nd Internacional Workshop on Middleware for Grid Computing. Canada, Oct. 2004. Poster.

[11] REAL, R. et al. Resource scheduling on grid: handling uncertainty. In: *4th International Workshop on Grid Computing*. Phoenix, Arizona, USA: IEEE/ACM, 2003.

Self-Organization and Autonomic Informatics (I)
H. Czap et al. (Eds.)
IOS Press, 2005

A Multi-Agent Approach for the Construction of a Peer-to-Peer Information System in Grids

Agostino FORESTIERO, Carlo MASTROIANNI and Giandomenico SPEZZANO
ICAR-CNR 87036 Rende (CS), Italy
{forestiero,mastroianni,spezzano}@icar.cnr.it

Abstract. A Grid information system should rely upon two basic features: the replication and dissemination of information about Grid services and resources, and an intelligent distribution of such information among Grid hosts. This paper examines an approach based on ant-based systems to replicate and map Grid services information on Grid hosts according to a given semantic classification of such services. Information is disseminated by agents (ants), which traverse the Grid by exploiting the P2P interconnections among Grid hosts. An entropy index is used to evaluate the performance of the proposed Ant-based Replication and MApping protocol (ARMAP), and control the dissemination of resource information. This approach enables the use of a semi-informed search algorithm which can drive query messages towards a cluster of peers having information about resources belonging to the requested class. A simulation analysis has been performed to evaluate the performance of the ARMAP protocol.

Keywords. Grid, P2P, information system, multi agent system, ant system, spatial entropy, resource mapping.

1. Introduction

The information system of a Grid framework provides resource discovery and browsing services which are invoked by Grid clients when they need to use hardware or software resources matching given criteria and characteristics. In currently deployed Grid systems, e.g. in the Web Services Resource Framework (WSRF) [20]), the information system is handled through a centralized or hierarchical approach. Nowadays the Grid community agrees that in large and highly heterogeneous Grids it is more efficient to devise scalable Grid information services based on a peer-to-peer (P2P) approach [9, 17].

In this paper, it is assumed that the resources offered by a service-oriented Grid are Grid services which are deployed and published by Grid Virtual Organizations; for example, within the WSRF framework, published Grid services are Web services having enriched functionalities such as state management. Furthermore, it is assumed that such Grid services can be semantically classified according to their features: a *class of resources* is defined as a set of Grid services matching specified properties. Generally, a query is not generally issued to search a single resource, but to collect information about resources belonging to a given class [4, 12]. After receiving a number of responses, a user or client can choose the resource which is more appropriate for their purposes.

This paper proposes a novel approach for the construction and management of a Grid information system which allows for an efficient search of resources. It is assumed that an underlying P2P infrastructure interconnects Grid nodes and can be used to explore the Grid. The proposed approach exploits the features of (i) epidemic mechanisms tailored to the dissemination of information in distributed systems [13] and (ii) self adaptive systems in which some sort of "swarm intelligence" emerges from the behaviour of a high number of agents which interact with the environment [3].

The rationale of using a replication approach is the following: even if a Grid service is provided by a particular Grid host, a number of information documents describing this service should be distributed on the Grid in order to facilitate discovery operations. An information document can be composed of a description of the service (i.e. a set of parameter/value pairs which specify the main characteristics of the service) and an URL reference to the WSDL interface of the service. For the sake of simplicity, in the following an information document describing a Grid service or resource will be referred to as a *Grid resource* or simply as a *resource*.

This paper proposes to disseminate information in a controlled way, with the purpose of maximizing the benefit of the replication mechanism and facilitate the discovery operations. Replicas are spatially sorted (or "mapped") on the Grid so that resources belonging to the same class are placed in nearby Grid hosts. The mapping of resources is managed through a multi agent approach, based on the model that was introduced in [6] to emulate the behavior of ants which cluster and map items within their environment. This paper proposes a variant of that model, in which items (in our case the Grid resources) are both *replicated* and *mapped*. A number of agents traverse the Grid via the underlying P2P interconnections and copy or move resources from one host to another, by means of *pick* and *drop* random functions. In particular, each agent is tailored to *pick* resources of a given class from a region in which such resources are scarcely present, and *drop* them in a region where they are already being accumulated.

An entropy function is defined for two main purposes: (i) to evaluate the effectiveness of the replication and mapping protocol for different values of network and protocol parameters; (ii) to choose the modality of mapping, between the *copy* modality and the *move* modality. In a first phase, the *copy* modality is used to generate an adequate number of resource replicas on the network. With this modality, when executing a pick operation, an agent does not remove resources from the current host: it generates a copy of the resources belonging to a given class, and takes such resources until it will leave them in another host. However, the copy modality cannot be maintained for a long time, since eventually every host would have a huge number of resources, thus weakening the efficacy of resource mapping. Accordingly, after a proper interval of time, the protocol should switch to the *move* modality: when an agent picks some resources, they are actually *removed* from the current host, thus preventing an excessive proliferation of replicas.

A *semi-informed* discovery protocol can efficiently exploit this form of resource mapping: if a number of resources of the same class are accumulated in a restricted region of Grid hosts, queries for such resources can be driven towards that region, in order to maximize the number of useful responses. A discovery operation can be performed in two phases. In the first phase, a query is forwarded through a *blind* mechanism. In the second phase, whenever a query gets close enough to a Grid region specialized in the needed class of resources, the search becomes *informed*: the query is driven towards the specialized Grid region and will easily discover a high number of useful resources.

This paper shows that the proposed protocol, namely the ARMAP protocol (Ant-based Replication and MApping Protocol), can be effectively used to build a Grid information system in which resources are properly replicated while keeping the overall entropy and the network load as low as possible.

The remainder of the paper is organized as follows. Section 1 introduces the ARMAP protocol, and discusses the random functions that drive the behaviour of mobile agents. Section 2 analyzes the performance of the proposed protocol by means of an event-driven simulation framework built upon the Swarm simulation environment [15]. Section 3 discusses related work and compares our approach to other ones proposed in the last years. Section 4 concludes the paper.

2. A Multi-Agent Protocol for Mapping Resources on the Grid

In this section the ARMAP protocol is defined and discussed. The aim of this protocol is to disseminate Grid resources and spatially map them on the Grid according to their semantic classification, in order to gather a consistent number of resources of the same class in a restricted region of the Grid. It is assumed that the resources have been previously classified into a number of classes `Nc`, according to their semantics and functionalities (see [4] and [12]).

The ARMAP protocol exploits the random movements and operations of a number of mobile agents that travel the Grid using the P2P interconnections. This approach is inspired by ant-based systems [3, 5, 6], in which swarm intelligence emerges from the collective behaviour of very simple mobile agents (ants), and a complex overall objective is achieved.

In ARMAP, each mobile agent can pick a number of resources on a Grid host, carry such resources while moving form host to host, and deposit them on another Grid host. Initially, it is assumed that each agent is "class-specific", i.e. it manages the resources of only one class. This assumption will be released later. The basic features of the ARMAP protocol (agent movements and pick and drop operations) are described in Section 1.1. Section 1.2 introduces the spatial entropy function used to evaluate the effectiveness of ARMAP and discusses a decentralized approach, based on ants' pheromone, that is used by a single agent to evaluate the correct time at which it should switch the protocol modality from *copy* to *move*. Section 1.3 discusses the role of the ARMAP protocol in the design of a Grid information system.

2.1 ARMAP Basic Operations

Agent Movement

Each agent travels over the Grid through the P2P interconnections among Grid hosts. For the sake of simplicity, the ARMAP protocol has been analyzed in a P2P network in which peers are arranged in a grid-like topology, as in the Swarm simulator [15]: each peer is connected to 8 neighbour peers, including horizontal, vertical and diagonal neighbours. At random times, each agent makes a random number of hops along the P2P network (the maximum number of hops Hmax is a protocol parameter), executes the agent's algorithm specified by the ARMAP protocol, and possibly performs a *pick* or *drop* operation.

Pick operation

Once an agent specialized in a class Ci gets to a Grid host, if it is currently unloaded (i.e. it is not taking resources of class Ci), it must decide whether or not to pick the resources of class Ci that are managed by the current host. The probability of picking the resources of class Ci is defined through by a *pick* random function; to favor the spatial mapping of resources, such probability must be inversely proportional to the number of resources of class Ci that are currently located in the local region of the Grid.

More precisely, the Ppick random function, defined in formula (1), is the product of two factors, which take into account, respectively, the relative accumulation of resources of a given class (with respect to the other classes), and their absolute accumulation (with respect to the initial number of resources of that class).

$$(1)\quad Ppick = \left(\frac{k1}{k1 + fr}\right)^2 \cdot \left(\frac{(fa)^2}{k2 + (fa)^2}\right)^2$$

The fr fraction can assume values comprised between 0 and 1 and is computed as the number of resources of class Ci accumulated in the peers located in the *visibility region* divided by the overall number of resources (of all classes) that are accumulated in the same region. The visibility region includes all the peers that are reachable from the current peer with a given number of hops (i.e. the peers located within the *visibility radius*). Here it is assumed that the visibility radius is 1, so that the *visibility region* is composed of 9 hosts, the current one included.

The fa fraction, which also can assume values comprised between 0 and 1, is computed as the number of resources of class Ci that are *owned* by the hosts located in the visibility region (i.e. resources that are directly published by such hosts) divided by the number of resources of the same class that are currently *maintained* by such hosts (i.e. including resources, published by other hosts, that have been previously deposited within the visibility region by mobile agents). Note that Ppick is directly proportional to the fraction fa, which in turns is inversely proportional to the extent to which the hosts within the visibility region have accumulated resources of class Ci so far. k1 and k2 are threshold constants which are both set to 0.1.

If the ARMAP protocol works in the *copy* modality, when an agent picks some resources of class Ci, it leaves a copy of them in the current host; otherwise, if the *move* modality is assumed, such resources are removed from the current host. In the latter case, the current host will only maintain the resources of class Ci that it directly owns, but it loses all the information about the resources of that class that are owned by other hosts.

Note that the ARMAP protocol assumes that each host is informed about the resources that are maintained by the hosts located within the visibility region. This assumption is not restrictive, since it is only required that a host periodically sends to the adjacent hosts a message containing information about the resources that it is currently maintaining. A soft state mechanism can be used to manage the possible disconnection of neighbors: if information about the resources maintained by a neighbor host is not refreshed after a proper amount of time, it must be deleted.

Drop operation

Whenever an agent specialized in a class `Ci` gets to a new Grid host, it must decide whether or not to drop the resources of class `Ci`, in the case that it is carrying any of them. As opposed to the pick operation, the dropping probability should be directly proportional to the relative and absolute accumulation of resources of class `Ci` in the visibility region. The `Pdrop` function is shown below.

$$(2)\quad Pdrop = \left(\frac{fr}{k3 + fr}\right)^2 \cdot \left(\frac{k4}{k4 + (fa)^2}\right)^2$$

In (2), the threshold constants `k3` and `k4` are set to 0.3 and 0.1, respectively.

A first set of experiments (not reported here) was performed to evaluate the impact of the threshold constants on ARMAP performance and to choose a proper set of values for such parameters. However, this setting is not very critical, since it was noted that a different set of values can only affect the rapidity of the mapping procedure, but has no remarkable effect on the qualitative behavior of ARMAP.

A high-level description of the ARMAP algorithm executed by each agent is given in Fig. 1: the different behaviour of an agent with the copy and move modalities can be noted.

```
// Na = number of agents: each one is specialized in a class of
        resources
// Hmax = max number of P2P hops that an agent can perform between two
//        successive operations
// mod = ARMAP modality (copy or move)
For each agent a (specialized in class Ci) do forever {
  Compute integer number h between 1 and Hmax;
  a makes h P2P hops;
  if (a is unloaded) {
    compute Ppick;
    draw random real number r between 0 and 1;
    if (r<=Ppick) then {
      pick resources of class Ci from current host;
      if (mod == move)
        remove resources of class Ci from current host;
    }
  }
  else {
    compute Pdrop;
    draw random real number r between 0 and 1;
    if (r<=Pdrop) then
      drop resources of class Ci into current host;
  }
}
```

Fig. 1. High-level description of the ARMAP algorithm

To tune the number of agents Na, it is supposed that each peer, at the time it joins the Grid, generates an agent with a given probability Pgen. For example, Pgen is set to 0.5 if it is wanted Na to be approximately equal to half the number of peers (Np/2).

So far, only class-specific agents were considered: with this assumption, each peer casually selects the class of resources in which the generated agent will be specialized. However, to improve performance, it is also possible to generate *generic* agents, which are able to pick and drop resources belonging to all the resource classes or a subset of them. In such a case, the algorithm shown in Fig.1 is slightly modified: the agent computes the pick and drop random functions separately for each class it can manage. This way an agent may pick the resources of class Ci from a Grid host, and drop the resources of another class Cj into the same host. The performance increase obtained by using generic agents will be shown in Section 2.2.

2.2 Entropy Function and Pheromone Mechanism

An entropy function is defined to evaluate the effectiveness of the ARMAP protocol. For each peer p, the entropy Ep gives an estimation of the extent to which the visibility region centred in p has accumulated resources belonging to one class, thus giving a contribute in the sorting process. As shown in formula (3), the overall spatial entropy E is defined as the average of the entropy values Ep computed at all Grid hosts. In (3), fr(i) is the fraction fr, as defined in Section 1.1, evaluated for the class of resources Ci. Note that Ep has been normalized, so that its value is comprised between 0 and 1.

$$(3)\quad E_p = \frac{\sum_{i=1..Nc} fr(i)\cdot \log_2 \frac{1}{fr(i)}}{\log_2 Nc}, \qquad E = \frac{\sum_{p\,\varepsilon\,Grid} Ep}{Np}$$

Simulation runs were executed to evaluate the correct time at which the modality switch (from *copy* to *move*) should be performed in order to minimize the entropy function. The assumption here is that each agent knows the value of the entropy function at every instant of time. Results are given in Section 2.

However, in the real world an agent has only a local view of the environment and cannot determine its behaviour on the basis of global system parameters such as the overall system entropy. Therefore, a method is introduced with the purpose of enabling a single agent to perform the modality switch only on the basis of local information. Such a method is based on the observation that an increase of the overall entropy value corresponds to a significant decrease of the mean level of activity of agents, i.e. of the frequency of pick and drop operations that are performed by agents. Indeed a low agent activity is a clue that a high degree of spatial sorting among resources has already been achieved.

Accordingly, a single agent can evaluate its own level of activity by using a pheromone mechanism, as suggested in [7]. In particular, at given time intervals, each agent counts up the number of successful and unsuccessful pick and drop operations (a pick or drop operation attempt is considered *successful* when the operation actually takes place – i.e. when the random number extracted is lower than the operation probability function, see Fig.1). At the end of each time interval, the agent makes a deposit into its pheromone base, by adding a pheromone amount equal to the fraction of unsuccessful operations with respect to the total number of operation attempts. An evaporation mechanism is also used to give a higher weigh to the recent behaviour of the agent.

The pheromone level at the end of the i-th time interval, Φi, is computed with formula (4).

$$(4)\quad \Phi_i = E \cdot \Phi_{i-1} + \varphi_i$$

The evaporation rate E is set to 0.9, and φi is the fraction of unsuccessful operations performed in the last time interval. As soon as the pheromone level exceeds a given threshold Tf, the agent can conclude that the frequency of pick and drop operations has remarkably reduced, and switches its protocol modality from *copy* to *move*. The value of Tf is set by observing the global system behaviour, as explained in Section 2.2. Note that the choice of updating the pheromone level every time interval, instead of every single operation, has been made to fuse multiple observations into a single variable, so giving a higher strength to agents' decisions. The length of the time interval must be long enough to collect a significant number of observations and short enough to give the right weight to pheromone evaporation; a value of 2000 sec was found to be a good compromise.

2.3 Role of the ARMAP Protocol for the Design Of P2P Information Systems in Grids

The ARMAP protocol can be a significant step towards the design and construction of a P2P-based information system in a Grid environment. However, to better understand its role, it is necessary to discuss how ARMAP can be related to the overall information system design process, which could be composed of the following three components:

(1) classification or clustering of Grid resources;
(2) replication and mapping of resources with the ARMAP protocol;
(3) discovery service.

The first component allows users to identify the features and functionalities of the resources they need (i.e. a particular resource class). Classification of resources can be performed with different techniques, as discussed in the Section 3.

The ARMAP protocol assumes that resources have already been classified.

The third component, the discovery service, assumes that the resources have been re-organized through the ARMAP protocol, or at least that ARMAP is working while discovery requests are being forwarded. The use of ARMAP permits to handle resource discovery by combining the flexible and scalable features of a *blind* approach with the efficiency and fastness of an *informed* approach. A possible discovery protocol that takes full advantage of the ARMAP work is briefly described in the following. A query message first travels the Grid network with a blind/random mechanism; however, the search procedure is turned into an informed one as soon as the query approaches a low-entropy region, i.e. a region which has gathered resources belonging to one particular class. In each low-entropy region a peer - for example the peer that collects the maximum number of resources belonging to the class in which such a region is specialized - is elected as a *representative peer*. It is possible to devise a procedure that allows representative peers to exchange information with each other and allows for the construction of an overlay network connecting representative peers. This way, a query can be routed towards the nearest representative peer, regardless of the class of resources in which it is specialized. Then, this representative peer could re-route the query to another representative peer which is specialized in the class of resources under interest, or simply send an informative message to the requesting peer. When a query finally gets to the proper representative peer, it will easily find a high number of useful results.

This paper is focused on the performance of the ARMAP protocol; the resource discovery protocol described above will be evaluated in future work.

3. Simulation Analysis

In this section we introduce and discuss the parameters and performance indices used to evaluate the ARMAP protocol, than we report and discuss some relevant simulation results which demonstrate the protocol effectiveness in a Grid environment.

3.1 Simulation Parameters and Performance Indices

Simulation runs were performed on a square toroidal grid of peers, by exploiting the software architecture and the visual facilities offered by the Swarm environment [15]. When an agent moves to a destination peer, it performs the algorithm shown in Fig. 1, possibly picks and/or drops a number of resources, and finally moves to another peer. Table 1 and Table 2 report, respectively, the simulation parameters and the performance indices used in our analysis. The number of peers Np (or Grid size) was varied from 225 (a 15x15 grid) to 10000 (a 100x100 grid). Each peer generates one or more agents with a given probability: by modulating this probability, the overall number of agents Na was varied from $Np/4$ to $2Np$. Both class-specific and generic agents were considered in the simulations.

The number of resources (Grid services) owned and provided by a single peer is determined through a gamma stochastic function having an average value equal to 15 (see [9]). Grid resources are assumed to be classified in a number of classes varying from 3 to 7; the class to which each resource belongs is selected by the simulator with a uniform random function. The average time between two successive agent movements is set to 60 sec. To move towards a remote host, an agent exploits the P2P interconnections between Grid hosts. The number of P2P hops that are performed within a single agent movement is also a random function. The maximum number of hops, referred to as $Hmax$, is varied from 1 to $D/2$, where D is equal to the square root of Np. The visibility radius, defined in Section 1.1, is set to 1.

The overall entropy E, defined in Section 1.2, is used to estimate the effectiveness of the ARMAP protocol in the reorganization of the resources. The $Nrep$ index is defined as the mean number of replicas that are generated for each resource: note that new replicas are only generated when the ARMAP protocol works with the *copy* modality. Fop is the frequency of operations (pick and drop) that are performed by each agent; this index gives estimation of agents' activeness and system stability, since such operations are less frequent when a low level of entropy has already been achieved. Finally, the traffic load L is defined as the number of hops per second that are performed by all the active agents. This index is used to evaluate the communication costs produced by ARMAP.

Table 1. Simulation parameters

Parameter	**Value**
Grid size (number of peer), **Np**	225 to 10000
Number of agents (class-specific or generic), **Na**	Np/4 to 2Np
Mean number of resources provided by a peer	15
Number of classes of resources, **Nc**	3 to 7
Mean amount of time between two successive movements of an agent	60 s
Maximum number of hops, **Hmax**	1 to D/2
Visibility radius, **Rv**	1

Table 2. Performance indices

Performance Index	**Definition**
Mean Spatial Entropy, **E**	Defined in Section 1.2
Mean number of replicas, **Nrep**	Mean number of replicas of a generic resource (included the original copy) which have been disseminated in the Grid
Mean frequency of operations, **Fop**	Mean number of successful operations - pick or drop - that are performed by a single agent per unit time (operations/s)
Traffic load, **L**	Mean number of hops that are performed by all the agents of the Grid per unit time (hops/s)

3.2 *Simulation Results*

Simulation results shown in this section are relative to simulations performed for a network with 2500 hosts that provide resources belonging to 5 different classes. In particular, results shown in Figures 2-6 are obtained by setting the number of agents Na to Np/2 (i.e. each peer generates an agent with probability 0.5), and the maximum number of hops Hmax to 3. The impact of these two parameters is discussed later. All agents are supposed to be generic, i.e. they can pick and drop resources belonging to every class.

Performance measures are reported versus time to illustrate the effect of the ARMAP protocol in the reorganization and mapping of resources. Figure 2 shows that the exclusive use of the copy modality is not effective: the system entropy decreases very quickly in a first phase, but increases again when the agents create an excessive number of replicas (Figure 3). The reason of this behaviour is the following: if agents continue to create new replicas, eventually all peers will possess all the resources, thus completely undoing the mapping operation. The curve labeled as "copy/move" in Figure 2 is obtained by switching the protocol modality from *copy* to *move* when it is observed that the entropy function increases two times in succession (entropy is calculated every 2000 sec). The effect of the modality switch is that the system entropy not only ceases to increase but decreases to much lower values. This is due to the action of agents that do not generate further replicas, as shown in Figure 3, but continue their work in creating low-entropy regions specialized in particular classes of resources.

Note that in these it is assumed that a *global* view of the system – the value of the overall system entropy - is known by all the agents. In the following this approach will be referred to as the *global* one.

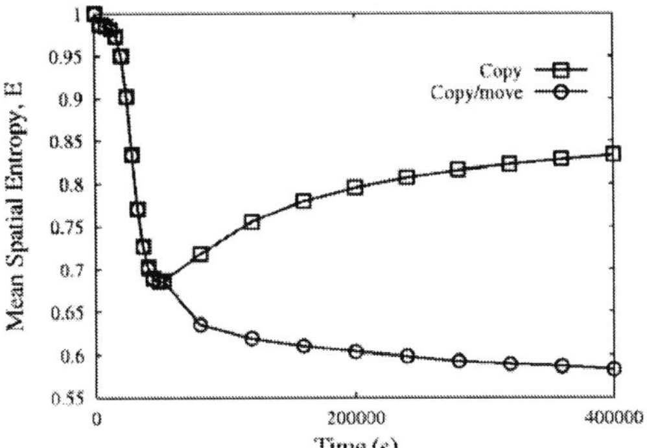

Fig. 2. Mean spatial entropy vs. time; comparison between ARMAP used with the copy modality and ARMAP with the copy/move modality switch.

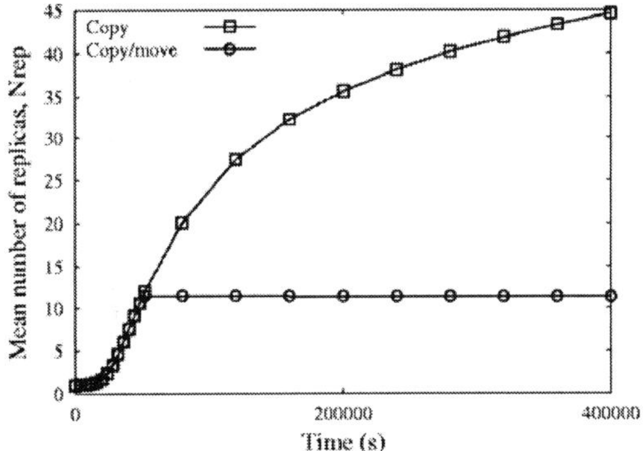

Fig. 3. Mean number of replicas vs. time; comparison between ARMAP used with the copy modality and ARMAP with the copy/move modality switch.

Figure 2 and 4 show that, with the *copy* modality, at the same time as the system entropy begins to increase, the frequency of operations of a single agent begins to decrease. This experimental result can be exploited to define a *local* approach that allows agents to perform the modality switch on their own. To this aim, the pheromone mechanism explained in Section 1.2 is used, but it is necessary to set a proper value of the pheromone threshold `Tf` beyond which an agent must perform the modality switch. This value was set with the following method: by running simulations with the *global* approach, we calculated the mean pheromone value of a generic agent at the time at which the system entropy begins to increase. This value was used as the pheromone threshold of an agent in the *local* approach. Figure 5 compares the curves of system entropy obtained with the *global* and *local* approaches. It is confirmed that the local approach is effective, since it approximates the global one very strictly, by creating a slightly higher number of replicas, as depicted in Figure 6.

Another interesting result of simulations (not shown here) is that the behavior of ARMAP does depend neither on the Grid size, nor on the mean number of resources owned by a generic peer. This assures that each agent can ignore such global parameters and confirms the effectiveness of the local approach based on the pheromone mechanism. The value of the pheromone threshold `Tf` only depends on the number of agents per peer, i.e. the ratio `Na/Np`; this parameter can be easily known by the agents since it is equal to the probability that a peer generates and forwards an agent.

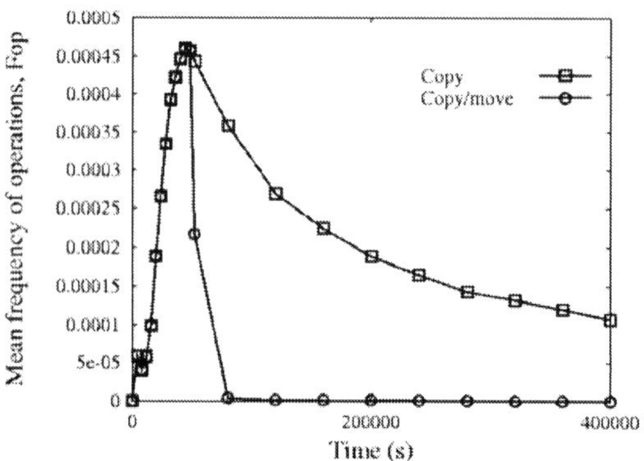

Fig. 4. Mean frequency of operations vs. time; comparison between ARMAP used with the copy modality and ARMAP with the copy/move modality switch.

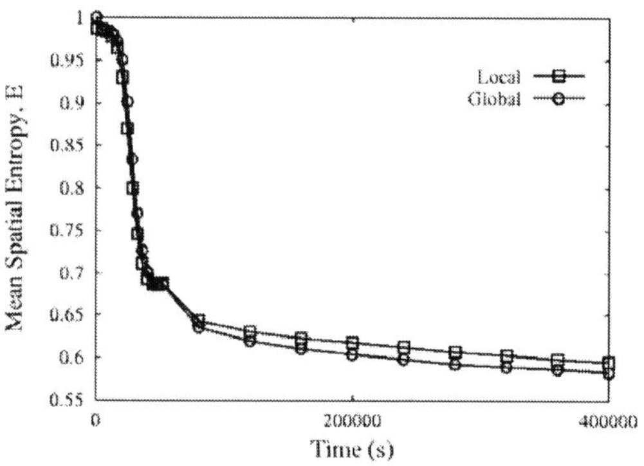

Fig. 5. Mean spatial entropy vs. time; comparison between global and local approaches.

Table 3 shows, for different values of the ratio `Na/Np`, the values of the time at which the switch modality should be performed (calculated with the *global* approach), and the mean pheromone level of a generic agent at that time. Such a pheromone level, as explained above, is the threshold at which agents must switch to the move modality when the *local* approach is used. It can be observed that, as the number of agents increases, the time at which such agents must stop creating new replicas decreases, since the overall activity of agents is higher. As a consequence, the threshold pheromone level decreases as well: a single agent will reach the pheromone threshold after a shorter interval of time.

Figure 7 reports the trend of the overall system entropy obtained with different numbers of agents: the local approach is used and the pheromone thresholds are set to the corresponding values shown in Table 3. An increase in the number of agents makes the system entropy decrease faster and reach lower values. However, a higher activity of agents also causes an increase in the traffic load (Figure 8). A correct setting of the ratio Na/Np should take into account the trend of these performance indices and in general should depend on system features and requirements, for example on the system capacity of sustaining a high traffic load.

Figure 9 compares the trend of the overall system entropy obtained with generic and specialized (class-specific) agents: in both cases the number of agents is set to Np/2. The performance increase achieved with the use of generic agents is confirmed, since they permit to obtain much lower values of the system entropy.

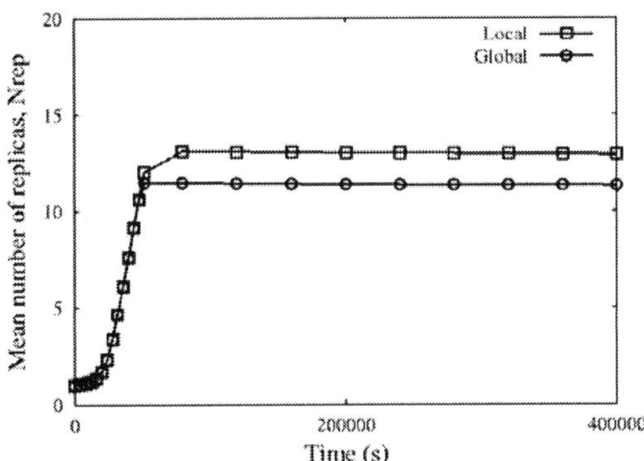

Fig. 6. Mean number of replicas vs. time; comparison between global and local approaches.

Table 3. Modality switch time and pheromone level at switch time with different numbers of agents.

Number of agents Na	Modality switch time (s)	Pheromone level at switch time
Np/4	116,000	9.44
Np/2	52,000	8.94
Np	28,000	7.41
2Np	18,000	5.89

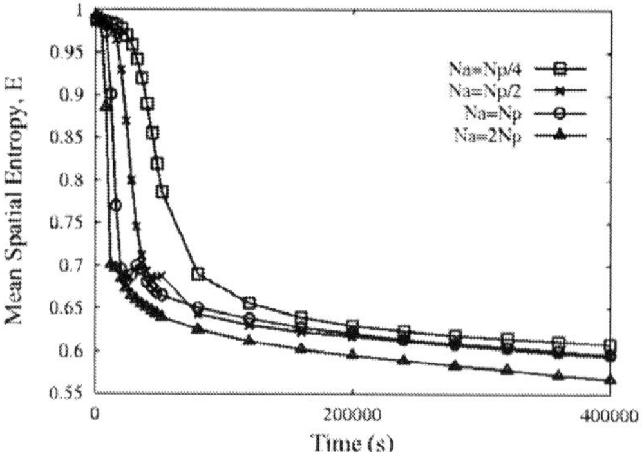

Fig. 7. Mean spatial entropy vs. time, with different numbers of agents.

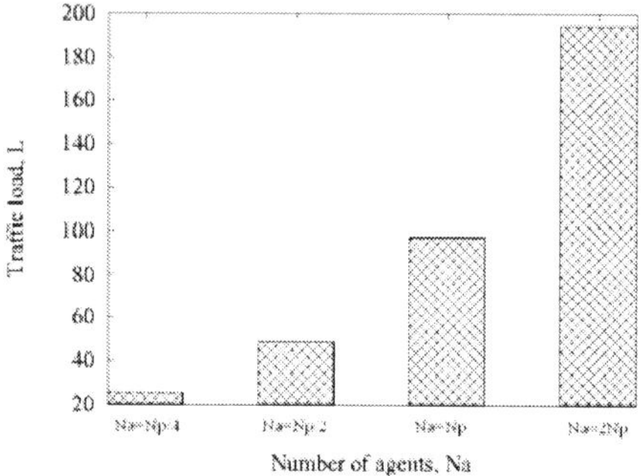

Fig. 8. Mean traffic load (hops/s), with different numbers of agents.

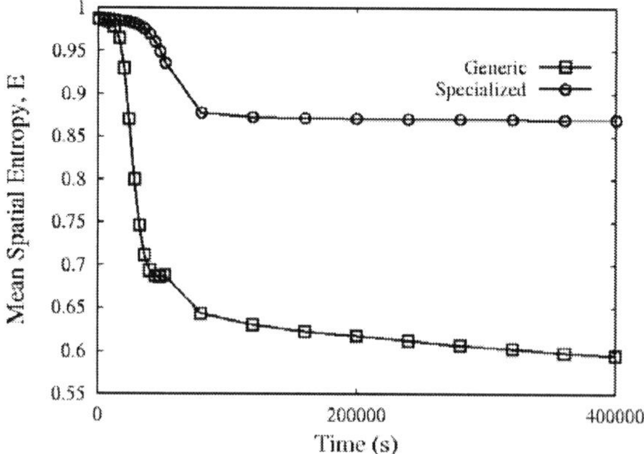

Fig. 9. Mean spatial entropy vs. time. Comparison between generic and specialized agents.

Finally, the effect of the parameter Hmax is analyzed in Figure 10-11. Here, the number of (generic) agents Na is set to Np/2. An increase of Hmax speeds up the sorting of resources (Figure 10), since an agent can move resource between more distant peers, but causes a significant increase in the traffic load (Figure 11). Again, a correct setting of Hmax should take into consideration the network requirements. We chose to set Hmax to 3, because this is the minimum value that permits to move an agent between two visibility regions that are completely disjoint, given that the visibility radius is set to 1.

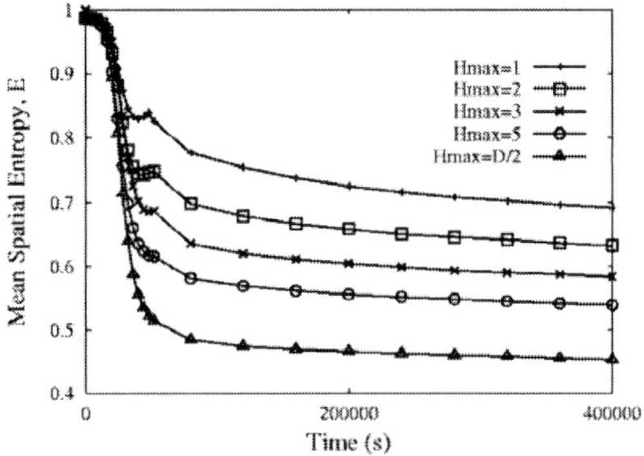

Fig. 10. Mean spatial entropy vs. time, with different values of Hmax.

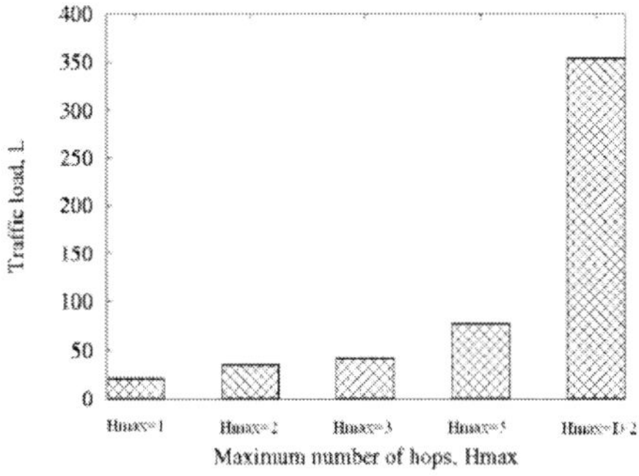

Fig. 11. Mean traffic load (hops/s), with different values of Hmax.

4. Related Work

Since Grid hosts provide a large set of distributed and heterogeneous resources, an efficient Grid information service is a pillar component of a Grid. Current Grid information services offer centralized or hierarchical information services, but this kind of approach is going to be replaced by a decentralized one, supported by P2P interconnection among Grid hosts.

In the last years, a number of P2P techniques and protocols have been proposed to deploy Grid information services: for example, super-peer networks [12, 21] achieve a balance between the inherent efficiency of centralized search, and the autonomy, load balancing and fault-tolerant features offered by distributed search.

P2P search methods can be categorized as structured or unstructured. The structured approach assumes that hosts and resources are made available on the network with a global overlay planning. In Grids, users do not usually search for single resources (e.g. MP3 or MPEG files), but for software or hardware resources that match an extensible set of resource descriptions. Accordingly, while structured protocols, e.g. based on highly structured overlays and Distributed Hash Tables (e.g. Chord [14]), are usually very efficient in file sharing P2P networks, unstructured or hybrid protocols seem to be preferable in largely heterogeneous Grids. Unstructured search methods can be further classified as blind or informed [18]. In a blind search (e.g. using flooding or random walks [11]), nodes hold no information that relates to resource locations, while in informed methods (e.g. routing indices [4] and adaptive probabilistic search [19]), there exists a centralized or distributed information service that drives the search for the requested resources. As discussed in Section 1.3, the approach presented in this paper aims to combine the flexible and scalable features of a blind approach with the efficiency and rapidity of an informed approach.

The ARMAP protocol introduced in this paper is based on the features of Multi-Agent Systems (MAS), and in particular of ant-based algorithms. A MAS can be defined as a loosely coupled network of problem solvers (agents) that interact to solve problems that are beyond the individual capabilities or knowledge of each problem solver [16]. Research in MASs is concerned with the study, behaviour, and construction of a collection of autonomous agents that interact with each other and the environment. Ant-based algorithms are a subclass of agent systems which aim to solve very complex problems by imitating the behaviour of some species of ants [3].

The Anthill system [2] is a framework that supports the design, implementation and evaluation of P2P applications based on multi-agent and evolutionary programming. In Anthill, societies of adaptive agents travel through the network, interacting with nodes and cooperating with other agents in order to solve complex problems. Reference [5] introduces an approach based on ant behaviour and genetic algorithms to search resources on a P2P network. A sub-optimal route of query messages is learnt by using positive and negative pheromone feedbacks and a genetic method that combines and improves the routes discovered by different ants. Whereas in [5] the approach is tailored to improve search routes with a given distribution of resources in the network, this paper proposes to reorganize and replicate the resources in order to decrease the intrinsic complexity of discovery operations. Instead of directly using ant-based algorithms to search resources, the ARMAP protocol exploits an ant-based replication and mapping algorithm to replicate and reorganize resources according to their categorization.

Our protocol is a variant of the basic sorting algorithm proposed in [6]. However, the latter assumes that only one item can be placed in a cell of a toroidal grid, and items can only be moved form one cell to another. Conversely, the ARMAP protocol assumes that a cell (i.e. a Grid host) can store several items (i.e. Grid resources) and an agent can create many replicas of the same item.

The problem of driving the behaviour of a single agent, which should autonomously be able to take actions without having an overall view of the system, has been discussed in [7]. There, a decentralized scheme, inspired by insect pheromone, is used to limit the activity of a single agent when it is no more concurring to accomplish the system goal. In this paper, a similar approach is used to drive the behaviour of an agent, in particular to evaluate the correct time at which it should switch from the copy to the move modality.

Information dissemination is a fundamental and frequently occurring problem in large, dynamic, distributed systems, since it consents to lower query response times and increase system reliability. Reference [8] proposes to disseminate information selectively to groups of users with common interests, so that data is sent only to where it is wanted. In this paper, instead of classifying users, it is proposed to exploit a given classification of resources: resources are replicated and disseminated with the purpose of creating low-entropy Grid regions that are specialized in specific resource classes. The so obtained information system allows for the definition and usage of a semi-informed search method, as explained in Section 1.3.

The ARMAP protocol assumes that a classification of resources has already been performed. This assumption is common in similar works: in [4], performance of a discovery technique is evaluated by assuming that resources have been previously classified in 4 disjoint classes. Classification can be done by characterizing the resources with a set of parameters that can have discrete or continuous values. Classes can be determined with the use of Hilbert curves that represent the different parameters on one dimension [1]; alternatively, an n-dimension distance metric can be used to determine the similarity among resources [10].

5. Conclusions

This paper introduces an approach based on multi agent systems to manage resources and construct an efficient information system in Grids. The presented ARMAP protocol accomplishes two main purposes: it replicates resource information on Grid hosts, and gathers information related to similar resources in restricted regions of the Grid. The work is performed by ant-like agents who travel over the network by exploiting P2P connections. A simulation study allowed for a deep analysis of the ARMAP protocol for different values of network and protocol parameters. The controlled dissemination of information assured by ARMAP guarantees a high availability of resources and favours an efficient management of the information system, since different clusters of peers can become specialized in different resource classes. Furthermore, with the use of ARMAP it is possible to devise a semi-informed discovery protocol that maximises the success of a query request by routing it towards a cluster of peers which are specialized in the resource class specified in the query. Further work is currently focused on the performance analysis of such a discovery protocol.

Acknowledgements

This work has been partially supported by the Italian MIUR FIRB Grid.it project RBNE01KNFP on High Performance Grid Platforms and Tools and by the project KMS-Plus funded by the Italian Ministry of Productive Activities.

References

[1] Andrzejak, A., Xu, Z.: Scalable, efficient range queries for grid information services, Proceedings of the Second IEEE International Conference on Peer-to-Peer Computing, Linkping University, Sweden (2002).
[2] Babaoglu, O., Meling, H., Montresor, A.: Anthill: A Framework for the Development of Agent-Based Peer-to-Peer Systems, Proceedings of the 22th International Conference on Distributed Computing Systems (ICDCS '02), Vienna, Austria (2002).
[3] Bonabeau, E, Dorigo, M., Theraulaz, G.: Swarm Intelligence: From Natural to Artificial Systems, Oxford University Press, Santa Fe Institute Studies in the Sciences of Complexity 1999.
[4] Crespo, A., Garcia-Molina, H.: Routing indices for peer-to-peer systems. In: 22 nd International Conference on Distributed Computing Systems (ICDCS'02), Vienna, Austria (2002), pp.23-33.
[5] Dasgupta, P.: Intelligent Agent Enabled P2P Search Using Ant Algorithms, Proceedings of the 8th International Conference on Artificial Intelligence, Las Vegas, NV, 2004, pp. 751-757.
[6] Deneubourg, J. L., Goss, S., Franks, S., Sendova-Franks, A., Detrain, C., Chrétien, L.: The dynamics of collective sorting: robot-like ants and ant-like robots, Proceedings of the first international conference on simulation of adaptive behaviour, From animals to animats, February 1991, Paris, France, pp. 356-365.
[7] Van Dyke Parunak, H., Brueckner, S. A., Matthews, R., Sauter, J., Pheromone Learning for Self-Organizing Agents, IEEE Transactions on Systems, Man, and Cybernetics, Part A: Systems and Humans, vol. 35, no. 3, May 2005.
[8] Iamnitchi, A., Foster, I., Interest-Aware Information Dissemination in Small-World Communities, IEEE International Symposium on High Performance Distributed Computing (HPDC 2005), Research Triangle Park, NC, July 2005.
[9] Iamnitchi, A., Foster, I., Weglarz, J., Nabrzyski, J., Schopf, J., Stroinski, M.: A Peer-to-Peer Approach to Resource Location in Grid Environments, eds. Grid Resource Management, Kluwer Publishing (2003)
[10] Kronfol, A. Z.: FASD: A Fault-tolerant, Adaptive, Scalable, Distributed Search Engine, PhD thesis at Princeton Univerity, May 2002.
[11] Lv, C., Cao, P., Cohen, E., Li, K., Shenker, S.: Search and replication in unstructured peer-to-peer networks, ACM, Sigmetrics (2002)
[12] Mastroianni, C., Talia, D., Verta, O., A Super-Peer Model for Resource Discovery Services in Large-Scale Grids, Future Generation Computer Systems, Elsevier Science, Elsevier Science, Vol. 21, No. 8 (October 2005), pp. 1235-1456.
[13] Petersen, K., Spreitzer, M., Terry, D., Theimer, M., Demers, A.: Flexible Update Propagation for Wakly Consistent Replication, Proc. of the 16th Symposium on Operating System Principles, ACM, 1997, pp. 288-301.
[14] Stoica, I., Morris, R., Karger, D., Kaashoek, M. F., Balakrishnan, H.: Chord: a scalable peer-to-peer lookup service for internet applications, Proc. of ACM SIGCOMM, San Diego, CA, USA (2001)
[15] The SwarmWiki environment, Center for the Study of Complex Systems, the University of Michigan, http://www.swarm.org/wiki.
[16] Sycara, K.: Multiagent systems, Artificial Inteligence Magazine, vol. 19, no. 2, pp. 79–92, 1998.
[17] Talia, D., Trunfio, P.: Towards a Synergy between P2P and Grids, IEEE Internet Computing 7(4) (2003) 94-96
[18] Tsoumakos, D., Roussopoulos, N.: A Comparison of Peer-to-Peer Search Methods. Proc. of the Sixth International Workshop on the Web and Databases (WebDB), San Diego, CA, 2003, pp.61-66.
[19] Tsoumakos, D., Roussopoulos, N.: Adaptive probabilistic search for peer-to-peer networks. In: Third International Conference on Peer-to-Peer Computing (P2P'03), Linkoping, Sweden (2003), pp. 102-110
[20] The Web Services Resource Framework, http://www.globus.org/wsrf/
[21] Yang, B., Garcia-Molina, H.: Designing a Super-Peer Network, 19th Int'l Conf. on Data Engineering, IEEE Computer Society Press, Los Alamitos, CA, USA (2003)

Self-Organization and Autonomic Informatics (I)
H. Czap et al. (Eds.)
IOS Press, 2005

Resource Discovery in Computational Grids: State-of-the-art and Current Challenges

Aisha NASEER and Lampros STERGIOULAS
Brunel University, School of Information Systems, Computing and Mathematics
Brunel University, Uxbridge UB8 3PH, United Kingdom
{aisha.naseer, lampros.stergioulas}@brunel.ac.uk

Abstract. The emerging Grid technologies hold out the promise of a global information channel that is far more powerful and uniquely distinct from the existing internet framework. This paper aims to provide a state of the art of work addressing the need of resource discovery in computational Grids. The various resource discovery models, techniques and approaches are discussed along with their pros & cons, and recommendations are made with respect to practical implementation and directions of future research in Grid resource discovery.

Keywords. Computational Grid, infrastructure, resource discovery, taxonomy

1. Introduction

The emerging Grid technologies hold out the promise of a global information channel that is far more powerful and uniquely distinct form the existing internet framework. The Future Interconnection Environment [2] is expected to usher in an era of intelligent interconnectivity by deploying *Intelligent Computational Grid (ICG) Infrastructure* that would ultimately lead to globalization, where man, machines[1], programs and processes act like *communicational agents* and each plays a vital role remotely according to its own semantics and render its services as intelligent resource.

Cross-communication among the various types of agents is needed on a large scale in order to enhance the capabilities and capacities of ICGs. Such agents or resources are capable of sending and receiving requests/commands and act as independent intelligent communicators. In a Grid of computers, the resources could be networks, clusters of computers offering information related to various fields, memory space or storage capacity, CPU time, CPU cycles, computational power, data repositories, files, attached peripheral devices, sensors, software applications, online instruments and data, all connected through the Internet and a middleware software layer that provides basic services for security, monitoring, resource management, and so forth [4].

Computational Grids are Multi-Peer to Multi-Peer network architectures in which all the units (agents, nodes, resources, etc.) are integrated and interconnected in such a way that no unit is stand-alone or solitary. If given privilege, a unit can remotely access any other Grid unit and be accessed by another depending on authorization.

---

[1] The term machines encapsulate all types of computer systems and attached peripheral devices.

In order to be able to access certain resources, a request is made. In order to get the request fulfilled and for the successful completion of the required tasks, it is necessary to find the right resource. If the resource you are targeting or sending request to is not the right one, then the job cannot be properly completed.

On a Computational Grid, multiple resources are dispersed/scattered across different regions. Their disparate geographical locations, heterogeneous platforms and diverse dynamic statuses make these resources rather specialized in nature and while offering much desired versatility, they are difficult to manage. This difficulty in tracking down the right resource in a vast interconnectivity environment raises the issue of Resource Discovery, namely gaining access to resources for successful completion of the job at hand. For this purpose, there must exist a consistent and controlled relationship among the various resources defined on a computational Grid. The cost and time taken to complete a job vary tremendously therefore, it is necessary to monitor the job consumption rate. Successful allocation, aggregation, management, selection and utilization of autonomous, versatile and distributed resources operating under different authentication policies on an intelligent computational Grid are issues not resolved yet. This paper intends to provide a conceptual view of the past and ongoing efforts to resolve the issue of resource discovery in computational Grids. A new concept for a taxonomy of various resource discovery methods along with their approaches is discussed and eventually recommendations are made with respect to practical implementation and directions of future research in Grid resource discovery.

2. Intelligent Computational Grid Environments – A Revolutionary Interconnectivity Infrastructure

The concept of Computational Grids or simply Grids emerged in late 90's when Ian Foster [1] introduced the computational Grid as: "A computational Grid is a hardware and software infrastructure that provides dependable, consistent, pervasive, and inexpensive access to high-end computational capabilities".

Since then many definitions and terminologies have been given to this relatively new concept. In a broader view, Computational Grid can be viewed as a paradigm, an emerging technology that aims to change the perspective of not only today's computer usage but also of computer resources and computer networks. Here, we attempt to offer a new, comprehensive definition of a Computational Grid:

"Computational Grid is a large-scale, high-performance, always-on and geographically distributed network system architecture that comprises and unifies a variety of autonomous, heterogeneous components such as processes, resources, layers, interfaces, protocols and services, etc. with strong, consistent and controlled relationships among them."

Grid is a type of parallel and distributed system that enables the sharing, selection and aggregation of geographically distributed 'autonomous' resources dynamically at run time depending on their availability, capability, performance, cost, policies or rules which govern them and user's quality-of-service requirements. This sharing is, necessarily, highly controlled, with resource providers and consumers defining clearly and carefully what exactly is to be shared, who is allowed to partake this "sharing", and the conditions for sharing. A set of individuals and/or institutions defined by such sharing rules is called a *virtual organization* (VO) [3].

It would not be far off the mark to describe the Computational Grid as a vast network, but calling it merely a computational network, will not do it justice in the semantic sense, rather it would be more fitting to view a computer network or a cluster of computers as a Grid resource (if its registered on a Grid). The Grid infrastructure, being distributed in nature, allows for high variability in user & resource participation. It deploys a decentralized computing environment, which compiles with all sorts of network architectures. In spite of being heterogeneous and distributed in terms of resources, the Computational Grid system differs quite a lot from the conventional distributed systems and resource sharing environments, such as P2P networks and clusters. It provides abstraction at both the user and resource level, which is transparent to the user and relies on a standard-based service infrastructure to share computers, storage space, sensors, software applications and data, etc. across organizational boundaries. Computational Grids provide a platform to support various (distributed) applications in resource sharing.

Compared to Grid environments, P2P systems provide limited, specialized functionality to larger and less homogeneous communities [24]. P2P systems are less reliable whereas Grids are ideal to cater for professional organizations and are more reliable. In clusters, the resource management is performed by a centralized resource manager and the nodes cooperatively work together as a single unified resource; whereas in Grids, each node has its own resource manager and does not aim at providing/supporting a single system view. Thus in computational Grids autonomous resources are managed by distributed resource managers. Each and every node on a computational Grid is considered to act both as client and server simultaneously.

Intelligent Computational Grids (ICGs) are the next generation of Computational Grids that aim to provide an "Intelligent Interconnectivity Environment". This is a global network infrastructure that contains several other networks (local or wide) and exhibits extensively high synergy among all its units (humans, networks, programs, processes and services, etc). In addition to providing a decentralized environment and possessing features of a distributed computing model, an ICG has the capability to implement intelligent interconnectivity and support diverse services such as Intelligent Resource Discovery. ICGs are expected to possess extra capabilities & features as: -

a. Automatic upgrading of resources – (Self-Managed)
b. Automatic configuration of settings – (Self-Configured)
c. Automatic scheduling – (Self-Motivated)
d. Automatic expansion (addition of new nodes) – (Self-Developed Growth)
e. Automatic repairing/elimination of distortion among resources/nodes
f. Intelligent synthesis of knowledge & creation of security models for P2P Grid
g. Intelligent security system generating automatic alerts throughout the Grid
h. Automatic tracking of resources (resource discovery)
i. Automatic intelligent distributed resource management
j. Automatic resolution of semantic compatibility issues

Intelligent Resource Discovery is considered as one of the most important aspects of an ICG. The effective and efficient discovery of resources is one of the most critical issues in the implementation of intelligent Grids. It involves not only finding but also retrieving resources that are relevant to the user (of the system) who placed request for the discovery of a specific resource.

3. Resource Discovery in Computational Grids

On a computational Grid various resources are dispersed or scattered in different regions. Their heterogeneous geographical locations, different platforms and dynamic status make these resources versatile in nature and therefore difficult to manage. It is not easy to track or locate the right resource in a vast interconnected environment. The mechanism of resource discovery can be viewed through different lenses in various domains; it is a multi-disciplinary task and is one of the most important issues to be dealt with in the future Computational Grid Technology. For the successful deployment of a computational Grid infrastructure, it is essential to access and make maximum use of resources that are available on the Grid and this is only possible if the resources are tracked effectively and efficiently. A new comprehensive definition of Resource Discovery can be introduced here as follows:

"The operation of tracking, accessing, matching, selecting and eventually requesting the right or the most accurately suitable resource for successful accomplishment of the desired job is known as Resource Discovery."

For each job, the expected cost and time consumed by the job has to be taken into consideration and be monitored. Each resource has to comply to standard connectivity protocols for communication and security, and other resource-specific protocols for enquiry, allocation, and management [3]. Techniques adopted for resource discovery should be both location and platform independent. Upon selection of the requested resource its availability status is checked and if available, the desired task is performed.

4. Types of Resources on a Computational Grid

A resource can be any real or conceptual object that's needed to be accessed by other entities, such as human users of the system or programs that generate requests for accessing particular resources.

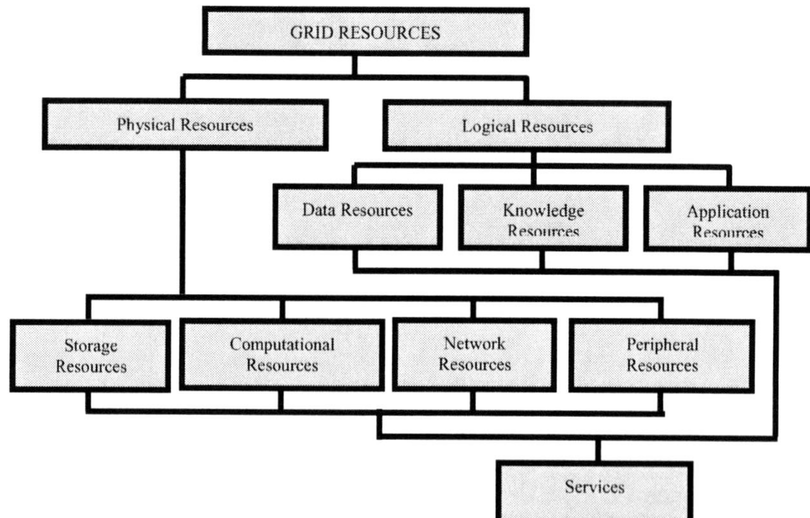

Figure 1. A Hierarchy of Grid Resources

The types of resources available on a computational Grid are generally more powerful, more diverse, and better connected than the typical P2P resources. A Grid resource might be a cluster, storage system, database, or scientific instrument of considerable processing/computation value that is administered in an organized fashion according to some well-defined policy [5]. Figure 1 depicts the categorization of Grid resources in a simple hierarchical model where the two major categories are Physical and Logical resources. Resources provide services. All the Grid resources mentioned above are utilized and accessed to get some sort of specific services. This service could be in the form of data manipulation (storage/retrieval) using a storage resource, getting solutions to complex problems, calculations via computational resources, or accessing and using a hardware device, such as a printer, etc. Service providers are entities that provide access and retrieval of resources to various users.

Both the physical and logical resources are strongly integrated and support each other in carrying out the various operations on a computational Grid. Each of these resources has individual availability status and is responsible for performing some specific tasks. Resources are located in geographical regions belonging to different time-zones and are heterogeneous as their properties, capabilities, configurations and their status change over time.

5. Why is Resource Discovery Needed?

In the face of recent technological advancements in different fields of computer science, there is a need for an infrastructure that would assist society to cope with and make maximum use of these rapid advancements. The infrastructure known as Computational Grid opts to fulfill these expectations and provides a platform for carrying out remote processing through intensive communicational ability. For successful deployment of a computational Grid model, it is necessary to discover the right resources available on it.

The challenging issues for on-demand applications derive primarily from the dynamic nature of resource requirements and the potentially large populations of users and resources. These issues include resource location, scheduling, code management, configuration, fault tolerance, security, and payment mechanisms [1]. Grid's resource discovery really is an enabler for bringing idle system resources to use.

5.1 Grid Resource Discovery Issues

Issues like geographical dispersion, heterogeneity, large number of users/requesters; dynamic nature and status of resources make resource discovery a challenge in the deployment of computational Grid systems. Resources of the same type can be highly heterogeneous [34], such as computers with different operating systems, number of CPUs and speed, datasets of various types and sizes, services of various sorts, etc. Heterogeneity in itself is not a small issue; it encompasses all aspects of compatibility conflicts, such as differences in operating systems, platforms, domains and protocols, etc. Moreover, it is a complicated task to achieve efficient job execution in a Grid environment that is constrained by deadlines and budget. Moreover, a representative taxonomy of resource descriptions might prove to be a powerful tool for engineering resource discovery solutions.

5.2 Technical Limitations

The need for resource discovery mechanisms on Grids emerged from technical limitations such as autonomous & heterogeneous resources, dynamic nature & status of resources, geographical dispersion of resources, large number of users and large distributed networks, different operating systems/platforms, different administrative domains, lack of portability, availability status of resources and different policies, etc.

To resolve all these issues, computational Grids need a consistent, efficient, time-saving and cost-effective resource discovery mechanism. To make discovery of resources more powerful, various resource discovery techniques and approaches have been proposed and applied to different settings, but no generic/comprehensive solutions to this problem has been yet available. What is still needed is a powerful, platform-independent and multi-user handling resource discovery infrastructure to support computational Grid environments.

5.3 Expected Benefits from Resource Discovery in Computational Grids

Some expected benefits incurred from resource discovery solutions include efficient resource allocation, optimal distribution of "Grid Power", maximization of usage of resources, increased usefulness of Grid and successful deployment of an infrastructure more powerful than the Internet.

From the above, there is a clear need for authentic resource discovery mechanism.

6. Taxonomy of Resource Discovery Methods

To resolve the issue of Resource Discovery in Computational Grids, different methods have been devised and models proposed for Grid taxonomy as in [21], [22] and [7]. Here in this paper we propose a new taxonomy for resource discovery models in a Grid environment.

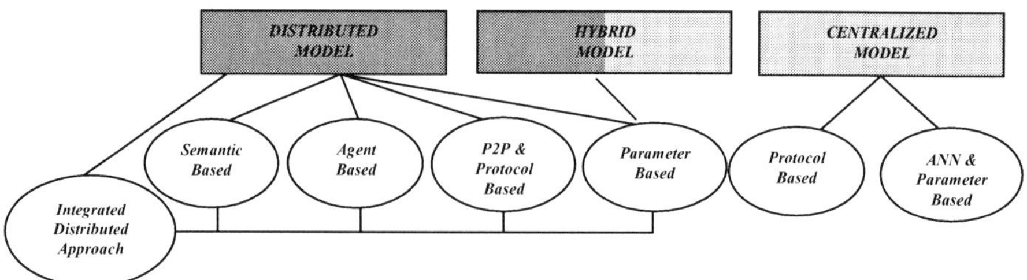

Figure 2. Taxonomy of Resource Discovery Models

This section examines three resource discovery approaches based on centralized, distributed and hybrid architectural models, which have been adopted so far for developing resource discovery solutions. A concept of these three models along with their applications is discussed further in the section. Figure 2 depicts a taxonomy of resource discovery methods in computational Grids, where the main models - Centralized, Distributed and Hybrid - are further sub-categorized.

6.1 The Centralized Resource Discovery Model

In centralized resource discovery, there is a single hosting site which acts as a central repository for hosting complete information about the entire Grid resources. This hosting site could be a single computer or a cluster of computers collectively operating as a central server. The resource information and various sharing policies reside on this centralized point. Whenever a new resource is added or an existing resource is deleted/removed or modified, information on the central server's resource directory is updated. The client nodes or units, which send their requests for some particular resource, can access this information. Since there is centralized control, the entire network is dependent on a central site whose failure will inevitably cause the entire network to crash. This model has been used to develop Grid-enabled resource discovery systems employing various approaches such as the Artificial Neural Network (ANN) & Parameter Based approach [13] and the Protocol Based approach [9,34,25].

6.2 The Distributed Resource Discovery Model

In distributed resource discovery, the resource information is dispersed across different sites. These sites could be a single computer or peer, each operating as server or a cluster of peers collectively operating as server. Each peer is hosting the directory of its local resources and an index or link to the resource registry of other peers. Whenever demand to discover a specific resource arises, the search query is sent to the immediate peer, and if match is not found then it is forwarded to the second nearest peer - if still not found then again to the next peer and so on. So each peer can send a resource request query and each is "surfed" and checked for resource availability. Since there is no centralized control, failure of any peer(s) does not affect the network in a catastrophic way. This model has been employed to develop Grid-enabled resource discovery systems by using various approaches such as P2P & Protocol based approach [23,10,34,31,29,11,8,30,19,16,6,26,27,28], Parameter-based approach [20], Agent-based approach [35,14], Semantic-based approach [33,18] and Integrated Distributed approach [32,15].

6.3 The Hybrid Resource Discovery Model

The hybrid resource discovery model integrates centralized and distributed models into a formalized broker system which maintains the resource directory and registers each resource on the Grid. The broker is responsible for matching or assigning the right resource to the request query for resource discovery [12,17,36].

7. Concluding Remarks

Various approaches are available to resolve the problem of resource discovery in computational Grids, based on three basic resource discovery models: centralized, distributed and hybrid.

The centralized resource discovery model is the simplest and a centralized architecture is easy to design. Moreover, data management is easy since the entire data is hosted at a single point. However, it is not likely to be recommended as a resource discovery solution in general purpose computational Grids, since a centralized architecture is entirely different from computational Grid architecture, which is multi-peer to multi-peer and hence would be an inappropriate route to follow in the case of ever-growing networks, due to the major issue of poor scalability. Furthermore, poor security and reliability are serious disadvantages of the centralized model.

The distributed resource discovery model can address this problem to some extent, but is not a perfect solution, since even though being distributed in nature, if the network grows, it becomes difficult to maintain and track the huge reservoir of resources, reducing the efficiency of resource discovery mechanism. The issues of scalability and architectural compatibility arise in both the centralized and distributed models.

The hybrid resource discovery model can provide the best option for creating resource/request brokering systems and designing related middleware packages, since overall it seems to be more reliable. However, this model also has some limitations, such as complexity, time, costs and difficulty of managing/maintaining, etc. In order to provide optimal service, such systems need to be easily configurable, flexible and generic. Moreover, hybrid network architecture should be modeled in a sophisticated manner, so as to address the scalability issue properly, so that its effectiveness and efficiency should remain constant regardless of the number of nodes or peers or resources added or removed from the network. It must also possess load-balancing and fault-tolerance features.

Thus it currently seems that using a hybrid approach can help resolve the problem of resource discovery in computational Grids to some extent. However, and although the potential and promise is there, it is clear that further advances are needed in this field in order to provide a satisfactory, all-round solution framework.

References

[1] I. Foster and C. Kesselman, "The Grid: Blueprint for a New Computing Infrastructure". Morgan-Kaufmann, 1998.

[2] H. Zhuge, "Semantics, Resource and Grid, Future Generation Computer Systems". Editorial of the special issue Semantic Grid and Knowledge Grid: the Next-Generation Web. 20(1):1-5, 2004.

[3] I. Foster, C. Kesselman and S. Tuecke, "The Anatomy of the Grid: Enabling Scalable Virtual Organizations". International Journal on Super Computing Applications, 2001.

[4] I. Foster, C. Kesselman, J. Nick and S. Tuecke, "The Physiology of the Grid: An Open Grid Services Architecture for Distributed Systems Integration". Globus Project, 2002.

[5] I. Foster and A. Iamnitchi, "On death, taxes, and the convergence of peer-to-peer and Grid computing". Proceedings of the 2nd International Workshop on Peer-to-Peer Systems, Berkeley, CA, 2003.

[6] N. Antonopoulos and F. Salter, "Efficient Resource Discovery in Grids and P2P Networks". Internet Research-Electronic Networking Applications and Policy, 14(5):339-346, 2004.

[7] H. Zhuge, "Resource Space Model, Its Design and Applications". Journal of Systems and Software, 72(1):71-81, 2004.

[8] I. Stoica, R. Morris, D. Karger, M.F. Kaashoek and H. Balakrishnan, "Chord: A Scalable Peer-To-Peer Lookup Service for Internet Applications". IEEE/ACM Transactions on Networking, 11(1):17-32, February 2003.

[9] S. Saroiu, P.K. Gummadi and S.D. Gribble, "A Measurement Study of Peer-To-Peer File Sharing Systems". Proceedings of Multimedia Computing and Networking 2002 (MMCN'02), San Jose, CA, 18-25 January, 2002.

[10] M. Ripeanu, "Peer-to-Peer Architecture Case Study: Gnutella Network". University of Chicago Technical Report TR-2001-26, University of Chicago, Chicago, IL, 2001.

[11] I. Clarke, O. Sandberg, B. Wiley and T.W. Hong, "Freenet: A Distributed Anonymous Information Storage and Retrieval System". Lecture Notes in Computer Science, Vol. 2009, 46-66, 2002.
[12] R. Buyya, D. Abramson and J. Giddy, "Nimrod/G: An Architecture for a Resource Management and Scheduling System in a Global Computational Grid". Proceedings of the HPC ASIA'2000, The 4th International Conference on High Performance Computing in Asia-Pacific Region (HPC Asia 2000), Beijing, China. IEEE Computer Society Press, USA, 2000.
[13] H.P. Chen, J.W. Jiang and B.W. Zhang, "Design of an Artificial-Neural-Network-Based Application-Oriented Grid Resource Discovery Service". Proceedings of the Third International Conference on Machine Learning and Cybernetics, Shanghai, 26-29, August, 2004.
[14] R. Aversa, B.D. Marino, N. Mazzocca, and S. Venticinque, "Terminal-Aware Grid Resource and Service Discovery and Access Based on Mobile Agents Technology". Proceedings of the 12th Euromicro Conference on Parallel, Distributed and Network-Based Processing (EUROMICRO-PDP'04), 1066-6192, 2004.
[15] F. Heine, M. Hovestadt, and O. Kao, "Towards Ontology-Driven P2P Grid Resource Discovery". Proceedings of the Fifth IEEE/ACM International Workshop on Grid Computing , 1550-5510, 2004.
[16] V. Iyengar, S. Tilak, S., M.J. Lewis and N.B. Abu-Ghazaleh, "Non-Uniform Information Dissemination for Dynamic Grid Resource Discovery". Proceedings of the Third EEE International Symposium on Network Computing and Applications (NCA'04), 0-7695-2242-4, 2004.
[17] J. Brooke, D. Fellows, K. Garwood and C. Goble, "Semantic Matching of Grid Resource Descriptions". 2nd European Across-Grids Conference, 2004.
[18] H. Tangmunarunkit, S. Decker and C. Kesselman, "Ontology-Based Resource Matching in the Grid - The Grid Meets the Semantic Web". Proceedings of SemPGRID '03, 2003.
[19] C. Zhu, Z. Liu, W. Zhang, W. Xiao, Z. Xu and D. Yang, "Decentralized Grid Resource Discovery Based on Resource Information Community". Journal of Grid Computing, (2): 261-277, 2004.
[20] M. Maheswaran and K. Krauter, "A Parameter-Based Approach to Resource Discovery in Grid Computing System". GRID, 181–190, 2000.
[21] R. Buyya, S. Chapin and D. DiNucci, "Architectural Models for Resource Management in the Grid". The First IEEE/ACM International Workshop on Grid Computing (GRID 2000), Springer Verlag LNCS Series, vol. 1971, 18-35, 2000.
[22] K. Krauter, R. Buyya and M. Maheswaran, "A Taxonomy and Survey of Grid Resource Management Systems". Software Practice and Experience, 32: 135-164, 2000.
[23] M. Harchol-Balter, T. Leighton and D. Lewin, "Resource Discovery in Distributed Networks". 18th ACM Symposium on Principles of Distributed Computing, 229-237, 1999.
[24] A. Iamnitchi, I. Foster and D. Nurmi, "A Peer-to-Peer Approach to Resource Location in Grid Environments". Proceedings of the 11th Symposium on High Performance Distributed Computing, Edinburgh, UK, August 2002
[25] I. Foster and C. Kesselman, "Globus: A Metacomputing Infrastructure Toolkit". International Journal of Supercomputing Applications, 11(2):115-128, 1997.
[26] S. Ratnasamy, P. Francis, M. Handley, R. Karp and S. Shenker, "A Scalable Content Address Network". SIGCOMM (San Diego USA), 2001.
[27] A.I.T. Rowstron and P. Druschel, "Pastry: Scalable, Decentralized Object Location, and Routing for Large-Scale Peer-to-Peer Systems". Middleware, pp. 329-350, 2001.
[28] B.Y. Zhao, J.D. Kubiatowicz and A.D. Joseph, "Tapestry: An Infrastructure for Fault-Tolerant Wide-Area Location and Routing". Technical Report CSD-01-1141, Berkley, 2001.
[29] W. Li, Z. Xu, F. Dong and J. Zhang, "A Grid Resource Discovery Model Based on the Routing-Transferring Method". Proceeding of 3rd International Workshop on Grid Computing, 2002.
[30] L. Huang, Z. Wu and Y. Pan, "Virtual and Dynamic Hierarchical Architecture for e-Science Grid". Proceedings of International Conference on Computational Science, pp. 316–329, 2003.
[31] A. Chander, S. Dawson, P. Lincoln and D.S. Calvert, "NEVRLATE: Scalable Resource Discovery". Proceedings of CCGrid, 2002.
[32] C. Mastroianni, D. Talia and P. Trunfio, "Metadata for Managing Grid Resources in Data Mining Applications". Journal of Grid Computing, (2): 85-102, 2004.
[33] S.A. Ludwig and P. Santen, "A Grid Service Discovery Matchmaker based on Ontology Description". Euroweb 2002-The Web and the GRID: from e-science to e-business, 2002.
[34] R. Raman, M. Livny and M. Solomon, "Matchmaking: Distributed Resource Management for High Throughput Computing". 7th IEEE International Symposium on High Performance Distributed Computing, 1998.
[35] K. Jun, L. Bolon, K. Palacz and D. Marinescu, "Agent-based Resource Discovery". Proceedings of IEEE Heterogeneous Computing Workshop, 2000. (HCW 2000), pp: 43 -52, 2000.
[36] E. Afgan, "Role of the Resource Broker in the Grid". Proceedings of the 42nd Annual Southeast Regional Conference, April, 2004.

Self-Organization and Autonomic Informatics (I)
H. Czap et al. (Eds.)
IOS Press, 2005

Providing Reliable Distributed Grid Services in Mobile Environments

T KIRKHAM, J GALLOP, S LAMBERT, B MATTHEWS, D Mac RANDAL, B RITCHIE
CCLRC Rutherford Appleton Laboratory, Didcot, UK

Abstract. This paper presents a high level model for the reliable use of mobile distributed services within Grid applications arising from research within the European research project Akogrimo. In the paper we discuss reliability and mobility in respect to static and mobile Grid services. We then introduce current research from the project into a high level blueprint/model for the architecture surrounding the provision of reliable distributed Grid services in a mobile environment.

Keywords: Grid, Mobility, Reliability, Context, SLA, Middleware, Web Services.

1. Introduction

Mobility has become part of many people's everyday lives and popular wireless and portable computing technologies are now common at home and in the workplace. This popularity is driving the development of mobile computing applications and the creation of powerful mobile computing devices, such as PDA's, laptops and mobile phones. At the same time the Grid is developing from a network of servers geared toward scientific processing, to one where Grid applications are delivering consumer services on PC's. This development is enabled by web services that are rapidly becoming part of Grid computing [1].

These two driving forces in computing will increasingly merge, leading to a powerful new generation of computing applications, for example location based services delivered to an individual's mobile phone. A new computing environment supporting semantics and collaboration will be needed to support these applications. In order to support this goal, the European sponsored project Akogrimo is developing a middleware prototype for the mobile Grid computing environment [2]. Central to this middleware development is the need for the design to be as robust and reliable as wired applications. In this paper an initial blueprint for this middleware (based on elements of research from the Akogrimo project over the last twelve months) will be presented and discussed in relation to the provision of reliability.

2. Related Work

Mobility has been a focus of previous European projects such as Moby Dick [3] and Daidalos [4]. These two projects have focused on the network layer and the creation of network interfaces and infrastructure capable of supporting a mobile set of Grid services. This work can be seen as a forerunner for the development of a Grid middleware capable of supporting mobility. Akogrimo is linked to these projects and aims to build on this infrastructure and design a Grid middleware infrastructure capable of supporting mobile services and mobility within Grid applications based on the OGSA model [5].

In Japan the 6Grid [6] project was an early project aimed at developing Mobile Internet Protocol Version Six (MIPV6) support for the Grid. (MIPv6 being a version of IPv6 standard where a mobile node is always identified by its home address, regardless of its current point of attachment to the Internet.). 6Grid's efforts focused particularly at development around the Globus toolkit [7]. Grid toolkits will provide the tools behind the creation of a mobile Grid middleware. The fruit of the 6Grid project can be seen in the latest release of Globus (GT4) that now has MIPv6 support. Aside from Globus there are also a number of implementations of Grid toolkits at various stages of mobility support. As of yet the main rival to Globus, WSRF.NET is at a disadvantage in its support for mobility as it stems from the Microsoft platform which as yet does not support MIPv6 [8].

Concerning the area of middleware development, in Korea the K*Grid project funded by the Korean Ministry of Information and Communication is developing along similar lines to Akogrimo, though mobility in K*Grid is part of a larger project to develop Korea's Grid infrastructure [9]. The main mobility focus of the project is investigating mobility in order to use idle resources from mobile devices. K*Grid aims to develop a mobile Grid platform to support this.

Whilst projects at academic and government level can be seen to be addressing the wider middleware issues, commercial companies can be seen to be working to improve their technologies to cater for specific elements/user requirements of mobility. For example, research into the provision for interfaces for specific mobile devices. Currently Oracle [10] and Sun's [11] latest releases of their Grid compatible technology claim to have incorporated mobility to some degree. However it is fair to say that no-one as yet has presented a mobile Grid middleware capable of running on a device such as a mobile phone, both in the commercial and public spheres.

The use of such Grid middleware frameworks on mobile devices that have few computing resources is generally regarded as unlikely. Current commercial and academic / governmental research is focused on the development of agents on mobile devices which can communicate with a central Grid gateway [12]. The development of such a model is the focus in this paper, where we present a central Grid middleware that can support service instance mobility.

3. Reliability

3.1 Challenges

Mobility in any form of computing has an impact on the reliability of applications which require a network connection to access external resources. Within Grid computing this strain on reliability can be seen as compounded by the applications' dependence on multiple distributed resources. From a technical point of view, these issues are often manifest in connection related problems, and are being addressed at both the lower network layers and within the middleware layer. At the network layer the problems are often approached in terms of developing QOS solutions [13]. Although they occur at the network layer, problems with the signal (for example via movement causing loss of signal in tunnels or remote areas) cannot be fully solved via network innovation and the solution to this problem has to therefore involve innovation in the middleware layer.

Within Grid middleware reliability can be measured and presented in various ways. Services can be designed to advertise their reliability metrics (i.e. up-time and average connection speed) within their Web Service Description Language (WSDL) [14]. This would aid the discovery of services for a job based on reliability criteria. In addition a Grid application could be created to store performance details of services it has used in the past, thus creating an internal reliability index to be used during service selection. However these types of approach using past data can only be seen as appropriate in Grid systems where the job that requires the services is neither urgent nor critical, as despite the provision of these reliability metrics there is nothing to 100% guarantee that the service will not fail.

The most popular form of reliability guarantee in computing applications from Grids to e-mail is the use of Service Level Agreements (SLA). Within the majority of Grid computing environments, as in many other types of service provision environments, the required level of reliability is associated with some pre-defined metrics that need to be kept above certain thresholds. Using a contractual agreement between user and provider ensures that service owners aim to maintain a service to a specific level of quality; as it would be in their best interests to fulfil the contracts. It also it allows the user to define what they mean and want in terms of reliability when negotiating this agreement. Reliability therefore becomes a metrical approach like QOS on a network, but is defined by both the supplier and user of a service.

Reliability expectations can be formed and expressed within service level agreements (SLA), which also include penalties for the provider or user to accept if the SLA is broken. The use of SLA and negotiation would encourage a marketplace within the Grid environment that would lead to greater effort to meet agreements and compete on terms. It is this negotiation and contract provision coupled with robust context aware middleware design that can be seen as the key elements in the provision of reliability in the Akogrimo mobile Grid middleware.

In the case of virtual organizations (VO) using mobile service instances to deliver business applications, it is clear that the use of SLA only encourages service providers to increase their own (bilateral) reliability. In an application exploiting many services, the overall reliability (multilateral) will be at best that of the weakest link, and generally will be far below that. To improve on this, the application requires middleware support to detect and adapt to changes in the connection to, performance of and context of the underlying services. Adaptation may be as simple as switching to an alternative network provider or as radical as changing the application to achieve the business goals in a different way. The middleware must be aware of and provide support for such changes, for example by providing solutions for handling data in an application when these changes occur.

3.2 Reliable Virtual Organisations

Later we will discuss how SLA and monitoring of workflow and service invocation is used to aid reliability in our model, but before that we will present the use of Virtual Organizations (VO's) in the Akogrimo system. Virtual Organisations are at the centre of most Grid computing architectures. During research into the provision of mobile Grid services we have developed several extensions to the ways virtual organizations are used in the execution of a job.

Grids which do not support mobility can be seen to build stable virtual organizations consisting of statically connected services geared to the execution of a specific task [15]. Within Akogrimo the services needed in order to execute a task are often mobile and as previously discussed are likely to have greater potential to experience difficulties in maintaining reliability. This introduces the need for dynamic service discovery during the execution of an application running on a mobile Grid, which in turn affects how we structure our VO.

Essentially, mobile services have the potential to drop in and out of the Akogrimo Virtual Organization after they have registered. The only time they need to guarantee availability is when they are needed for execution. Thus the discovery process must ensure only services that are available are chosen; these services therefore cannot be seen to be part of a traditional (static) VO where associated services are permanently connected and continuously available to the VO. The provision of the network connection to the VO from a mobile device may be far more costly than a traditional wired one. Thus it is also in the VO's interest to ensure the minimal use of a service's time. Akogrimo has introduced the concept of the Operative VO (Operational VO) to ensure effective use of mobile services keeping connection time to the VO at a minimal and in turn aiding the provision of service reliability.

As Figure 1 illustrates, within our design mobile Grid services are registered to a Base VO. The Base VO has two types of service. One (permanent) set are Base VO specific and reside within the boundary of the Base VO in order to aid its functionality. The other (variable) sets of services are not Base VO specific and relate to other applications. They are members of the Base VO but often are not located as permanent fixtures inside the Base VO boundary. These application services are instantiated via the Base VO in order to become a part of a Base VO application.

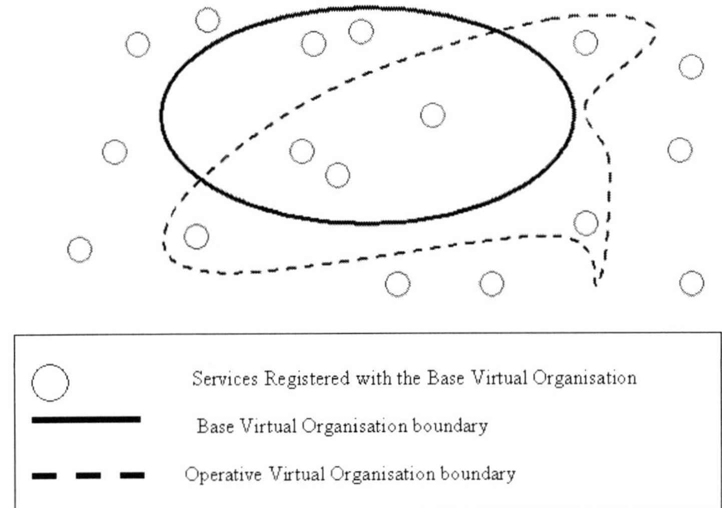

Figure 1: Services, an Operative Virtual Organisation and a Base Virtual Organisation

Application invocation for a Base VO application involves the component services becoming part of an Operational VO (see Figure 1), created and owned by a Base VO but extending out of the Base VO boundary. The Operational VO also includes and is managed by Base VO services. The Operational VO structure is unique, in the sense that it is temporary in nature and when the application has finished the Base VO dissolves it. In other words, the Base VO is a permanent entity with the power and privilege to form Operative VO's, which then provide an environment in which to execute services for the Base VO applications. The services external to the Base VO, although known and registered to it, are not owned or executed by the Base VO owner.

This VO structure is designed to suit the reliable execution of mobile services, the Operational VO's temporary nature ensures the use of services when needed thus keeping connection time and the margins for error to a minimum. Reliability in application execution inside an Operational VO is provided by Base VO services. We will now take a look at how these core VO services provide this reliability.

4 Services

4.1 Core VO Services and the Base VO

As mentioned previously, within the Base VO there exist key permanent components that provide the core functionality of the Base VO and which are essential to the provision of reliability in the execution of mobile Grid applications. These main core application services in the Base VO can be seen in figure 2.

The Base VO Manager resides at the head of the Base VO core application service hierarchy. The Base VO Manager is the initial point of contact for requests for application delivery from the VO, if the request is accepted the Base VO manager starts an application execution process interacting initially with the main services in Figure 2.

The Base VO Manager is the key locus of authentication and authorization within the VO, working with both the Participants' registry and Policy Manager to achieve this. The Base VO Manager then uses the Grid Service Discovery Service (GSDS) to find Services that can deliver the requested application functionality. These applications (for example emergency response) are typically workflows which are executed by the Workflow manager under supervision by the SLA Manager. An Operational VO Broker is created by the Base VO Manager to find services matching Workflow and SLA requirements.

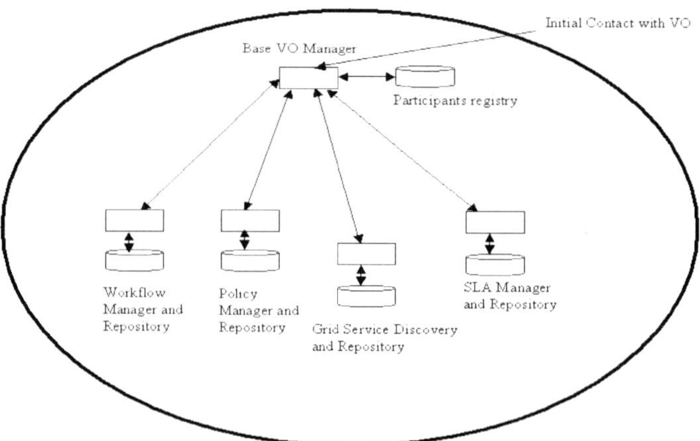

Figure 2: Core application service components of the Base VO

In order to maintain reliability, it is envisaged that the Base VO services would need to be hosted in a reliable environment, or one that is resilient keeping network downtime of the core services to a minimum. A feasible option would be to have replication of core services within the Base VO to provide back up services in case of problems, make possible a reliable mobile Base VO that could make decisions based on its current mobility, i.e., connection and location. However the focus of our current research is on non-mobile Base VOs and context and mobility is only expected to come from associated services external to the Base VO.

4.2 Mobile Services and the Operational VO

Mobility in the model is achieved by the use of mobile Grid Service instances that are registered with the Base VO, but are located in separate hosting environments and only become active within Base VO infrastructure when they are incorporated in to an Operational VO. The Operational VO instantiates registered services as part of the larger Grid application run from the Base VO. During this execution process the core Base VO services work together to ensure reliable instantiation and execution of services.

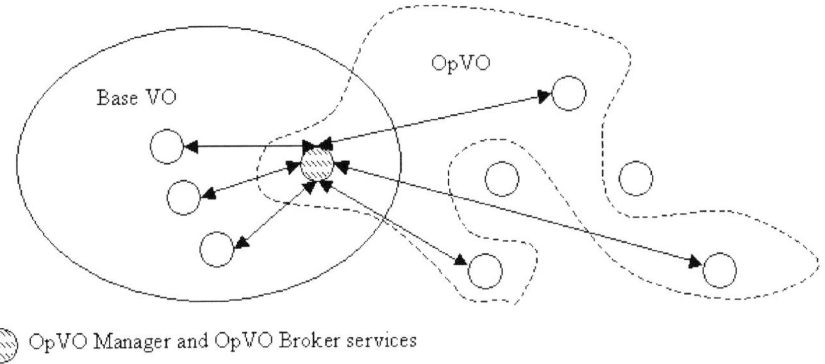

Figure 3: Operational VO and Base VO linkage

Key to the creation of the Operational VO is the use of an Operational VO Broker during the service instantiation phase. Created by the Base VO manager, this negotiates between the workflow and SLA managers in respect to which service instances should make up the membership of the Operational VO. A separate VO Manager for the Operational VO is also created by the Base VO manager. These Operational VO managers along with the Operational VO Broker are services that act as a bridge between invoked Grid service instances in the Operational VO and the core Grid services in the Base VO. As Figure 3 illustrates, the Operational VO Manager and Broker both reside in the Base VO and Operational VO.

The Operational VO and Broker are temporary services of the Base VO and are also key services of the Operational VO. The Operational VO Manager and Broker communicate directly with the services invoked in the Operational VO and the core services in the Base VO. For example the services in the Base VO use the Operational VO Manager and Broker to aid context management and workflow implementation in the Operational VO while services in the Operational VO use the Operational VO

Manager and Broker for communication with the Base VO services.

In terms of reliability the Operational VO can be seen as an innovation ensuring the reliability of execution of services orchestrated and chosen by Base VO core services. The Operational VO often contains multiple services from various distributed hosting environments. The services are executed via workflows from the workflow manager via the Operational VO Manager and are monitored during this process for errors as well as changes in context. The workflow Manager can then enhance reliability by rollback or compensation followed by the use of alternative services if there were changes in service context or other execution problems.

The Base VO has the ability to run multiple Operative VO's at the same time. This allows the processing of several workflow threads simultaneously in the case of aggregated services. This use of Operational VO's has the potential to influence positively the reliability and efficiency of the application. Furthermore the separation of the Operational VO service execution from the Base VO core services hosting environments helps ensure that any problems with the Operational VO affecting the services hosting environments would not likely affect the reliability of the Base VO; examples here could include security problems via rogue services or interception of wireless data packets.

4.3 Service Selection

What we have discussed so far is how Virtual Organizations in Akogrimo are used to establish a base architecture for the reliable use of mobile Grid services. Building on this we will now consider our provisions for the reliable behavior of services within the VO's. In order to do this we will first discuss how we aim to provide reliable service selection, and then to provide reliable service execution.

Service level agreements (SLA's) are central to the process of choosing a reliable Grid service. As discussed, metrics on service reliability can be taken into account and can help predict behaviour. However, the process of defining an SLA agreement between all parties ensures an agreement is met concerning what is expected in terms of reliability of services in the application. SLA Management is a core service residing in the Base VO.

Within Akogrimo, SLA management is based around two main roles; the first role is that of Service Provider (SP) and is occupied by the owner of the services made available to the Base VO for use in applications. When these services are registered, the SP must provide a set of contract templates specifying details such as QOS and penalties if this is not met. The other role is that of Service Consumer (SC); this role is held by the requestor of a service, for example a user who wishes to access an Akogrimo location-based service provided by several service providers. The SC also must complete a series of SLA contract templates when joining in order to outline their requirements in relation to their QOS requirements and how much they are prepared to pay for the service.

These Quality of Service agreements and requirements are outlined in SLA contracts held by the SLA Manager. When a workflow is being executed and services are being discovered, service SLA contracts are matched by the SLA Manager against the service requester's SLA requirements. The SLA manager also negotiates with the Policy Manager and Workflow Manager via the Operational VO Broker to ensure specific targets (as specified in SLA contracts) are met when the execution of a set of services occurs. The reservation phase then takes place as the services Execution Management Service (EMS) are contacted to ensure availability of the service during the time that the workflow is due to be executed. This can be seen in step 4 in Figure 4.

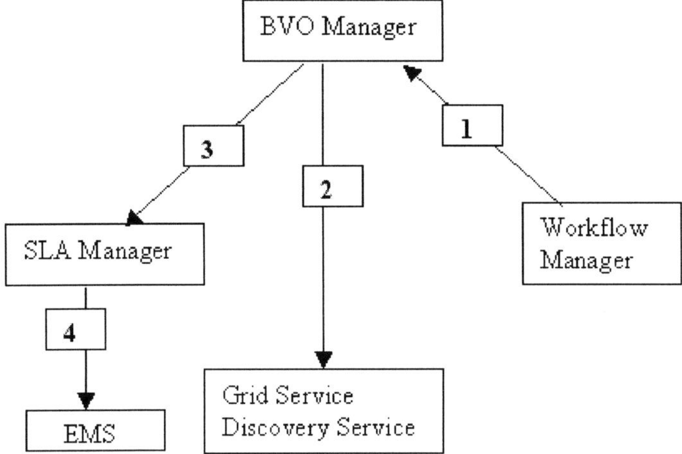

Figure 4: Grid Service Reservation Process.

In figure 4 the key three steps of the reservation process are shown. Here the initial workflow is sent to the Base VO Manager for approval. If this is accepted (subject to privileges and security policy), the Grid Service Discovery Service GSDS then selects all services appropriate for the workflow. This selection is returned to the SLA Manager who then negotiates with the services to find a best fit for the SLA requirements of the requester of the workflow. Once this is completed the services are reserved by the service provider's Execution management service (EMS).

Although this process offers no guarantees of reliability, it is the main point in the model where the standards reflecting service reliability are agreed. The process reserves the appropriate services for the workflow based on reliability statements from the service providers, and the reliability requirements specified in the SLA by the service consumer.

4.4 Service Execution

The previous work on negotiation is put to the test in the execution phase of the model. In an ideal case the workflow should be able to execute the pre-reserved services within the Operational VO with no problems or breach of SLA. In cases where this does not happen the model has to consider provision for various potential problems in order to ensure that the workflow can still be completed, thus ensuring the reliability of the framework. This area is handled by the monitoring and orchestration services within Akogrimo.

Reliability in workflow execution is achieved by the use of a hierarchy of workflow-related components around the Workflow Manager. The Workflow Manager sits at the top of the hierarchy and can start and stop workflows as well as submit and discover workflows from the Workflow Repository, see Figure 5. During the execution of a workflow the Workflow Enactment Engine and Monitoring Daemon work side by side. The latter informs the Workflow manager of the progress of the Enactment Engine, and therefore influences decisions relating to the control of the workflow, and possible corrections to the workflow in cases of error to ensure reliable execution.

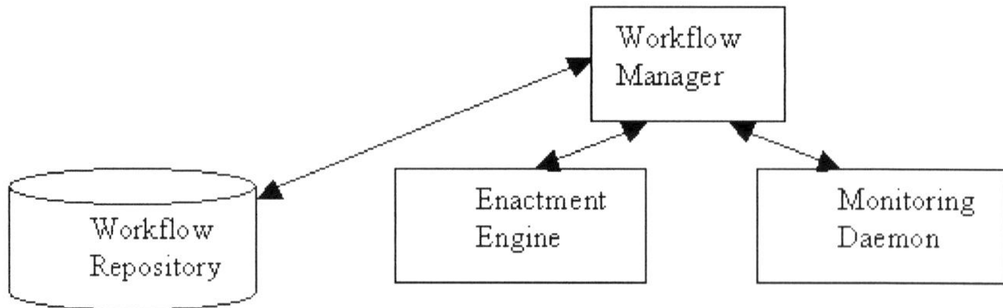

Figure 5: The Workflow Manager and workflow components.

In the case of monitoring of service context, this is achieved by context management services initiated from the service instantiation process of the Operational VO. For example the movement of a mobile device from one urban area with good network coverage to a rural area with poor coverage may cause issues effecting reliability, to the extent that a backup service instance from another provider may need to be invoked.

Therefore in respect to reliable service instance execution, the workflow management and context management partnership aids the reliable use of service instances in relation to workflow execution. The Enactment Engine's close involvement with the Operational VO Broker and Manager in the processing of services in the Operative VO is central to this. New services can be discovered and workflow blocks re-executed in cases of error picked up by the Monitoring Daemon.

5. Conclusion.

Central to the concept of reliability in Akogrimo is the process of its definition between customers and providers during SLA registration. The central control of the Base VO and its main services aids this. The SLA details are used to ensure that appropriate services are selected in order to support an application in respect of the customer's reliability expectations. The reliable use of these services in Akogrimo is supported by the concept of an Operational Virtual Organisation and by the Akogrimo Orchestration and Context Management services.

6. Future Work

In the next phase we intend to progress the Akogrimo architecture from conceptual diagrams to implemented technology. We aim to do this by developing several test beds from the architecture. As this development occurs it is likely that we will refine our initial plans for the reliability provisions in the system

We also aim to investigate the possibility of enriching the service selection and execution process. We hope to incorporate more contextual information during service discovery and execution. This would require modification of technologies such as

BPEL and possibly WSDL. Requirements to support location-specific information in a mobile environment may also lead to modification of existing standards.

7. Acknowledgements

We wish to acknowledge the support from the European Commission's IST program under the Akogrimo project.

8. References

[1] I Foster, C Kesselman, JM Nick, S Tuecke : Grid services for distributed system integration: *Computer, 2002*
[2] Akogrimo home page: www.akogrimo.org
[3] Moby Dick University of Stuttgart home page: www-ks.rus.uni stuttgart.de/Events/030516_MobyDick-Summit/Demonstrations.html
[4] Daidalos home page: www.ist-daidalos.org/
[5] I. Foster, C. Kesselman, J. Nick, S. Tuecke,: The Physiology of the Grid: An Open Grid Services Architecture for Distributed Systems Integration: *Open Grid Service Infrastructure WG, Global Grid Forum, June 22, 2002*
[6] 6Grid Project homepage: www.biogrid.jp/e/research_work/gro1/6Grid/products_Grid.html
[7] Globus home page: www.globus.org
[8] WSRF.NET home page: www.cs.virginia.edu/~gsw2c/wsrf.net.html
[9] K*Grid home page: http://Gridcenter.or.kr/
[10] Oracle Mobile Access 10g release preview: www.oracle.com/technology/products/owireless/pdfs/ocs10g_mobile_preview.pdf
[11] Sun N1 Grid Engine 6 home page: http://www.sun.com/software/Gridware/datasheet.xml
[12] Antonios Litke Dimitrios Skoutas Theodora Varvarigou: Mobile Grid Computing: Changes and Challenges of Resource Management in a Mobile Grid Environment: *5th International Conference on Practical Aspects of Knowledge Management, University of Vienna, 2004*
[13] Beaujean, Ch., Chaher, N., et al.: Implementation and Evaluation of an End-to-End IP QoS Architecture for Networks Beyond 3rd Generation. *Proc. of 12th IST Summit on Mobile and Wireless Communications, Aveiro, Portugal, June 2003*
[14] Foster (ed), J. Frey (ed), S. Graham (ed), S. Tuecke (ed), K. Czajkowski, D. Ferguson, F. Leymann, M. Nally, I. Sedukhin, D. Snelling, T. Storey, W. Vambenepe, S. Weerawarana: Modeling Stateful Resources with Web Services v. 1.1:. I*Globus Alliance, 2004*
[15] I. Foster, C. Kesselman, S. Tuecke: The Anatomy of the Grid: Enabling Scalable Virtual Organizations: *International J. Supercomputer Applications, 15(3), 2001.*

Self-Organization and Autonomic Informatics (I)
H. Czap et al. (Eds.)
IOS Press, 2005

Mobile Agent Implemented Grid Resource Reservation[1]

Wandan ZENG[a], Guiran CHANG[a], Dengke ZHANG[a] and Yu GUO[b]
[a]*Computing Center of Northeastern University, Shenyang Liaoning, China, 110004*
[b]*School of Computing, Robert Gordon University, Aberdeen, Scotland*

Abstract. It is a challenge to achieve the end-to-end QoS assurance for Grid applications in the dynamic and complex Grid environment. A Grid resource reservation model is proposed in this paper. The mobile agent implemented method of the model is given. The simulation environment of it is designed and implemented under RSVP/ns on linux. The result of simulations shows that it can flexibly adapt to the changing of the environment and achieve better end-to-end QoS assurance both of computing resources and network resources in Grid than the traditional way.

Keywords. Grid, QoS, resource reservation, mobile agent, RSVP/ns

1. Introduction

Grid makes users transparently visit the computing and storage resources possible and at the same time assures some levels of QoS [1]. Grid QoS is different from the traditional network QoS. Grid runs on top of the networks. In order to satisfy the end-to-end QoS assurance of the high performance Grid applications, it needs QoS implementations in Grid besides on the network. But because of the heterogeneity, complexity, and autonomy of the widely spread Grid resources, the assurance of the end-to-end Grid QoS becomes extremely difficult.

One of the three standards for judging a Grid proposed by Ian Foster is to supply the outstanding quality of service[2]. As an important research area of Grid, QoS gradually becomes a focus.

2. Related Research

GARA is a Grid QoS architecture developed in Argonne National Lab[3,4,5]. It first proposed the unified API to reserve and mange the heterogeneous resources and to construct higher level services. At the same time because of the hierarchical structure it can be easily extended. When a new technique of QoS appears it can be embedded into the lower level and needn't consider about higher levels. GARA uses the Globus security framework to realize the end-to-end QoS. G-QoSM is a service-oriented

---

[1] This work was support the National research Foundation for the Doctoral Program of Higher Education of China under Grant No.20030145017.

platform and attaches more importance to the implementation of the Grid QoS infrastructure[6,7]. It developed the methods of supplying and managing the Grid QoS services such as the registration and discovery of the Grid services, and so on. G-QoSM can be implemented on top of GARA.

A Grid resource reservation Model implemented by mobile agent is proposed in this paper. The model is constructed on the current Grid supporting environment. TCP/IP RSVP protocol is also the network transfer standard for it to fulfill the reservation of resources. It is easy to implement and can flexibly adapt to the dynamic Grid environment to achieve both the computing resources and network resources reservation for Grid applications.

3. G-RSVPM: Grid Resource Reservation Model

RSVP protocol of TCP/IP achieves the network resource reservation. There are two kinds of RSVP packets, PATH and RESV[10]. The process of the network RSVP is as the two following steps. First the PATH packet of RSVP with the full transmission path from the receiver to the user is sent back to the user. Then the user sends his RESV packet of the network resource reservation information on the way of PATH.

In the Grid resource reservation model, PATH packet is also used for the end-to-end address searching for the computing resource reservations.

But the end-to-end QoS negotiation in Grid is different from the traditional network. First, the reservation resource in Grid is not only the network resources but also Grid computing resources. The computing resource reservation information is set in job RSL in GRAM It will be unchanged parsed by the GRAM in the other QoS end. The network resource reservation request of the job is in the RESV packet by RSVP. Second, the PATH packet of RSVP is also used to return the end-to-end negotiation path. But because of the dynamical changing Grid environment the repeated QoS re-negotiation is inevitable and this instability makes the end-to-end QoS assurance in Grid extremely difficult. In order to resolve the difficulties mobile agents are used in the implementation of G-RSVPM. See it in section 4.

The Grid Resource Allocation Manager(GRAM) component of Grid supporting environment is installed in every Grid resource environment. GRAM is responsible for the local resource management and reflects the dynamic resource states to the Meta-computing Directory Service component(MDS) in Grid server. The Grid supporting components, MDS and DUROC are installed in the server. MDS is a Grid supporting component which can take responsibility of the resource finding, registration, resource query and getting the changing information of resources and obtain the update state of Grid resources. DUROC(Dynamically–Updated Request Online Coallocator) can dynamically co-allocate the resource owners in Grid and execute the tasks among them.

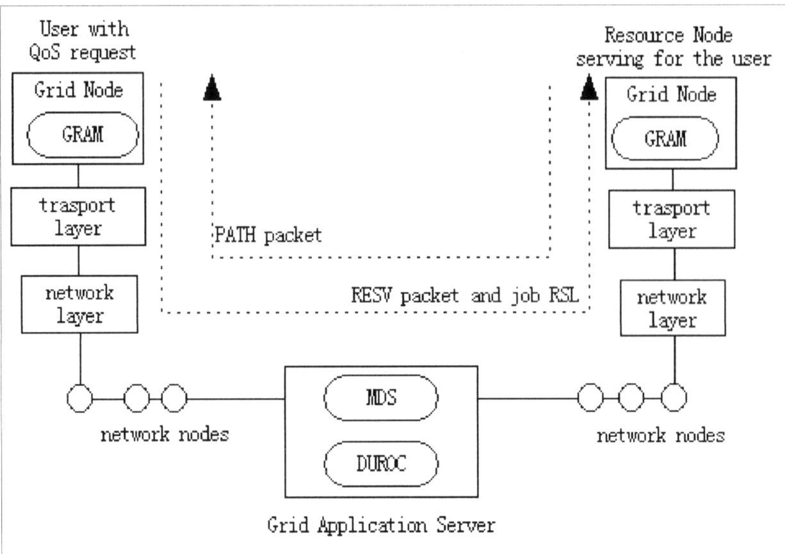

Figure 1. The Overview of the G-RSVPM

4. Mobile Agent Implemented Grid Resource Reservation Model

4.1. Using Mobile Agents to Resolve the Problems in the Grid Resource Reservation

Grid resources are fully autonomous and they can randomly leave out or suspend services running on them. When the job comes to the pre-allocated resources, the resources may have already changed and can't satisfy the QoS requests. Secondly, the PATH in RSVP may change in the dynamic environment so the QoS re-negotiation would be taken repeatedly.

In order to overcome these difficulties mobile agents are used to implement the system. Mobile agents can automatically monitor the changing environment and get the most fresh resource information timely. The QoS negotiation is primarily taken by mobile agents so that avoids the repeated transmission of jobs. So that reduces the workload burden greatly.

Mobile agents can supply dynamic run-time adaptability in Grid environment to insure the end-to-end QoS. The QoS re-negotiation by mobile agents shields the changing of the environment, makes the resource reservation and the QoS re-negotiation easier.

Certainly, this approach needs techniques supporting ,such as the interaction between mobile agents and the local Grid resource management system, the security of Grid-based mobile agents. Some researches have done to resolve these problems such as research [8,10,14] and so on. These will also be further researched in our work. The assumed precondition in this paper is that the authentication and interaction for mobile agents to the Grid are all in ideal conditions.

4.2. Mobile Agent Implemented Topology

The mobile agent management architecture in[13] is adopted in the implementation of our model. The agent system is composed of three parts and they are LQoSA, NA, GQoSA. The GQoSA is management station and it dispatches LQoSA and NA. Every Grid node is taken as a managed node and the mobile agent supporting environment is installed in every node.

The mobile agents implemented Grid Resource Reservation Model topology is as Figure 2.

Some numbers of Grid nodes in a geographical near area or in an administration domain are looked as one Grid area. A local QoS management agent(LQoSA) in each Grid area manages the QoS requests from local nodes, obtains the newest local resource information from local PBS or LSF. It can automatically forward the changing resource information to the GQoSA. LQoSA communicates with the NA or the GQoSA to do the QoS negotiation.

The global QoS management agent (GQoSA) is set in the Grid server. GQoSA is responsible for the global Grid QoS control. It can interact with the MDS component to obtain the resource information in the whole Grid. Its resource information also can be refreshed by the coming LQoSA with the new information state. Once the LQoSA monitors the information changes in local nodes they will report to the GQoSA. At the same time, the GQoSA tracks the LQoSA and prepare for the coming QoS requirements from other LQoSA. GQoSA communicates with the NA to monitor the current network reservation state.

Network Agent (NA) is set in passing networks. The main function of it is to assure the network resource reservation. There may be several NAs through the path. NA interacts with the administration of networks collecting the changed network information, reserves network resources and monitors the QoS. When network congestion happens, NA starts up the QoS adaptation. In RSVP-unsupported tunnels NA is responsible for the information collection of routers in the tunnel. NA gets network QoS-related parameters by SNMP, such as the loss rate and the queue length, etc, and sends them to the NA on other edge of the network to take the QoS re-negotiation.

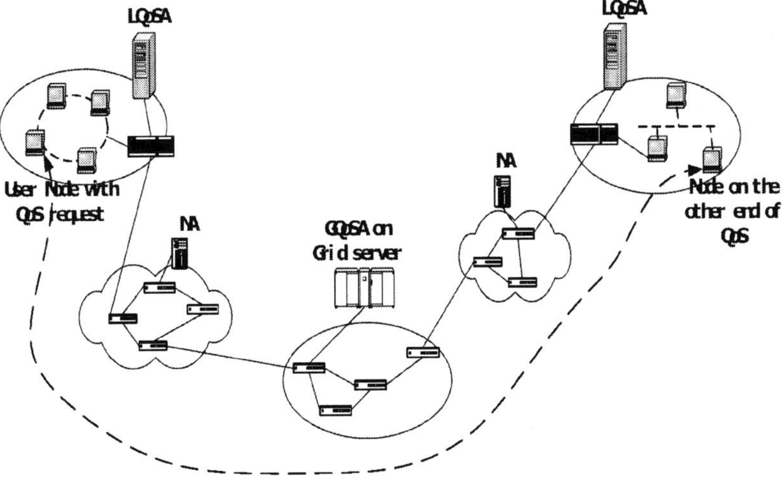

Figure 2. The Topology of the Model

QoS negotiation or re-negotiation is taken primarily by mobile agents so that to overcome difficulties in Grid end-to-end QoS assurance.

Figure 3 shows the end-to-end QoS negotiation levels in the system.

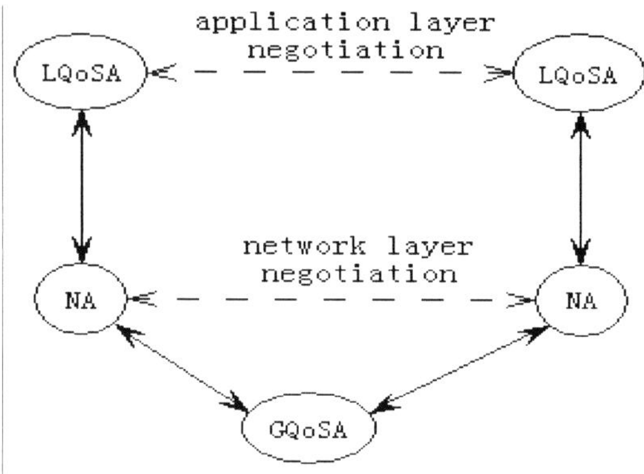

Figure 3. The End-to-End QoS Negotiation Levels

5. Simulation of the Model

5.1. Simulation Design

The mobile agent based G-RSVPM is simulated using Updated version of Marc Greis' RSVP/ns environment on Linux[11][12].

In order to simulate Grid nodes, network nodes in NS-2 are extended. Nodes in NS-2 are attached a mechanism, which can simulate GRAM in G-RSVPM to obtain the local computing resource information timely. A dynamic changing table of resource states is used to simulate the local PBS. One of the nodes in the simulation topology is taken as the server so that to simulate the Grid server in G-RSVPM. The server in NS-2 is attached a mechanism which can simulate MDS in G-RSVPM to obtain the global computing resource information dynamically.

Agents are attached to every node. The mobile agent attached to the node can get information from local PBS. This simulates the LQoSA in G-RSVPM. The agent attached to the server can obtain the dynamic resource information from MDS. This simulates the GQoSA in G-RSVPM. The mobile agent can automatically forward its resource changing information to the agent on the server. The automatism behavior of the LQoSA to GQoSA is simulated by this way.

5.2. Simulation Results

The first set of simulations (Figure 4.) shows that it can achieve better computing resource reservation performance in the G-RSVPM than without it.

The second set of simulations (Figure 5.) shows the comparison of the reservation success ratio in G-RSVPM with mobile agents and without mobile agents.

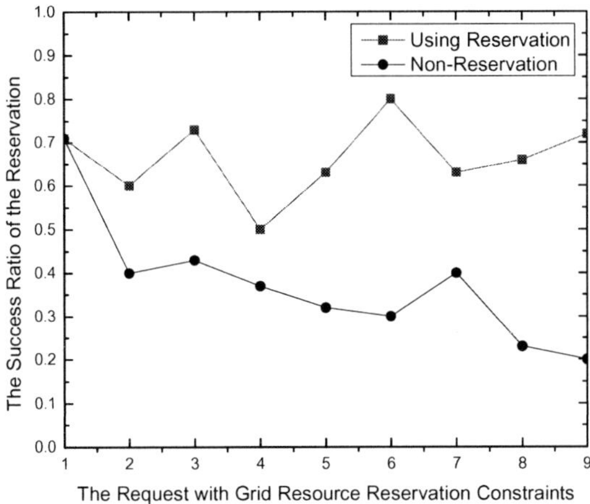

Figure 4. The Success Ratio of the Request with Reservation and without Reservation

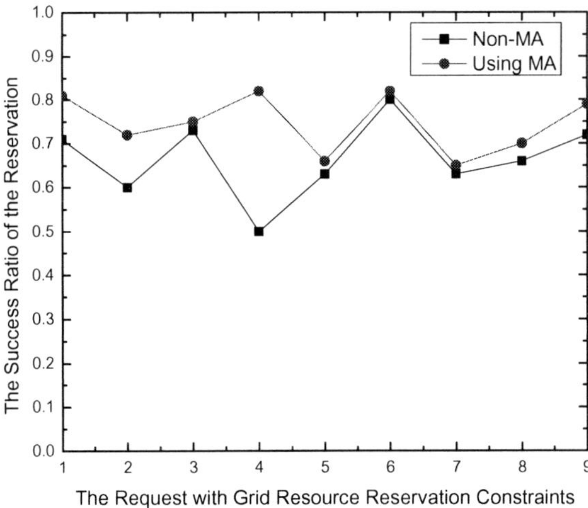

Figure 5. The Success Ratio of the Request with Mobile Agents and without Mobile agents

The third set of simulations (Figure 6.) shows the statistical re-negotiation times of the end-to-end resource reservation in G-RSVPM with mobile agents and without mobile agents.

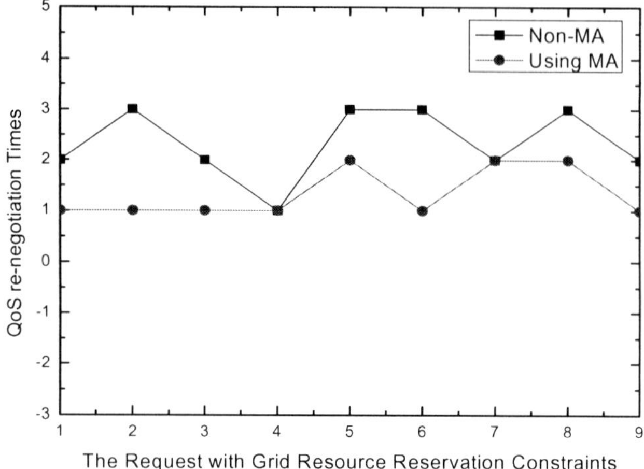

Figure 6. The QoS Re-Negotiation Times

The simulations show that G-RSVPM implemented by mobile agents can achieve computing resource reservation QoS besides the network resource reservation. It can also adapt to the dynamic environment to get much better QoS performance than the traditional way.

6. Conclusion

The mobile agent implemented Grid resource reservation model proposed in this paper can flexibly adapts to the dynamic and complex Grid environment to achieve the end-to-end QoS assurance. But because the simulation is taken under the ideal circumstance, the model should be testified and optimized in the real Grid applications in the future. It also needed further work on the quantitative performance evaluation of the QoS in this model.

References

[1] Dario Bruneo, et al. Communication Paradigms for Mobile Grid Users. Proceedings of the 3rd IEEE/ACM International Symposium on Cluster Computing and the Grid. 2003 IEEE.
[2] Ian Foster. What is The Grid? A Three Point Checklist. http://www-fp.mcs.anl.gov/`foster
[3] Ian Foster, Carl Kesselman, et al. A Distributed Resource Management Architecture that Supports Advance Reservation and Co-Allocation. Proceedings of the International Workshop on Quality of Servicce(IWQoS), 1999:27-36
[4] Ian Foster, Alain. Roy, et al. A quality of service architecture that combines resource reservation and application adaptation. In International Workshop on Quality of Service, pages 181-188, 2000.
[5] Alain. Roy. End-to-End Quality of Service for High-End Applications.PhD Dissertation. The University of Chicago, 2001.

[6] Rashid Al-Ali, Kaizar Amin, Gregor von Laszewski, Omer Rana and David Walker. An OGSA-Based Quality of Service Framework. Proceedings of the Second International Workshop on Grid and Cooperative Computing (GCC2003), Shanghai, China, December 2003.
[7] Rashid Al-Ali, Omer Rana and David Walker. G-QoSM: A Framework for Quality of Service. In e-Science AHM03 Proceedings, Nottingham UK, September 2003.
[8] Ma Tianchi, Li Shanping. An Instance-Oriented Mechanism in Grid-based Mobile Agent System.Proceedings of the 2th GCC. ShangHai, China. 2003,12.
[9] Lin Chuang, Shan Zhiguang, Ren Fengyuan. Quality of Service of Computing Networks. Beijing:Tsinghua University Press.2004,4.
[10] Ma Tianchi. Securing Mobile Processes in Grid Environment. PhD Dissertation. Zhe Jiang Univercity, China. 2004.
[11] http://www.cse.unsw.edu.au/~mamalik/rsvponns2.html
[12] Marc Greis. RSVP/ns: An Implementation of RSVP for the Network Simulator ns-2.
[13] Gavalas D, Greenwood D, et al. An infrastructure for distributed and dynamic network management based on mobile agent technology [J]. Proceedings of the IEEE International Conference on Communications(ICC'99), p1362-1366. Vancouver, June 6-10, 1999.
[14] Dhruba Raj Ghimire. Design of a Grid-based Geo-Service Architecture.

Part IV

Autonomic Computing in General

Self-Organization and Autonomic Informatics (I)
H. Czap et al. (Eds.)
IOS Press, 2005

Achieving Self-Healing in Autonomic Software Systems: a Case-Based Reasoning Approach

Cosimo Anglano, Stefania Montani
Dipartimento di Informatica
Università del Piemonte Orientale
Via Bellini 25/g, Alessandria, Italy.
email: {cosimo.anglano, stefania.montani}@unipmn.it

Abstract. Self-healing, one of the four key properties characterizing Autonomic Systems, can enable large-scale software systems delivering complex services on a 24/7 basis to meet their goals without requiring any human intervention. In this paper we present a self-healing methodology for software systems based on Case-Based Reasoning, a reasoning paradigm that exploits the unformalized knowledge embedded into already solved instances of problems, able to take advantage from the fact that in software systems most errors are just re-occurrences of known problems. We demonstrate the practical applicability of our methodology by showing how it can be used to obtain a self-healing software system delivering large-scale Internet services.

1. Introduction

As pointed out by Ganek and Corbi [14], "the computer industry has spent decades creating systems of marvelous and ever-increasing complexity, but today complexity itself is the problem." By looking at today's large-scale networked applications and services, whose growth has been quite substantial in the recent years, we can indeed observe that their inherent complexity, heterogeneity, and dynamism makes inappropriate, if not impossible, the traditional human-centered approach to system administration. As a result, the attention of the industrial and academic communities has been driven towards novel solutions allowing the design and the implementation of self-managing computer systems. The *Autonomic Computing paradigm* [14,21,31,44], inspired by the human autonomic nervous system, has been recently proposed as an approach for the development of computer and software systems and applications that can manage themselves in accordance with high-level guidance from humans. An Autonomic Computing Systems (ACS) is composed of *managed elements*, whose behavior is controlled by *autonomic managers* that apply suitable policies in order to automate the process of system management. Autonomic managers behave according to the so-called *autonomic cycle* [21,31,44], schematically depicted in Fig. 1, that encompasses

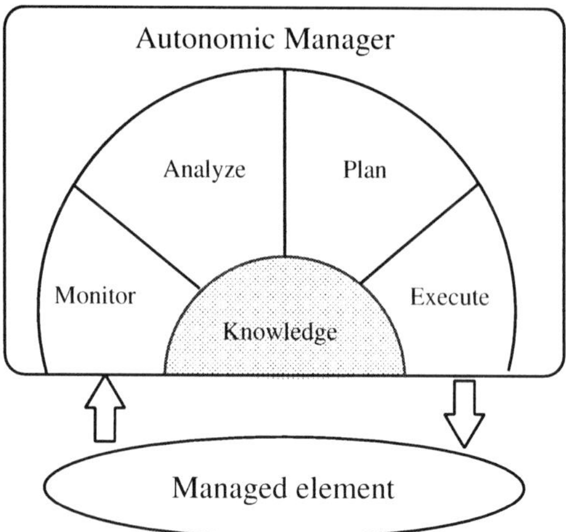

Figure 1. The architecture of an Autonomic Manager, illustrating the *autonomic cycle.*

four distinct steps. More specifically, an autonomic manager continuously monitors the corresponding managed element (in the *Monitoring* step) in order to collect information concerning its state and behavior, analyzes this information (in the *Analysis* step) to determine possible deviations from the correct or intended behavior, devises proper corrective actions (in the *Plan* step) to bring the system back into normal behavior, and implements these plans (in the *Execution* step) by exploiting suitable *actuation mechanisms* that must be provided by the managed element.

In order to be able self-managing, an ACS should exhibit the so called *Self-** properties, that is it should be *Self-Configuring* (i.e., able to (re)configure itself under varying and unpredictable conditions), *Self-Optimizing* (i.e., able to detect suboptimal behavior and to optimize itself to improve its execution), *Self-Protecting* (i.e., able to protect itself from both external and internal attacks), and *Self-Healing* (i.e., able to detect and recover from problems and/or failures). Obtaining these properties is the goal of autonomic managers, whose activity is driven by both human-specified policies and knowledge that is acquired and updated during the operation of the system.

The Autonomic Computing paradigm is particularly attractive for large-scale software systems aimed at delivering on-line services on a 24/7 basis, as for instance those described in [9,29]. The very large size of these systems (that may typically include from hundreds to thousands machines), and the adoption of customized application software and middleware, makes at the same time service failures relatively frequent and human-centered system administration very hard, if not impossible. Consequently, self-managing capabilities, and especially self-healing, represent a very attractive solution for the management problems of these systems. In this paper we focus on the problem of achieving self-healing in (large-scale) software systems, that can be defined as the ability of a system to repair itself after the occurrence of a *fault* (in one or more of its constituent components) that led to an *error*, that in turn caused a *service failure* (that is, a deviation from the correct or intended behavior of a system delivering a service) [1]. As prescribed by the autonomic cycle, a self-healing software system must be able to *monitor* its own behavior, in order to identify service failures, to *analyze* these failures with the aim of diagnosing the faults causing them, to *devise* a suitable repair plan able to fix the faults, and to *execute* the above plans by means of suitable mechanisms.

While monitoring [41,42] and reconfiguration[12,17,26] are certainly important, the ability to correctly diagnose the faults causing a service disruption, and to devise the corresponding repair strategy, is crucial to the achievement of self-healing. As a matter of fact, if a service failure is ascribed to the wrong cause, an ineffective (or even wrong) solution is proposed, with the consequence that other undesired behavior may be introduced into the system. In this paper we address the problem of devising fault diagnosis and remediation techniques that can be used by an autonomic manager of an Autonomic Software System to identify both the causes and the possible solutions of a service failure. More specifically, we propose a methodology supporting autonomous *service failure diagnosis and remediation* (as opposed to fault diagnosis and remediation pursued in traditional approaches) that exploits Case-Based Reasoning (CBR), a reasoning paradigm able to exploit the unformalized knowledge embedded into already solved instances of problems [2,23].

The rest of the paper is organized as follows. In Section 2 we present the related work, while in Section 3 we briefly review the main features of CBR. In Section 4 we discuss how CBR can be used for service failure diagnosis and remediation in Autonomic Software Systems, and in Section 5 we demonstrate our technique by using the failure and repair data of some real, large-scale Internet Service systems reported in [28]. Finally, Section 6 concludes the paper and outlines future research work.

2. Related Work

Various approaches to fault diagnosis and remediation have been proposed in the literature. Sterrit [40] proposes an approach to fault diagnosis based on *event correlation*, where various *symptoms* of system malfunctions (represented by alarms triggered by the various system components that are collected during the monitoring phase) are correlated in order to determine the (set of) fault(s) that have occurred. An alternative approach is proposed by Garlan and Schmerl [15], where fault diagnosis is performed by means of a suitable set of models. Brodie *et al.* [10] propose a technique in which recurring software errors are identified by means of the call stack generated by the faulty program, that are stored into a database and are used for fault diagnosis. Littman *et al.* [24] propose *cost-sensitive fault remediation*, a planning-based technique aimed at determining the most cost-effective system reconfiguration able to bring the system back to full functionality. Planning represents also the basis for the fault remediation strategy proposed by Arshad *et al.* [5].

The main drawback of these approaches is that they require the availability of formalized and widely recognized background knowledge (henceforth referred to as *structured knowledge*) on the structure and/or the behavior of the system. For instance, planning-based techniques require a description of the domain, the states, and the correct configurations of the system, while event correlation requires the availability of a model describing how the various system components interact among them. Unfortunately, significant efforts are usually required to build, maintain, and use structured knowledge, with the consequence that its applicability to large-size systems, exhibiting complex behaviors and interactions among their components, may be problematic. Another drawback of these approaches is their "fault orientation", that is they are triggered by individual component faults. Consequently, they attempt to correct a fault as soon as it is diagnosed, even if it is not (yet) causing any service disruption because it is *dormant fault* [1], or a fault that has been masked by the fault-tolerance techniques embedded into the system. Devising a repair plan for a fault that can be masked by the system is a waste of resources, and the same holds true for a dormant fault if, when it occurs, can be masked as well. Furthermore, from the perspective of service delivery, a dormant or a masked fault has little or no importance until it causes a service failure (i.e., it becomes *active* [1]). However, while pro-active repair of dormant faults can be important in physical systems, for software systems it is much less important, as an unnoticed bug or a misconfiguration (that are, by definition, dormant faults) may never turn into an active fault causing a service failure. For instance, an unnoticed bug may be corrected as a side effect of a software update performed to fix another problem. Moreover, the correction of dormant or masked faults (*preventive maintenance* [1]) requires the availability of a model of the system, that brings us back to the problem of structured knowledge mentioned before. Finally, the proposals discussed before either address fault diagnosis or fault remediation, but none of them addresses both issues at the same time.

The CBR-based methodology we present in this paper does not suffer from the problems mentioned above. It is indeed particularly suited to those domains in which a formalized and widely recognized background knowledge is not available. This is often the case when designing an Autonomic Computing System for which structured knowledge is too hard to collect and maintain, or when "retrofitting" self-healing capabilities into existing legacy applications [19]. Moreover, CBR seems particularly appropriate to failure diagnosis and remediation in software systems, as in this domain most errors are re-occurrence of known problems [5,10,20,28,39,46], and provides a unique framework in which failure diagnosis and remediation are performed jointly. Nevertheless, rather interestingly, very few proposals of adoption of CBR in this field can be found. To the best of our knowledge, the only proposal resembling ours has been published in [10], where a Case-Based retrieval system for discovering software problems without requiring human intervention is presented. Despite this contribution represents a first concrete step in the direction of relying on CBR for Autonomic Computing, the approach is still quite limited, as it consists in a pure retrieval systems, in which the other steps of the CBR cycle (see section 3) are ignored, and the problem solution is not provided.

3. Case-Based Reasoning

Case-Based Reasoning is a reasoning paradigm able to exploit the unformalized knowledge embedded into already solved instances of problems [2,23], called *cases*. In some sense, CBR is able to mimic human experts' analogical reasoning, by remembering solutions to similar problems adopted in the past, and by adapting them to the current situation. The problem solving experience gained in the past is explicitly taken into account by storing past cases in a library (the *case base*), and by suitably retrieving them when a new problem has to be dealt with. A case consists of the following information:

- the *problem description*, i.e. a collection of $\langle feature, value \rangle$ pairs able to summarize the problem at hand;
- the *case solution*, describing the solution adopted for solving the corresponding problem;
- the *case outcome*, describing the (positive or negative) result obtained by applying the solution itself.

Two basic possibilities exist for exploiting CBR for complex problems solving, namely:

- *Precedent Case-Based Reasoning*, where the emphasis is on retrieving past cases, and on using past solutions as a justification for the solution of the current problem, with almost no adaptation (e.g. legal reasoning);
- *Case-Based Problem Solving*, where retrieved solutions to previous similar cases need to be adapted to fit the current situation (e.g. planning, design, diagnosis, etc.).

Case-Based Problem Solving is the most general approach, and can be summarized by the following four basic steps, known as the *CBR cycle* [2]:

1. *Retrieve* the most similar case(s) from the case library;
2. *Reuse* them, and more precisely their solutions, to solve the new problem;
3. *Revise* the proposed new solution;
4. *Retain* the current case for future problem solving.

In the above cycle some steps may be missing or may be collapsed. For instance, it is quite common to view the *reuse* and *revise* steps as a single one, or to avoid the *retain* step if the current case is in some sense "covered" by other cases already stored in the library. These steps may be fully automated, although usually some user intervention is needed to perform adaptation and reuse.

A critical aspect affecting the efficiency of CBR is case retrieval, whose computational cost strongly depends on the organization of the case base. Various solutions have been proposed in the literature, ranging from *flat memories*, in which cases are stored sequentially as lists or feature vectors, to more structured organizations (like *shared features nets* and *discrimination nets* [23]). While flat memories are simple to update, but require clever strategies to avoid exhaustive search [32,36], more complex organizations are harder to organize and maintain, but allow for a faster search.

As anticipated in Section 2, CBR is particularly appealing in situations where the domain knowledge is poor and difficult to explicit (e.g., when a model of the system is not available), since the bottleneck of knowledge acquisition and representation is reduced, as new (unformalized) knowledge is automatically stored in the case base during the normal working process: no additional effort is required to the user and/or to the domain expert. As the case library grows, more and more representative examples can be retrieved, and it will become easier to find a proper solution to the problem at hand. CBR has been applied in several fields, mainly dealing with diagnostic problem solving [4,22] or planning, and has successfully been exploited also for industrial applications [34,33]. Moreover, CBR can be easily integrated with other knowledge sources (if available) and with other reasoning paradigms, making the methodology suitable also for of applications with a partially available background knowledge, or with a known problem-diagnosis model. The interest in multi modal approaches involving CBR is recently increasing through different application areas [3,13], from planning [7] to classification [43] and to diagnosis [25], and from legal [8,35] to medical decision support [6,27,37,47]. Different reasoning methods can be combined in the same application, or it can be possible to switch among alternative reasoning paradigms. CBR is particularly well suited for integration with Rule Based or Model Based systems[16].

4. A CBR-based Approach to Self-Healing

As discussed in the Introduction, a complete self-healing solution requires that all the steps of the autonomic cycle are carried out. Roughly speaking, there are two possible ways of doing so, namely (a) integrating the autonomic cycle into the system, thus in a certain sense embedding the autonomic manager into the managed system, and (b) "surrounding" the managed system by an external closed control loop (an approach named *externalization* in [15], where it has been proposed for the first time). While the former approach is undoubtedly more general, it requires innovative design and implementation techniques that, at the moment, are not mature enough. Furthermore, it cannot be used with existing applications whose size, complexity, and unavailability of source code may prevent any required modification. For these reasons, an externalization approach appears more appropriate.

Our proposal, schematically illustrated in Fig. 2, uses externalization, as done in Garlan and Schmerl's work [15], but relies on CBR, rather than resorting to a model of the system, so that the need of acquiring and maintaining structured knowledge about the system is avoided. The basic idea of externalization [15] is to treat the system as a "black box" surrounded by a set of external modules that form a "closed-loop" controlling the "health" of the system and performing proper repair actions in case of service failures. In our approach this closed loop includes, in addition to the `Running System` (corresponding to the *managed element* in the autonomic computing terminology), four external modules, that jointly act in such a way that self-healing is obtained. In particular, when the `Monitoring` module detects a service failure, it passes the set of *symptoms* of this misbehavior to the `Case Preparation` module, whose purpose is to assemble a case with the proper structure and passes it to the `CBR Analysis` module. The

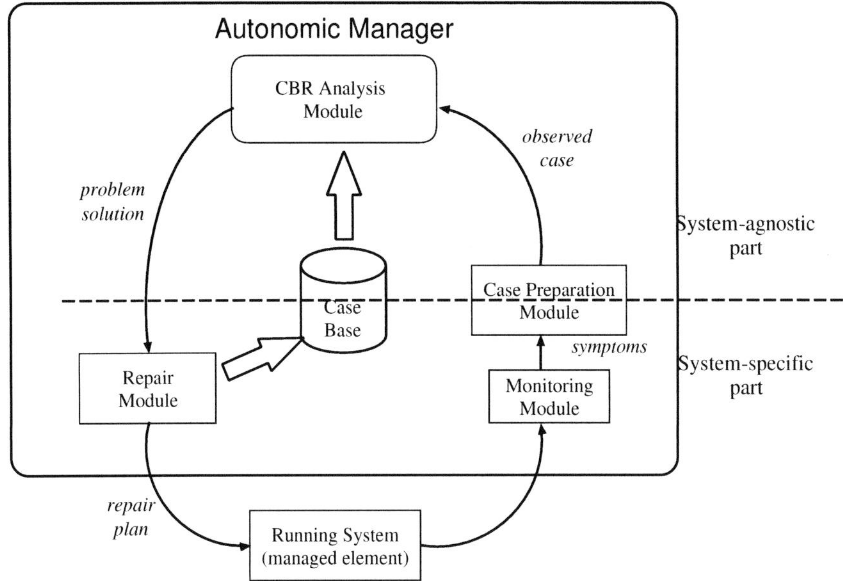

Figure 2. CBR-based Self-Healing

CBR Analysis module (which is the core of our architecture), upon receiving an *observed case* from the Case Preparation module, finds the solution to the case at hand (or, more precisely, the best solution among those available in the case base), and passes it to the Repair Module that, by using suitable system-specific mechanisms, executes the corresponding *repair plan*. Note that during the repair a new case, describing the new solution derived from a previously-stored one, can be generated and stored into the case base. As shown in Fig. 2, the external modules can be classified either into *System-specific* (i.e., that must be tailored to the specific characteristics of the running system) or *System-agnostic*. While the monitoring and repair activities require the availability and usage either of mechanisms provided by the running system or of adapted third-party components [18] (hence their classification as system-specific), the problem resolution activity (performed by the CBR Analysis module) does not rely upon any particular system feature (hence it classification as system-agnostic, although an adaptation strategy would require specific domain knowledge). The Case Preparation module has been instead drawn as crossing the dotted line separating system-specific from system-agnostic modules, as it interfaces the system-specific part of the system with the system-agnostic one.

As can be observed by comparing Fig. 2 with Fig. 1 (illustrating the autonomic cycle), our architecture closely matches the behavior of a self-healing autonomic manager. This is the consequence of the fact that the CBR cycle fits very well into the autonomic cycle, since it naturally covers the *Analysis* and *Planning* phases, while the other two phases (i.e., *Monitoring* and *Execution*) are covered by the Monitoring and Repair modules of our architecture. In particular, the *Retrieve* step of the CBR cycle corresponds to the *Analysis* step of the autonomic cycle, the *Reuse-Revise* steps of CBR, where past retrieved solutions are evaluated (and, if necessary, adapted) correspond to the *Planning* step, and the *Knowledge* used by the Autonomic Manager is contained in the `Case Base`.

In order to behave as a really self-healing system, our infrastructure must be able to work as much as possible in an autonomous way, i.e. without human intervention. However, this can happen only if a case base containing enough instances of solved cases is available, but in general it is not available when the system is initially put into operation. We therefore envision a *bootstrap* phase, enabling the collection of the initial cases, during which problem solution is performed by humans. However, as the system experiences failures that are solved by human operators, the case base grows to include the information concerning solved problems, so after a relatively limited amount of time problems solution can be performed autonomously. An obvious question that may arise is concerned with the size of the case base, that may potentially become very large if the number of stored cases keeps increasing. However, in these situations it is possible to keep the case base size within reasonable limits by resorting to technique already published in the literature [38], like for instance the definition of suitable "prototypes" able to summarize the information carried by the ground cases they represent, that are stored in place of the cases.

As a final consideration, we note that some aspects of the CBR implementation are strongly domain dependent; typically, the possibility of simply reusing a case solution vs. the need for adaptation and, in this situation, the details of the adaptation strategy, cannot be provided in a general fashion, but need to be tailored to the application under consideration. Nevertheless, the mapping of the CBR steps to the self-healing cycle appears to be general enough to be shared across different contexts.

5. Case Study: Self-Healing in Large-Scale Internet Services

In this Section we illustrate how our approach can be practically used to obtain self-healing in a computing system conceived to deliver large-scale Internet services. In order to make our discussion as realistic as possible, we use as a case study one of the systems described in [28,29], for which the data concerning some representative service failures and the corresponding repair actions are available.

The system chosen for our discussion is the one named *Online* in [28,29], whose purpose is to deliver an online service/Internet portal on a 24/7 fashion. The system is spread across two data centers, and includes about 500 machines altogether. Each data center hosts a subsystem using a three-tier architecture, in which a load-balancing tier distributes the load across a set of front-end servers, that in turn rely on a set of back-end servers to carry out their operation. The *Online* computing system uses both Sparc and x86 machines running the Solaris operating system, while the software and the middleware used to provide the service is customized, and is updated very frequently (typically every week).

The sheer size of the system makes the fault frequency relatively high, so proper fault-tolerant techniques and strategies have been incorporated into its design. As shown in [28], these techniques work reasonably well and are able to mask a large fraction of those faults that can be ascribed to node hardware/software problems (although some of them turn into service failures that must be properly handled), but are very often unable to mask those faults that are caused by operators' errors (typically software misconfigurations), that consequently turn into service failures.

In order to illustrate how our self-healing approach works, we consider three service failures (and the corresponding solutions), two of which have been taken from [28], while the third one has not been observed in the actual system, but is very plausible in practice. Starting from these failures, we describe a possible case structure, and we show how a service failure can be autonomously repaired by our self-healing infrastructure. The service failures we consider for our case study are the following:

- *Failure 1* (from [28]): the system was not delivering the email messages to its users. The failure was found to be caused by a software upgrade of the front-end daemon that handles username and alias lookup, that inadvertently changed the format of the string used by that daemon to query the back-end database server that stores usernames and aliases. The daemon continually retried all lookups because they were failing, eventually overloading the back database, causing also the risk of bringing down other services using it. The problem was finally fixed by rolling back the software upgrade and rebooting the database and front-end nodes.
- *Failure 2* (fictitious): the system was, again, not delivering the email messages to its users, but this time no increase in the username/alias lookup frequency occurred. The inspection of the logs of the email server however revealed a significant fraction of discarded messages. The failure was found to be caused by a misconfiguration of the email servers, that set an internal buffer to a too small value. The problem was fixed by increasing the buffer size and by restarting the email service on front-end nodes.
- *Failure 3* (from [28]): users' posting were (sometimes) not showing up on the service's newsgroup. The failure was due to an operator misconfiguration that had caused the newsgroup posting server not to run the daemon performing username/alias lookups. This server requires indeed that the message sender is a valid *Online* user, otherwise the message is silently dropped. The posted messages were being silently dropped since all the user authentication requests were failing because the username lookup daemon was not running. The failure was fixed by correcting the configuration file, and by restarting the newsgroup service on front-end nodes.

Let us now discuss a possible structure for the cases based on the information concerning the failures and their symptoms. As discussed in Section 3, a case consists in the *problem description* (i.e., a collection of <feature,value> pairs), the *case solution* (i.e., a description of the solution adopted to solve the corresponding problem) and the *case outcome* (describing the positive or negative result obtained by applying the solution). Let us start with problem description by giving the list of features used for problem description, reported in Table 1, that correspond to the symptoms of the failures. This list is derived from the three service failures discussed above and has been chosen in order to suit the example (of course, in general the cases may have a more complex structure). Each case will contain all

Table 1. Features used to describe cases

Feature	Possible Values	Description
email deliv.	*Yes/No*	indicates whether email messages are received (*Yes*) or not (*No*)
lookup number	*None/Low/ Normal/High*	indicates the number of username/alias lookups per unit of time
disc. msg.	*Yes/No*	indicates whether any service log reports discarded messages (*Yes*) or not (*No*)
news msg appear	*Yes/No*	indicates whether all posted messages appear in the newsgroup (*Yes*) or not (*No*)
lookup daemon running	*Yes/No*	indicates whether the username/alias look daemon is running (*Yes*) or not (*No*)

the features listed above, although some of them may have not been observed for a specific case, plus two fields corresponding to its solution and outcome. Table 2 reports the case base for our example.

Table 2. The case base corresponding to the three service failures. The *NULL* value is used to indicate that the corresponding feature has not been observed.

Case desc.	email deliv.	lookup number	disc. msg.	news msg appear	lookup daemon running	problem solution	case outcome
case 1	No	High	No	NULL	NULL	rollback to previous software version and reboot front-end nodes	positive
case 2	No	Normal	Yes	NULL	NULL	resize buffer and restart email service	positive
case 3	NULL	None	Yes	No	No	start lookup daemon and restart news service	positive

Let us now describe how our self-healing methodology would solve a new service failure, after it has been observed by the `Monitoring` module. We assume that the `Case Preparation` module (not described here because of its simplicity) formats the new case according to the structure described before, and passes it to the `CBR Analysis` module. For our example, let us assume that the new case has the structure reported in Fig. 3, that corresponds to an email messages delivery

email deliv.	lookup number	disc. msg.	news msg appear	lookup daemon running
No	NULL	Yes	NULL	Yes

Figure 3. Case corresponding to the new service failure that is being observed

failure, for which message discards are reported into the servers logs, and the lookup daemon is observed to be running.

Starting from this case, the `CBR Analysis` module retrieves the most similar past case chosen among those stored into the case base. In order to do so, it must have a way of measuring the similarity between cases, that in turn requires the definition of a measure of distance among cases. Generally speaking, the distance $d(c_i, c_j)$ between cases c_i and c_j can be computed as weighted average of the normalized distances among their various features, that is:

$$d(c_i, c_j) = \frac{\sum_{k=1}^{N} w_k \cdot d(c_i(k), c_j(k))}{N} \tag{1}$$

where $d(c_i(k), c_j(k))$ and w_k denote the normalized distance between feature k of cases c_i and c_j, and the weight associated with this feature, respectively. The features of the cases used in our example may take only either boolean or categorical values, for which two different measures of distance can be used. For boolean features we use to so-called *overlap distance* [45], that is defined as:

$$d(c_i(k), c_j(k)) = \begin{cases} 0 & \text{if } c_i(k) = c_j(k) \\ 1 & \text{otherwise} \end{cases}$$

For categorical features (only *lookup number*, in our case) we use a *similarity table* that explicitly lists the distance among all the pairs of possible values, defined as follows: [1]

$$d(c_i(k), c_j(k)) = \begin{cases} 0 & \text{if } c_i(k) = c_j(k) \\ 0.5 & \text{if } c_i(k) = None \land c_j(k) = Low \text{ or viceversa} \\ 0.5 & \text{if } c_i(k) = Low \land c_j(k) = Normal \text{ or viceversa} \\ 0.5 & \text{if } c_i(k) = Normal \land c_j(k) = High \text{ or viceversa} \\ 0.75 & \text{if } c_i(k) = None \land c_j(k) = Normal \text{ or viceversa} \\ 0.75 & \text{if } c_i(k) = Low \land c_j(k) = High \text{ or viceversa} \\ 1 & \text{otherwise} \end{cases}$$

By using the above distance measures, and by defining as '1' the distance between the value *NULL* and any other value, the distances between the new case (Fig. 3) and each case included in the case base (Table 2) are those reported in Table 3, from which we can conclude that the closest match to the new case is *case 2*.

Table 3. Distances between the new service failure and the cases in the case base

distance w.r.t.	email deliv.	lookup number	disc. msg.	news msg appear	lookup daemon	overall distance
case 1	0	1	1	1	1	$\frac{4}{5}$
case 2	0	1	0	1	1	$\frac{3}{5}$
case 3	1	1	0	1	1	$\frac{4}{5}$

Once the most similar case has been determined, the corresponding solution can be used to solve the problem corresponding to the new case. As already

[1] Of course, other distance values are possible, but for our purposes this simple definition is sufficient.

mentioned in Section 3, sometimes this solution may be readily used to solve the new problem (as in the example discussed in this section), while in other cases it may have to be adapted if it does not solve directly the new problem. Solution adaptation, however, cannot be always automated in a simple way, so in these cases some form of human intervention may be required. A similar consideration applies to the execution of the devised repair plan, that requires the availability of suitable reconfiguration/repair mechanisms, that sometimes may be hard to develop in a completely autonomous manner. In spite of that, we believe that our system represents a step towards the achievement of a full self-healing behavior. More precisely, we believe that our system fits between *Level 3* (*Predictive*) and *Level 4* (*Adaptive*) of the Ganek and Corbi's scale of autonomicity [14], depending on the amount of human intervention required to carry out the tasks mentioned above. As a final consideration, it is worth to point out that, although the example presented in this section is relatively simple, at the same time it is representative of a large class of real-world situations. Moreover, since the CBR methodology can handle exactly in the same way more complex cases, we believe that our approach can be practically used to achieve self-healing in real-world, large-scale software systems.

6. Conclusions

In this paper we have presented a CBR-based approach for the achievement of self-healing in software systems that, unlike alternative solutions, directly addresses service failures rather than individual component faults, so that unnecessary repair actions are avoided. Moreover, it does not require the availability of structured knowledge like, for instance, models of the behavior of the system, thus easing its applicability to large-scale, complex software systems. The suitability of this approach to real world applications has been exemplified by showing how it can be used to make a large-scale Internet service system able to self-heal. The relative simplicity of the case study chosen to illustrate our methodology is in no way a requirement for the applicability of CBR, that can handle exactly in the same way more complex cases. Therefore, we believe that our methodology represent a concrete step towards the achievement of self-healing for complex, large-scale software systems.

As future work, we plan to apply our methodology in practice by using it to provide self-healing in real-world large-scale systems, like PlanetLab [11] and OurGrid [30]. In order to do this, an implementation of both the system-agnostic and system-specific modules of our infrastructure need to be implemented and integrated.

References

[1] A. Avizienis and J. Laprie and B. Randell and C. Landwehr. Basic Concepts and Taxonomy of Dependable and Secure Computing. *IEEE Transactions on Dependable and Secure Computing*, 1(1), January-March 2004.

[2] A. Aamodt and E. Plaza. Case-Based Reasoning: foundational issues, methodological variations and systems approaches. *AI Communications*, 7(1):39–59, 1994.

[3] D. Aha and J. Daniels, editors. *Proc. AAAI Workshop on CBR Integrations*. AAAI Press, 1998.

[4] K.-D. Althoff and S. Wess. Case-Based Knowledge Acquisition, Learning, and Problem Solving in Diagnostic Real World Tasks. In *Proc. of Fifth European Knowledge Acquisition for Knowledge-Based Systems Workshop (EKAW '91)*, Crieff, Scotland, UK, Jun. 1991.

[5] N. Arshad, D. Heimbigner, and A. Wolf. A Planning Based Approach to Failure Recovery in Distributed Systems. In *Proc. of 2nd ACM Workshop on Self-Healing Systems (WOSS '04)*, Newport Beach, CA, USA, October 2004. ACM Press.

[6] I. Bichindaritz, E. Kansu, and K. Sullivan. Case-based reasoning in care-partner: Gathering evidence for evidence-based medical practice. In B. Smyth and P. Cunningham, editors, *Proc. 4th European Workshop on Case-Based Reasoning*, volume 1488 of *Lecture Notes in Computer Science*, pages 334–345, Dublin, Ireland, September 1998. Springer.

[7] P.P. Bonissone and S. Dutta. Integrating case-based and rule-based reasoning: the possibilistic connection. In *Proc. of 6th Conference on Uncertainty in Artificial Intelligence*, Cambridge, MA, USA, July 1990.

[8] L.K. Branting and B.W. Porter. Rules and precedents as complementary warrants. In *Proc. of 9th National Conference on Artificial Intelligence*, Anaheim, CA, USA, July 1991. AAAI Press.

[9] E. Brewer. Lessons from giant-scale services. *IEEE Internet Computing*, 5(4), 2001.

[10] M. Brodie, S. Ma, G. Lohman, T. Syeda-Mahmood, L. Mignet, N. Modani, J. Champlin, and P. Sohn. Quickly finding known software problems via automated symptom matching. In *Proc. of the 2nd International Conference on Autonomic Computing*, Seattle, WA, USA, June 2005.

[11] B. Chun, D. Culler, T. Roscoe, A. Bavier, L. Peterson, M. Wawrzoniak, and M. Bowman. PlanetLab: An Overlay Testbed for Broad-Coverage Services. *ACM SIGCOMM Computer Communications Review*, 33(3), July 2003.

[12] E. Dashofy, A. van der Hoek, and R. Taylor. Towards Architecture-based Self-Healing Systems. In *Proc. of 1st ACM Workshop on Self-Healing Systems (WOSS '02)*, Charleston, SC, USA, November 2002. ACM Press.

[13] E. Freuder, editor. *Proc. AAAI Spring Symposium on Multi-modal Reasoning*. AAAI Press, 1998.

[14] A.G. Ganek and T.A. Corbi. The dawning of the autonomic computing era. *IBM Systems Journal*, 42(1):5–18, 2003.

[15] D. Garlan and B. Schmerl. Model-based Adaptation for Self-Healing Systems. In *Proc. of 1st ACM Workshop on Self-Healing Systems (WOSS '02)*, Charleston, SC, USA, November 2002. ACM Press.

[16] K.J. Hammond. *Case-Based Planning: viewing planning as a memory task*. Academic Press, 1989.

[17] M. Hawthorne and D. Perry. Architectural Styles for Adaptable Self-Healing Dependable Systems. In *Proc. of 2005 ACM International Conference on Software Engineering (ICSE '05)*. ACM Press, May 2005.

[18] G. Kaiser, J. Parekh, P. Gross, and G. Valetto. Kenesthetics eXtreme: An External Infrastructure for Monitoring Distributed Legacy Systems. In *Proc. of 5th IEEE International Active Middleware Workshop*, Seattle, WA, USA, June 2003. IEEE CS Press.

[19] G. Kaiser, J. Parekh, P. Gross, and G. Valetto. Retrofitting Autonomic Capabilities onto Legacy Systems. Technical Report TR CUCS-026-03, Department of Computer Science, Columbia University, 2003.

[20] M. Kalyanakrishnam, Z. Kalbarczyk, and R. Iyer. Failure Data Analysis of a LAN of Windows NT Based Computers. In *Proc. of 18th IEEE Symposium on Reliable Distributed Systems*, Lausanne, Switzerland, October 1999. IEEE CS Press.
[21] J.O. Kephart and D.M. Chess. The vision of autonomic computing. *IEEE Computer*, January 2003.
[22] J. Kolodner and R. Kolodner. Using experience in clinical problem solving: introduction and framework. *IEEE Transactions on Systems, Man and Cybernetics*, 17:420–431, 1987.
[23] J.L. Kolodner. *Case-Based Reasoning*. Morgan Kaufmann, 1993.
[24] M. Littman, T. Nguyen, and H. Hirsh. Cost-Sensitive Fault Remediation for Autonomic Computing. In *Proc. of IJCAI Workshop on AI and Autonomic Computing: Developing a Research Agenda for Self-Managing Computer Systems*, Acapulco, Mexico, August 2003.
[25] D. Macchion and D. Vo. A hybrid knowledge-based system for technical diagnosis learning and assistance. In S. Wess, K. Althoff, and M. Richter, editors, *Proc. 1st European Workshop on Case-Based Reasoning*, volume 837 of *Lecture Notes in Computer Science*, pages 301–312, Kaiserslautern, Germany, November 1993. Springer.
[26] M. Mikic-Rakic, N. Mehta, and N. Medvidovic. Architectural Style Requirements for Self-Healing Systems. In *Proc. of 1st ACM Workshop on Self-Healing Systems (WOSS '02)*, Charleston, SC, USA, November 2002. ACM Press.
[27] S. Montani, L. Portinale, G. Leonardi, and R. Bellazzi. Case-based retrieval to support the treatment of end stage renal failure patients. in press.
[28] D. Oppenheimer, A. Ganapathi, and D. Patterson. Why do Internet services fail, and what can be done about it? In *Proc. of 4th Usenix Symposium on Internet Technologies and Systems (USITS '03)*, Seattle, WA, USA, March 2003.
[29] D. Oppenheimer and D. Patterson. Architecture and Dependability of Large-Scale Internet Services. *IEEE Internet Computing*, September-October 2002.
[30] The ourgrid home page. http://www.ourgrid.org. Accessed on August 2nd, 2005.
[31] M. Parashar and S. Hariri. Autonomic Computing: An Overview. In *Proc. of 2004 Workshop on Unconventional Programming Paradigms*, volume 3566 of *Lecture Notes in Computer Science*, Mont Saint-Michel, France, September 2004. Springer.
[32] L. Portinale, P. Torasso, and D. Magro. Selecting most adaptable diagnostic solutions through pivoting-based retrieval. In D. Leake and E. Plaza, editors, *Proc. of 2nd International Conference on Case-Based Reasoning*, volume 1266 of *Lecture Notes in Computer Science*, pages 393–402, Providence, RI, USA, July 1997. Springer.
[33] R. Bergmann. *Experience Management: Foundations, Development Methodology, and Internet-Based Applications*, volume 2432 of *Lecture Notes in Computer Science*. Springer, Berlin, 2002.
[34] R. Bergmann and K.-D. Althoff and S. Breen and M. Goker and M. Manago and R. Traphoner and S. Wess. *Developing Industrial Case-Based Reasoning Applications. The INRECA Methodology*, volume 1612 of *Lecture Notes in Artificial Intelligence*. Springer, Berlin, 2003.
[35] E. Rissland and D. Skalak. Combining case-based and rule-based reasoning: A heuristic approach. In N. S. Sridharan, editor, *Proc. of 11th International Joint Conference on Artificial Intelligence*, pages 524–530, 1989.
[36] J.W. Schaaf. Fish and shrink. a next step towards efficient case retrieval in large-scale case bases. In I. F. C. Smith and B. Faltings, editors, *Proc. 3rd European Workshop on Case-Based ReasoningEWCBR*, volume 1168 of *Lecture Notes in Computer Science*, pages 362–376, Lausanne, Switzerland, November 1996. Springer.
[37] R. Schmidt, S. Montani, R. Bellazzi, L. Portinale, and L. Gierl. Case-based reason-

ing for medical knowledge-based systems. *International Journal of Medical Informatics*, 64(2-3):355–367, 2001.

[38] R. Schmidt, S. Montani, R. Bellazzi, L. Portinale, and L. Gierl. Case-based reasoning for medical knowledge-based systems. *International Journal of Medical Informatics*, 2–3, 2001.

[39] C. Simache, M. Kaaniche, and A. Saidane. Event Log based Dependability Analysis of Windows NT and 2K Systems. In *Proc. of the 2002 Pacific Rim International Symposium on Dependable Computing (PRDC '02)*, Tsukuba, Japan, December 2002. IEEE CS Press.

[40] R. Sterrit. Autonomic networks: engineering the self-healing property. *Engineering Applications of Artificial Intelligence*, 17:727–739, October 2004.

[41] R. Sterrit and S. Chung. Personal Autonomic Computing Self-Healing Tool. In *Proc. of 11th International Conference and Workshop on the Engineering of Computer-Based Systems (ECBS '04)*, Brno, Czech Republic, May 2004. IEEE CS Press.

[42] R. Sterrit, D. Gunning, A. Meban, and P. Henning. Exploring Autonomic Options in an Unified Fault Management Architecture through Reflex Reactions via Pulse Monitoring. In *Proc. of 11th International Conference and Workshop on the Engineering of Computer-Based Systems (ECBS '04)*, Brno, Czech Republic, May 2004. IEEE CS Press.

[43] J. Surma and K. Vanhoof. Integration rules and cases for the classification task. In M. Veloso and A. Aamodt, editors, *Proc. 1st Int. Conference on Case-Based Reasoning*, volume 1010 of *Lecture Notes in Computer Science*, pages 325–334, Sesimbra, Portugal, October 1995. Springer.

[44] H. Tianfield and R. Unland. Towards autonomic computing systems. *Engineering Applications of Artificial Intelligence*, 17:689–699, 2004.

[45] D.R. Wilson and T.R. Martinez. Improved Heterogeneous Distance Functions. *Journal of Artificial Intelligence Research*, 6, 1997.

[46] J. Xu, Z. Kalbarczyk, and R. Iyer. Networked Windows NT System Field Failure Data Analysis. In *Proc. of the 1999 Pacific Rim International Symposium on Dependable Computing (PRDC '99)*, Hong Kong, China, December 1999. IEEE CS Press.

[47] L.D. Xu. An integrated rule- and case-based approach to AIDS initial assessment. *International Journal of Biomedical Computing*, 40:197–207, 1996.

Self-Organization and Autonomic Informatics (I)
H. Czap et al. (Eds.)
IOS Press, 2005

Why Applying Agent Technology to Autonomic Computing ?

Walid Chainbi [1]
ENIS, Département d'informatique, TUNISIA

Abstract. Recently, an interest has been witnessed in computing community to autonomic computing. Inspired by the functioning of the human nervous systems, autonomic computing is to design and build computing systems that posses inherent self-managing capabilities. While autonomic computing is a quite new revolutionary move to the discipline of computing and so far a holistic solution has not yet appeared, we argue in this paper that agent technology is already available for being integrated into the framework of autonomic computing. Starting with the high degree of match between agent systems and autonomic systems, an in-depth analysis is presented to denote that agent architectural concepts are crucially important.

Keywords. Autonomic systems, self-managing mechanisms, agent paradigm

1. Introduction

The managing of today's computing systems goes beyond the administration of individual software environments. The need to integrate several heterogeneous environments into corporate wide computing systems, and to extend that beyond company boundaries into the Internet, introduces new levels of complexity. Autonomic computing is considered as a promising solution to such problem [13].

Autonomic computing is a self-managing computing model named after, and patterned on, the human body's autonomic nervous system (see figure 1). It deals with the design and the construction of computing systems that posses inherent self-managing capabilities. Each autonomic system is a collection of autonomic elements - individual systems constituents that contain resources and deliver services to humans and other autonomic elements. It involves two parts : a managed system, and an autonomic manager. A managed system is what the autonomic manager is controlling. An autonomic manager is a component that endows the basic system with self-managing mechanisms. The managed system as well as the autonomic manager can be seen at different abstraction levels ranging from collections of components (autonomic managers and managed elements) to single conceptual units namely a managed system and an autonomic manager.

---

[1] Correspondence to: Walid Chainbi, ENIS, département d'informatique, B.P. W -3038- Sfax-Tunisia. Tel.: +216 74 274 088; Fax: +216 74 275 595; E-mail: Walid.Chainbi@lycos.com.

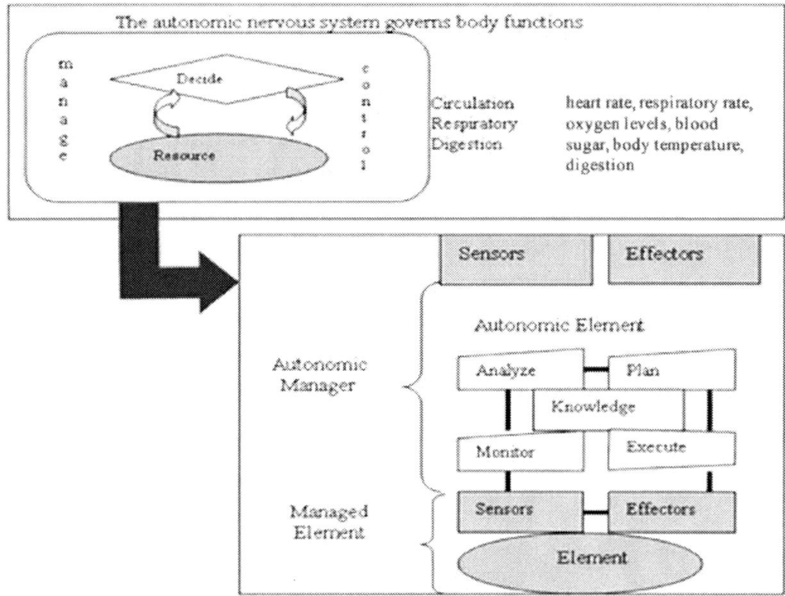

Figure 1. An autonomic computing architecture (drawn from [9])

Agent technology is one of the most prominent and attractive technologies in computer science at the beginning of the new millenium. It is not only a promising technology, but it is also emerging as a new way of thinking : a conceptual paradigm for analyzing problems and for designing systems, for dealing with complexity, distribution, and interactivity while providing a new perspective on computing and intelligence [12]. An agent is a computational entity that can be viewed as perceiving and acting upon its environment and that it is autonomous in that its behavior at least partially depends on its own experience.

Our approach, pragmatically motivated, has been to explore why agent technology can be applied to autonomic computing. The reminder of this paper is organized as follows. Section 2 motivates the key idea related to this study namely, the application of agent technology to autonomic computing. The sections 3 and 4 outline some properties of autonomic elements and explain why they meet agent properties. Section 5 discusses the merits of the presented study. Section 6 summarizes the appropriateness of the work done and gives directions for future work.

2. Motivations

Recently computing community has witnessed a great move to a new paradigm of computing namely autonomic computing. This latter denotes a move from the pursuit of high speed, powerful computing capacity to the pursuit of self-managing mechanisms of computing systems. Indeed, today's computing and information infrastructure have reached a level of complexity that is far beyond the capacity of human system administration [17].

For instance, follow the evolution of computers from single machines to modular systems to personal computers networked with larger machines. Along with that growth has come increasingly sophisticated architectures governed by software whose complexity now routinely demands tens of millions of lines of codes. The internet adds yet another layer of complexity by allowing us to connect this world of computers and computing systems with telecommunications networks. In the process, the systems have become increasingly difficult to manage, and ultimately, to use. Inspired by the functioning of the human nervous system which frees our conscious brain from the burden of dealing with some vital functions (such as governing our heart rate and body temperature), autonomic computing is considered as a promising solution to such problems.

As yet, however there is not a successful solution to autonomic computing which can be applied on a significant scale. So far a mature solution has not yet appeared. In part, this is due mainly to the youth of this paradigm and the absence of adequate tools, but our experience suggests that the absence of tools that allow system complexity to be effectively managed is a greater obstacle.

Agent paradigm has achieved a respectable degree of maturity and there is a widespread acceptance of its advantages : a relatively large community of computer-scientists which is familiar with its use now exists. A substantial progress has been made in recent years in providing a theoretical and practical understanding of many aspects of agents and multi-agent systems [5].

Managing complexity is perhaps the most challenging task that forced both academia and industries to re-think the fundamental issues of computing and information technology. It is now widely accepted that the construction of reliable, maintainable, and extensible systems that conform to their specifications requires the use of design methodologies and modeling techniques that support adequate mechanisms for managing their inherent complexity. Perhaps foremost amongst the paradigms used the last years for analyzing problems and for designing systems, for dealing with complexity, distribution and interactivity is agent paradigm, based upon the central notion of agents which are reactive, autonomous, internally-motivated entities embedded in changing uncertain worlds which they perceive and in which they can act. Besides, a confrontation of agent technology with autonomic computing has not yet been done. In this paper, we analyze in-depth the major points favoring the integration of agent technology into the framework of autonomic computing.

3. Properties' match

Autonomic systems are collections of autonomic elements -individual system constituents that contain resources and deliver services to humans and other autonomic elements.

3.1. Characteristics

Each autonomic element is responsible for managing its own internal state and behavior and for managing its interactions with an environment that consists largely of signals and messages from other elements and the external world [13].

An element may require assistance from other elements to achieve its goals. If so, it will be responsible for obtaining necessary resources from other elements and for dealing with exception cases, such as the failure of a required resource.

An element's internal behavior and its relationships with other elements will be driven by goals that its designer has embedded in it, by other elements that have authority over it, or by subcontracts to peer elements with its tacit or explicit consent.

Autonomy, proactivity, and goal-directed interactivity with their environment are distinguishing characteristics of software agents.

Learning in agency belongs to the characteristics which are of differing importance for different domains[1] [15]. Thus, for some applications, the ability of agents to learn from their experiences is of paramount importance ; for other applications, learning is not only unimportant, it is undesirable (e.g., Michael Georgeff, the main architect of the PRS system, gives the example of an air-traffic control system he developed ; the clients of the system would have been horrified at the prospect of such a system modifying its behavior at run time [20]). Learning is considered as important to self-managing mechanisms. Hence, learning in agency is of primary importance to deal with autonomic systems.

3.2. Behavior

Autonomic elements have complex life cycles, continually carrying on multiple threads of activity, and continually sensing and responding to the environment in which they are situated. Hence, a kind of control loop is created and dissected into three parts (see figure 2).

- the *measure* part provides the mechanisms that collect, aggregate, filter, and report details (e.g., metrics) collected from the controlled process.
- the *decision* part provides the mechanisms to produce the action needed to achieve goals and at the same time respect the constraints.
- the *actuation* part provide the mechanisms that control the execution of actions.

By measuring some activity in a process to be controlled, an autonomic element decides what needs to be done next and executes the required operations through a set of actuators. The process is then re-measured to determine whether the actions of the actuator had the desired effect. The whole routine of measure, decide, actuate is then continually repeated.

Modeling in terms of agents require the use of a methodology providing an iterative approach towards modeling and developing agent-based systems. Examples of methodologies include Gaia [21], MaSE (Mulitagent Systems Engineer-

[1] Actually, there is a general consensus that autonomy is central to the notion of agency, but there is little agreement beyond this.

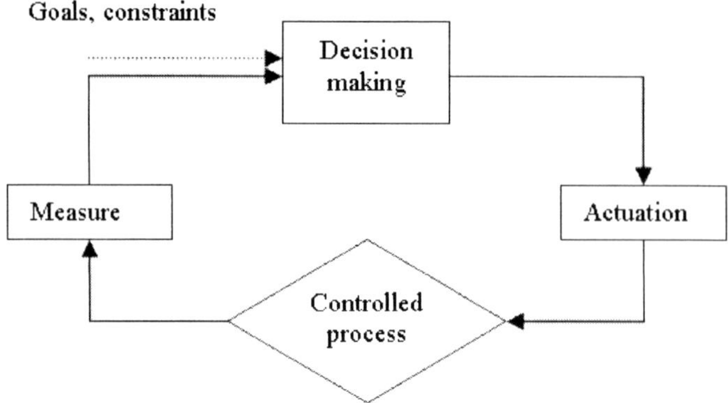

Figure 2. A control loop

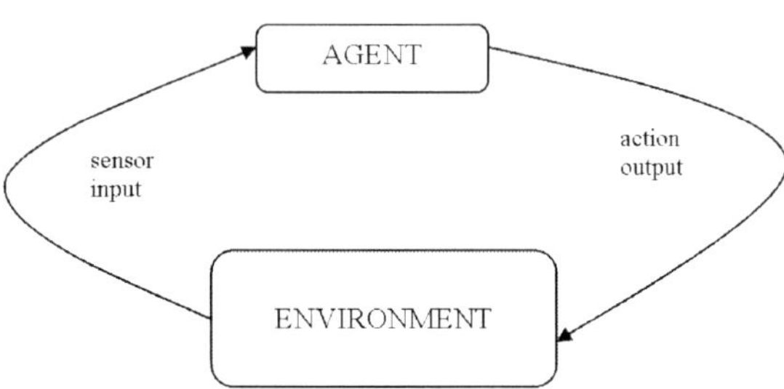

Figure 3. An ongoing interaction between an agent and his environment

ing Methodology) [19]. A survey of methodologies used within the agent oriented software community can be found in [18].

In general, an agent is an active computational entity that can perceive (through sensors), reason about, and initiate activities (through effectors) in its environment. The types of activities an agent can perform may be classified as either external (in which case the domain of the activity is the environment outside the agent i.e., the other agents and the spatial environment) or internal (in which case the domain of the activity is the agent himself).

An agent is able to measure some activity by his sensors and his ability to execute actions (either external or internal). An agent is also able to decide which action to produce in response to the goals and the state of the environment (see figure 3). Each produced action can be integrated into an interaction paradigm (for example an agenda). The interaction paradigm implicitly defines the metaphors used by the agents to interact and cooperate among each other [2].

The essence of autonomic computing systems is self-management for which four aspects are cited : *self-configuration*, *self-optimization*, *self-healing*, and *self-protection* [8][2].

[2]This instantiation of self-managing mechanisms is not unique (other instantiations exist - see [17] for instance)

- *Self-configuration* : adapting automatically to the dynamically changing environment.
- *Self-optimization* : continually seeking opportunities to improve performance and efficiency.
- *Self-healing* : discovering, diagnosing and reacting to disruptions.
- *Self-protection* : defending against malicious attacks or cascading failures.

Normally, each agent has a repertoire of actions available to him. This set of possible actions represents his ability to modify his environment. The types of actions an agent can perform at a point of time include *physical* actions (interactions between agents and the spatial environment), *communicative* actions (interactions between agents, which can be emission or reception actions), *private* actions (internal functions of an agent, which correspond to an agent exploiting his internal computational resources), and *decision* action.

An action can be classified either as adaptive, optimization, healing, or protection action. It depends on the reason of its execution. For example, increase (or decrease) the virtual memory space is a physical action which can be executed by an agent. This action can be an optimization action if it is intended to increase the performance of the system. It can be as well considered as a healing action if the system is unable to deal with many programs at a point of time, and the fact to try to execute a program is concomitant with an error occurrence.

4. Complexity's management

Previously, Frederick P. Brooks, Jr., observed "complexity is business we are in, and complexity is what limit us" [4]. The computer industry has spent decades creating systems of marvelous and ever-increasing complexity. But today, complexity itself is the problem [7]. Managing complex systems has grown too costly and prone to error. Indeed, people in charge of administering a myriad of system management details are under continuous pressure, which increase the potential of making mistakes and system outages. Besides, there continues to be a scarcity of highly skilled IT professionals to maintain these complex, heterogeneous systems.

Autonomic computing is considered as a promising solution to manage today's computing systems whose management goes beyond the administration of individual software environments [9]. Indeed, the need to integrate several heterogeneous environments into corporate-wide computing systems, and to extend that beyond company boundaries into the Internet, introduces new levels of complexity.

Agent-based techniques are well suited for tackling complex problems according to the high degree of match between two perspectives : on the one hand, the key characteristics of complex software systems and the well-known techniques for tackling complexity in software. On the other hand, the essential concepts and notions of agent-based paradigm [10]. Next, we outline the major underlying points of this match advocated by Jennings in favor of agent-based software engineering. Before, we briefly remind three techniques identified by Booch [3] for tackling complexity in Software :

- *Decomposition* : it consists in dividing large problems into smaller, more manageable chunks each of which can then be dealt with in relative isolation.
- *Abstraction* : the process of defining a simplified model of the system that emphasizes some of the details or properties, while suppressing others.
- *Organization* : the process of identifying and managing the interrelationships between the various problem solving components. This helps designers tackle complexity in two ways. Firstly, by enabling a number of basic components to be grouped together and treated as a higher-level unit of analysis (e.g., the individual components of a subsystem can be treated as a single coherent unit by the parent system). Secondly, by providing a means of describing the high-level relationships between various units (e.g., a number of components may cooperate to provide a particular functionality).

With this characterization in place, the match process argument in favor of agent-base software engineering can be expressed as follows.

- *Agent-oriented decomposition is an effective way of partitioning the problem space of a complex system* : The agent-oriented approach advocates decomposing problems in terms of autonomous agents that can engage in flexible, high-level interactions. Decomposing a problem in such a way helps the process of engineering complex systems in two main ways. Firstly, it is simply a natural representation for complex systems that are invariably distributed and that invariably have multiple loci of control. This decentralization, in turn, reduces the system's control complexity and results in a lower degree of coupling between components. The fact that agents are active entities means they know for themselves when they should be acting and when they should update their state. Such self-awareness reduces control complexity since the system's control know-how is taken from a centralized repository and localized inside each individual problem solving component [10]. Secondly, since decisions about what actions should be performed are devolved to autonomous entities, selection can be based on the local situation of the problem solver. This means that the agent can attempt to achieve its individual objectives without being forced to perform potentially distracting actions simply because they are requested by some external entity. The fact that agents make decision about the nature and scope of interactions at run-time makes the engineering of complex systems easier. Indeed, the system's inherent complexity means it is impossible to know a priori about all potential links : interactions will occur at unpredictable times, for unpredictable reasons, between unpredictable components. For this reason, it is futile to try and predict or analyze all the possibilities at design time. Rather, it is more realistic to endow the components with the ability to make decisions about the nature and scope of their interactions at run-time. Thus agents are specifically designed to deal with unanticipated requests and they can spontaneously generate requests for assistance whenever appropriate.

- *The key abstractions of agent-oriented mindset are a natural means of modeling complex systems* : In the case of a complex system, the problem to be characterized consists of subsystems, subsystems components, interactions and organizational relationships. Taking each in turn : firstly, there is a strong degree of correspondence between the notions of subsystems and agent organizations. They both involve a number of constituent components that act and interact according to their role within the larger enterprise. Secondly, the interplay between the subsystems and between their constituent components is most naturally viewed in terms of high level social interactions (e.g., agent systems are described in terms of "cooperating to achieve common objectives" or "negotiating to resolve conflicts"). Thirdly, complex systems involve changing webs of relationships between their various components. They also require collections of components to be treated as a single conceptual unit when viewed from a different level of abstraction. On both levels, the agent-oriented mindset again provides suitable abstractions. A rich set of structures is typically available for explicitly representing and managing organizational relationships such as roles (see [1], [6] for example). Interaction protocols exist for forming new groupings and disbanding unwanted ones (e.g., Sandholm's work [14]). Finally, structures are available for modeling collectives (e.g., teams [16]).
- *The agent-oriented philosophy for dealing with organizational relationships is appropriate for complex systems* : organizational constructs are first-class entities in agent systems. Thus explicit representations are made of organizational relationships and structures. Moreover, agent-based systems have the concomitant computational mechanisms for flexibly forming, maintaining and disbanding organizations. This representational power enables agent-oriented systems to exploit two facets of the nature of complex systems. Firstly, the notion of primitive component can be varied according to the needs of the observer. Thus, at one level, entire subsystems can be viewed as singletons, alternatively, teams or collections of agents can be viewed as primitive components, and so on until the system eventually bottoms out. Secondly, such structures provide a variety of stable intermediate forms, that, as already indicated, are essential for the rapid development of complex systems. Their availability means individual agents or organizational groupings can be developed in relative isolation and then added into the system in an incremental manner. This, in turn, ensures there is a smooth growth in functionality.

5. Discussion

Autonomic computing is being described as a new paradigm for computing [17]. But in order to become a new paradigm, robust and easy-to-use methodologies and tools have to be developed. Unfortunately, autonomic computing lacks mature tools that can be used on a significant scale. Accordingly, agent technology can be used with benefits to develop autonomic systems. The word development is sometimes interpreted as programming, but we use it here to include the full development process that covers requirement specification and design, in addition to the programming itself. This fact shouldn't be understood that we don't see anything novel in autonomic computing but should be seen as an endeavour to find an answer to the realisation of autonomic systems through agent technology.

Actually, the idea of the possibility to use agent technology within autonomic computing is not entirely new. Indeed, Kephart et al. have mentioned this eventuality but their analysis remains shallow (it was not the main subject of their study) [13].

An autonomic solution to a system requires that all steps of the autonomic cycle (see figure 2) are carried out. There are two possible ways of doing so, namely (a) integrating the autonomic cycle into the system, thus in a certain sense embedding the autonomic manager into the managed system, and (b)using agents as an external layer which provides the autonomic behavior. The latter approach treats the system to be managed as a "black box" surrounded by agents that form a control loop (or a hierarchy of control loops)[3] controlling the state of the system and performing actions each of which can be adaptive, optimization, healing, or protection action. This approach requires the development of specific interfaces facilitating the interaction between the agent-based manager part and the managed part of the system.

The former approach seems to be more appropriate. Indeed, adopting an agent solution for the autonomic computing system makes easier the interaction between the managed and the managing parts of the system. This is mainly due to the homogeneity of the adopted solution (an agent is the unit of design). Previously, Huaglory and Unland considered a major challenge for autonomic computing systems will be how to design explicitly the automation of system adaptation in addition to conventionally the design of the basic computing system[17]. We think that the former approach can be an answer to this question.

6. Conclusion

The presented study is an endeavour to find an answer to the possibility of integrating agent technology into the framework of autonomic computing.

Starting from the youth of autonomic computing and the lack of mature tools that can be used on a significant scale, we have analysed the high degree of match between the characteristics of agent systems and those of autonomic systems, as well as the ability of both systems to deal with complexity management.

In any design process, finding the right models for viewing the problem is a main concern. In general, there will be multiple candidates and the difficult task is picking the most appropriate one. When it comes to designing software, the most powerful abstractions are those that minimize the semantic distance between the units of analysis that are intuitively used to conceptualize the problem and the constructs present in the solution paradigm. Accordingly, an autonomic sys-

[3]Depending on the degree of changes in the environment, different strategies have been established to implement self-managing mechanisms within an autonomic computing system. These strategies range from a control loop to a hierarchy of control loops [17][9].

tem can be viewed as a multi-agent system whether its constituents (autonomic elements) are viewed as agents.

In future work, the following issues can be investigated. Most importantly, the *how* aspects should be treated i.e., how can the results related to agent technology be used in the development of autonomic computing systems. For example, developing elements that rarely fail is an important aspect of being autonomic. Accordingly, a major challenge will be to develop analysis tools and techniques for ensuring that autonomic systems will behave as we expect them to -or at least, will not behave in ways that are undesirable. A study is currently in progress to develop an agent-based formal framework to deal with such aspects.

References

[1] C. Baejis, Y. Demazeau, Organizations in multi-agent systems, Journées DAI, Toulouse, 1996.
[2] S. Bandinelli, E. Di Nitto and A. Fuggetta, Supporting Cooperation in the SPADE-1 Environment, in: IEEE Transactions on Software Engineering, vol. 22, no 12, 1996.
[3] G. Booch, Object-Oriented Analysis and Design with Applications, Addison-Wesley, Reading, MA, 1994.
[4] F. P. Brooks, The Mythical Man-Month : Essays on Software Engineering, Addison-Wesley Publishing Co., Reading, MA, 1995.
[5] M. d'inverno and M. Luck, Understanding Agent Systems (Second Edition), Springer, 2004.
[6] M.S. Fox, An organizational view of distributed systems, in: IEEE, Trans. SMC-11, p 70-80, 1981.
[7] A. G. Ganek and T.A. Corbi, The Dawning of the autonomic computing era,in: IBM Systems journal, vol 42, no 1, 2003.
[8] IBM, Autonomic Computing Manifesto, http://www.research.ibm.com/autonomic/manifesto, 2001.
[9] IBM, An architectural blueprint for autonomic computing, April 2003.
[10] N. Jennings, On agent-based software engineering, in: Artificial intelligence 117, 277-296, 2000.
[11] N. Jennings, K. Sycara and M. Wooldridge, A Roadmap of Agent Research and Development, Autonomous Agents and Multi-Agent Systems, 1998, 1, pp 7-38.
[12] N. Jennings and M. Wooldridge, Intelligent agents : theory and practice, The Knowledge engineering review, 1998, 10(2), pp 115-152.
[13] J. O. Kephart and D. M. Chess, The vision of autonomic computing, in: IEEE Computer, 2003, pp 41-50.
[14] T. Sandholm, Distributed rational decision making, Multi-Agent Systems, MIT Press, 201-258, 1985.
[15] J. Shavlik and T. Dietterich, Readings in Machine Learning, Morgan Kaufmann, San Mateo, California,1990.
[16] M. Tambe, Towards flexible teamwork, in: Artificial Intelligence 7, 83-124, 1997.
[17] H. Tianfield and R. Unland, Towards autonomic computing systems, in: Engineering Applications of Artificial Intelligence journal, Elsevier, 2004, 7, pp 689-699.
[18] A. Tveit, A Survey of Agent-Oriented Software Engineering. Proc. of the First NTNU CSGS conference, May, 2001.
[19] M.F. Wood and S.A. DeLoach, An Overview of the Multiagent Systems Engineering Methodology, AOSE'2000, 2000.
[20] M. Wooldridge, An Introduction to Multiagent Systems, Wiley, 2002.

[21] M. Wooldridge, N. Jennings and D. Kinny, The Gaia methodology for agent-oriented analysis and design. AAMAS'2000, September 2000, 3(3):285-312.

Self-Organization and Autonomic Informatics (I)
H. Czap et al. (Eds.)
IOS Press, 2005

A Programmatic Approach to Applying Sympathetic and Parasympathetic Autonomic Systems to Software Design

Philip MISELDINE [a] and Azzelarabe TALEB-BENDIAB [b]
[a,b] *Liverpool John Moores University, UK*
[a] *p.l.miseldine@2004.ljmu.ac.uk,* [b] *a.talebbendiab@ljmu.ac.uk*

Abstract: Research into solving the problems that the recent rapid evolution of hardware and software technology and increasing consumer requirements has yielded, produced several Autonomic Computing initiatives, spearheaded by IBM. This paper contends that whilst this research is inspired by the Autonomic Nervous System (ANS), current implementations have treated the ANS as a single system, rather than as two disparate, competing systems: the sympathetic and parasympathetic nervous systems. In this paper, the authors show how the current limitations in state-of-the-art autonomic system implementations that hinder the successful representation of such sub systems and their conflicts can be overcome through the specification, design and development of a runtime introspective programming language, Neptune. An introduction to the Neptune specific CA-SPA (Concept Aided Situation-Prediction-Action) methodology for policy definition is given, as is the Cloud framework that hosts Neptune objects in an autonomic grid framework An evaluation of the language is given using illustrative examples based on sympathetic and parasympathetic scenarios, concluding with a summary of the further aims of the research.

Keywords: autonomic behaviour, biological autonomic nervous systems, self management.

1. Introduction

Large distributed systems are now vital for organisations to effectively manage their own activities and to interact with others. Systems that were once singular in resource are now widely dispersed through interconnected networks that can typically span both internally and externally managed computer systems. With the layers of services and interwoven dependencies from which they are comprised, it is often the case that a failure in one service can produce failures in a set of services. This makes promoting reliability, robustness, and scalability difficult to design and implement. Thus, the need for software that intelligently reacts and adapts to the circumstances and situations within which it is executed, is paramount to future effective development of distributed software systems [1].

Such desires lead to the development of Autonomic Software [2], derived from the self-maintaining systems evolved in natural biology. In order to realise these autonomic specifications the system must have a notion of the context in which its actions and perceived events are taking place to achieve its self-governance, incorporating the context of the user. Also the method of interaction has moved away from just a desktop interface into the real world, a much more complex and dynamic place with widely varying contexts. It is for this reason that approaches tried so far have produced little in the way of scaleable autonomic systems [3]. However, research has often modelled biological autonomic systems as a singular intention system, rather than a more accurate model consisting of conflicting sub systems, principally the sympathetic and parasympathetic nervous systems [4].

In this paper, the authors show how the current limitations in state-of-the-art autonomic system implementations can be overcome through the specification of a runtime introspective programming language, Neptune that exposes policies and governance behaviour as compiled objects that can be inspected, modified, and executed at runtime. An evaluation of the language is given using illustrative examples based on sympathetic and parasympathetic scenarios, concluding with a summary of the further aims of the research.

2. Survey of Autonomic Systems

The human body represents a naturally evolved system of services that is regulated without direct intervention or planning. In this respect, the human body exhibits behaviour much sought after in software design. Research into replicating such abilities into artificial software systems has yielded several Autonomic Software initiatives. IBM has been at the forefront of recent research into autonomic system specification [2] defining the core self-management capabilities required by autonomic systems including: self-configuration, self-healing, self-optimisation, and self-protection. This approach is typified by DIOS++ [5] which provides an infrastructure that enables rule-based autonomic adaptation and control of distributed scientific applications.

Other research focused on self-healing software that modifies its architecture during the system's execution, commonly known as runtime dynamism [6]. Formally defined models of system architectures such as the Hirsh and colleagues approach [7], the Le Mètayer approach [8] and the CommUnity approach [9] allow operational changes of state to be modelled and analysed in the form "*given system x, what happens when change y occurs?*" [10].

The idea behind calling such software concepts *autonomic* is that they are inspired by the Autonomic Nervous System (ANS), part of the nervous system principally out of our control and awareness [11]. When light is shone directly into the human eye, it is an automatic response from our ANS that contracts the pupil to reduce the amount of light reaching the sensory cells in the retina of the eye thereby protecting the eye from a situational change. This response it is not a conscious reaction, rather an automatic response from the ANS [12].

Thus, it is foreseen that an appreciation and understanding of biological autonomic systems can yield insight into how artificial, software based autonomic systems should be constructed. There has been little research focus however to mimic the various subsystems of the autonomic nervous system, oftentimes, the ANS is treated as a single disparate system.

2.1. Biological Autonomic Nervous Systems

There are three main sections in ANS, each with different roles in regulating the health of the human body: the sympathetic nervous system, the parasympathetic nervous system, and the enteric nervous system [4]. For the purposes of applying the principles of these natural autonomic systems to software systems, the first two are of particular importance.

The sympathetic nervous system responds to situations that typically protect the systems within the body. These are often known as the "flight or fight" responses. If danger is sensed, energy and priorities within the body are altered so that blood pressure increases, the heart beats faster, and digestion slows so that the energy required to respond to the danger is available. The parasympathetic nervous system however, responds to situations that provide recuperation for the body, and are often called "rest and digest" responses. If the body is at rest, or in a period of sleep, the parasympathetic nervous system lowers blood pressure, lowers heartbeat, and starts digestion to increase energy reserved within the body system.

However, both systems remain in conflict at all times, each reconfiguring to achieve their distinctive operational goals [13, 14]. Whilst at rest, the body might be configured by the parasympathetic nervous system to reserve energy, however the sympathetic nervous system is still maintaining resources to enable alertness and responsiveness to detect danger. Whilst both systems are not achieving their own disparate configurations (energy is being used to maintain each system) the optimum operational configuration is achieved as both systems deliberate to reach equilibrium either in a synergistic or an antagonistic way [15].

The usual method of expressing deliberation within software has been via the Beliefs Desires and Intentions (BDI) [16] model. The beliefs represent what are known about the system, the desires represent the ideal state of the system, and the intentions are triggered by a conflict between the beliefs and the desires.

This however, does not take into account the normative intentions of software systems. For example, there is no allowance for cooperation or coordination with other system components who themselves are operating their own BDI model. This makes it unsuitable for distributed systems, where intersystem coordination is paramount (such as with effective modelling of the ANS). This has led to the proposal of improved models such as BOID [17] and EDA [18], which are more suited to cooperative deployment through distributed architectures.

2.2. Context Awareness

The context in which the ANS responds is primarily constructed from the transmissions sent by sympathetic and parasympathetic fibres, that consist of preganglionic and postganglionic nerve cells. Nervous impulses are then transmitted via the chemical acetylcholine, or "ACH" through the blood stream. The nerve cells used by the ANS are situated in the central nervous system, allowing the monitoring of all the major organs, the critical services of the human body system. The combination and construction of these various nerve impulses builds an operational context for the whole body, from which the ANS can react [19].

The context of the operating environment in which software operates is invariably affected by its host machine and the current processes being executed upon it. Detailed knowledge of the individual software components, their current and future status, and their capacity for work is essential for software to model this environment. In addition, networked distributed software systems require knowledge of all other systems from whom they use services, as they are as dependent on these services as they are on their own local processes. The production and distribution of this data has yielded much research, with software sensors [20] (synonymous with nerve impulses) producing a popular solution to the retrieval of operational-level data, and software grid architectures providing solutions to the distribution and management of sensor data [21] (synonymous with the chemical transmitters).

The analysis of this information to identify changes to the operational environment, and the actions that are needed to occur have often taken the form of policies that can be executed on specific operational triggers. For example, a service can have a policy that redirects requests made to it if its current load is over a certain value. These policies can be altered and modified at runtime by the system to provide a level of self-adaptive autonomy within the system. Such modelling is undertaken using logic or process calculi [22, 23, 24], where systems are modelled in the form of actions made to and by it, causing the transition between specified system states. This provides an explicit logical representation of how a system can modify its architecture to provide dynamic software evolution. Implementations exist [6, 25] that use graphical representations to process algebra and logical formalisms to provide basic support for self-managing architectures. An assessment and comparison of 14 state-of-the-art implementations is given in [26]. This found the major problem with many of these implementations is that the management logic is currently centralised and explicitly formalised within the architecture, making scalability an issue for complex systems. In addition, the flexibility to express the types of changes that yield operation changes are limited to the configuration operations that are available at the architectural level [27].

Whilst the research highlighted has concentrated on providing self-managing autonomic behaviour through static rule and goal orientated approaches that respond to operational triggers, the system itself has little knowledge of its own execution and composition at runtime. In paradigms, where this knowledge exists, (such as in generative software [25]) the autonomic features of the system are defined at a higher level than is suitable for runtime reasoning and adaptation of governance concerns.

2.3. Survey Conclusions

To conclude, the biologically derived ANS has often been considered as a single system in software, however it does in fact consist of several sub-systems that remain in conflict to provide optimum operational equilibrium. This itself is an intriguing concept for an autonomic software system: two competing subsystems, one responsible for reacting to crisis, such as attacks or failures (sympathetic), and one responsible for providing services and bringing the system back to a more optimal configuration (parasympathetic).

To fully appreciate and successfully attempt to apply sympathetic and parasympathetic nervous system models to autonomous software, a suitable expressive language is required to enable runtime deliberation, adaptation and introspection of the varying intentions defined by each model. In addition, a suitable architecture that can represent such intentions and allow their execution to propagate out onto a distributed system is required to overcome the problems highlighted in this survey, whilst producing a truer model of an artificial biological autonomic system.

3. The Neptune Language

To enable the required intention and system state representation within an adjustable framework at runtime, the Neptune Scripting Language [28] enables axioms, normative intentions, and governance rules to be symbolised within an introspective object framework, such that the individual constructs that comprise a logical statement or assignment can be both inspected and modified without recompilation at runtime. Rules collectively assemble a profile of accepted behaviour for the system they describe by associating logical conditions (or Concepts) with a set of actions to take place upon their success or failure. In the context of system design, actions can consist of both direct alteration to the system state or calls to facilitate change within the system. Concepts however, represent a logical condition whose value is calculated based on the context of where it is being executed.

As an illustrative example of Concepts, Figure 1 shows a Neptune script providing a logical statement that determines the availability of a software component by inspecting and deliberating upon metrics obtained from the component to be tested:

```
concept responsive as boolean
{
   if (service.roundtripTime > 50)
         return true
   else
         return false
}
```

Figure 1: Concept encoded in Neptune

In this example, the metric "roundtripTime" is compared to a static value "50 (as seen on the second line of the code). If the metric is above this threshold, the function returns true, otherwise, the function returns false. In essence, a set of logical constructs are encapsulated within a Concept, here defined as "responsive". Accordingly, alteration of the underlying logical constructs allows the behaviour of the concept to be altered without affecting the structural signature of the concept when it is referenced elsewhere within the system.

The second stage of a Neptune Script abstraction is within its process model. To assist structural fluidity in the decision process, Neptune provides a range of objects that can be joined to provide workflow modelling. These objects include data input objects, which prompts the executor of the Neptune script to obtain a series of data. With data access abstracted into a composition object, Neptune objects, as well as providing introspective logical interpretation at runtime, provide introspective and adaptable data interfaces. Figure 2 shows a UML representation of a Neptune object.

The actual progression of the decision process is provided by the logic represented in the Neptune language. Thus, systems that are dependent on Neptune Script implementations for their self deliberation can adapt and inspect both their architecture and reasoning models during the system's execution, making it well suited to defining concepts subject to wide distribution, as with modelling sympathetic and parasympathetic behaviour.

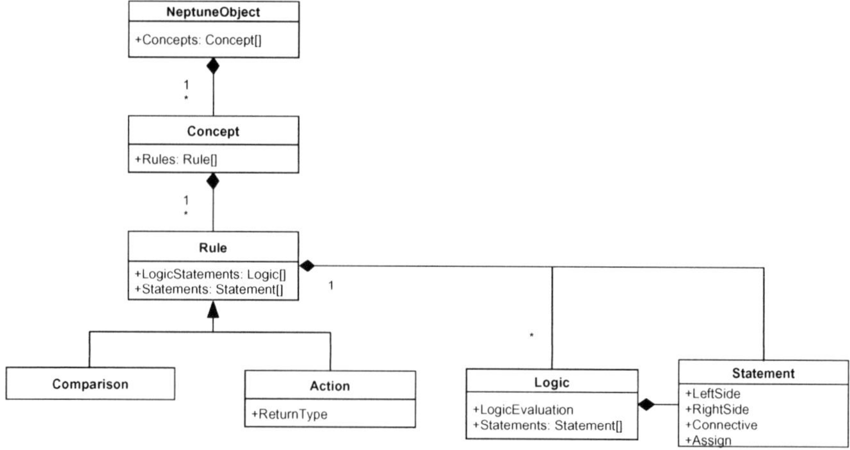

Figure 2: UML representation of a Neptune object

4. Co-ordination Framework: Clouds

The notion of a Cloud is a system with loose boundaries, which can interact and merge with other such systems [29]. A cloud can be thought of as a federation of services and resources controlled by a system controller and discovered through a system space. In a distributed system, oftentimes services and dependencies can overlap with different configurations on different systems. Systems based on the Cloud framework can interact with each other, sharing and pooling resources for greater efficiency over a large deployment such as an enterprise. Brief descriptions of relevant aspects of the Cloud Framework are given in this section.

4.1. Shared Memory: System Space

A System Space, a service that controls access to a distributed data store, provides persistent data storage for service registration and state data within a Cloud. Building on recognised technologies such as JavaSpace and TupleSpace, the System Space allows data to be stored in a generic format through a unified data structure. The structure of storage is also well suited to storing object definitions with the XML format capable of accepting serialised objects.

4.2. Service and Resource Meta-Objects

Each service and resource when it first registers itself to the Cloud sends a meta-object serialized from an XML definition file. This meta-object contains the properties and state data of the service it is describing and is stored within the System Space at registration. Each service maintains its own meta-object and updates the System Space when changes in state occur. Due to the generic form of these meta objects, they provide an attribute system for services that can be queried inside Neptune scripts just as an object can be queried through its published properties in traditional object orientated software. The XML definition file contains all information required for the

Cloud to query the service status through the published state properties contained within the state element. As the meta-objects are stored within the System Space, they enjoy system wide scope, allowing any other service or resources to request read and write access to the properties of a service through the system controller.

4.3. CA-SPA

CA-SPA (Concept Aided Situation Predicted Action) is a componentised form of system policy documentation. Using system defined concepts and actions, higher order policies can be built that automatically adapt on changes to the base components. A CA-SPA document contains 3 sections that determine its policy role:

- *Situation*: A formalisation of a state of a system component.
- *Prediction*: A formalisation of the preferred state of a system component.
- *Action*: An ontological view of the changes that need to be made to bring about the system change.

Each section is comprised from an aggregation of concepts and actions that encapsulate levels of system representation and system action in a modular form. The advantage in this approach over traditional reasoning models is that a change in one concept or action is propagated throughout the whole system. This results in a modular architecture for policy definition.

5. Evaluation

To evaluate the effectiveness and usability of Neptune and the Cloud Framework to address the limitations highlighted in Section 2, suitable scenarios that can exhibit sympathetic and parasympathetic behaviours from the system need to be determined.

A sympathetic behaviour can be categorised as a reaction to crisis. In distributed systems, such crises can be, for example the failure of a service by an abnormally high volume of requests (such as with a Denial of Service attack). It is proposed that parasympathetic behaviour in software can be categorised as a *reaction to equilibrium*. In this way, when the system is not in an adverse conditional state (as defined by the sympathetic model), parasympathetic behaviour can occur. Thus, there is a direct relationship between the two models. If the system is in a state defined by the sympathetic model, parasympathetic behaviour cannot occur, though when transition is made into a state not defined by the sympathetic model, the parasympathetic model can execute its own behavioural intentions.
Mathematically, this relationship be expressed as shown in Figure 3:

$$A \cap B = \varnothing$$

where A = *{all defined sympathetic states}*
B = *{all defined parasympathetic states}*

Figure 3: State Ownership Relationship

Which yields the implication shown in Figure 4, given state S:

$$(S \notin A) \Leftrightarrow (S \in B)$$
$$\text{where} \quad A = \{\text{all defined sympathetic states}\}$$
$$B = \{\text{all defined parasympathetic states}\}$$

Figure 4: State Ownership Implication

The relationship in figure 4 shows that the parasympathetic model requires knowledge of the composition of the sympathetic model, though the sympathetic model does not require knowledge of the parasympathetic model.

5.1. Software Implementation

The composition of a state, as with the biological ANS, consists of a series of sensed data that are connected via logical constructs to produce a condition that defines the boundaries of a state. In software, this sensed data is resolved from data exposed through meta-objects (see Section 4). Such conditional states are represented within the CA-SPA policy document format, allowing componentisation of states.

Accordingly, defining a state as sympathetic, and providing an action ontology to move the system out of the sensed state can mimic sympathetic behaviour mimicked within the system. As an example used in Section 2 to illustrate sympathetic responses, the human ANS contracts the pupil to protect it when a light source is shone directly upon it. This would translate into the high level Neptune CA-SPA representation shown in Figure 5:

```
situation s {
   if (Eye.LightLevel > CriticalLightLevel)
   {
       return true;
   }
}

action a as s {
  Pupil.Contraction--; //reduce contraction
}

prediction p to a {
   if (Eye.LightLevel <= SafeLightLevel)
   {
       return true;
   }
}
```

Figure 5: CA-SPA Policy

Thus, when the light level is over a critical level (as defined in the situational state) the pupil is contracted until the safe light level is reached (defined within the action ontology). The predicted state, within the context of sympathetic systems, determines the desired resolution of any sympathetic behaviour. Accordingly, predicted states can be considered as the states that determine the resolution to equilibrium making them suitable for representing parasympathetic behaviour as situational states.

However, determining whether a system has entered a parasympathetic state is dependent on the whole collective sympathetic model, as stated in the relationship given in Figure 4. As CA-SPA states are modularised by being represented in Neptune script, the CA-SPA documents that collectively provide a sympathetic behavioural model can be joined using this modular representation to provide a federated state representing the whole sympathetic model. Changes made to the individual models are then propagated to the sympathetic model, making it well suited for delivering adjustable autonomy. This enables a single access point to determine whether a system is in a parasympathetic state, making the whole approach scalable, a quality essential for distributed system design. For example, two CA-SPA representations of sympathetic behaviour are shown in Figure 6:

```
situation UnderHeavyLoad {
   if (Service.CPUUsage > 80%)
   {
       return true;
   }
}

situation IncreasedUsage {
  if (Service.RequestsPerSecond>CriticalRequestsPerSecond)
   {
       return true;
   }
}
```

Figure 6: CA-SPA Situational States

Thus we can produce a single condition representing whether the system has entered any sympathetic state, as shown in Figure 7:

```
if (State.UnderHeavyLoad || State.IncreasedUsage) {
        return true;
}
```

Figure 7: Federated Sympathetic State

On successful resolution of this state, the individual sympathetic states can be inspected to ascertain the action ontology to take. Otherwise, the system is able to accept parasympathetic behaviour, which again is determined by action ontology's, though using the predicted states of the sympathetic CA-SPA documents.

6. Further Aims of this Research

This paper has introduced the theoretical usage of the Cloud Framework and Neptune language to facilitate a more natural modelling of a biological autonomic nervous system. Several relationships that hold for such modelling have also been defined by this paper, and have been used to simplify the structural architecture of the proposed implementation, promoting scalability whilst maintaining adjustable deliberation.

Future directions of this research include building a grid based deployment to further analyse the concepts introduced in this paper, which can yield a better understanding of the tangible benefits of applying sympathetic and parasympathetic behaviour to autonomic system design.

References

[1] P. Horn, Autonomic Computing: IBM's Perspective on the State of Information Technology, October 15, 2001

[2] Jeffrey O. Kephart and David M. Chess. The vision of autonomic computing. *Computer*, 36(1):41–52, 2003.

[3] M Shackleton, F. Saffre, R Tateson, E Bonsma;,C Roadknight "Autonomic Computing for Pervasive ICT — A Whole-System Perspective", BT Technology Journal, Vol.22 No.3 pp 191-199, July 2004

[4] Shields RW. Functional anatomy of the autonomic nervous system. J Clin Neurophysiol. 1993;10:2-13.

[5] H. Liu and M. Parashar. "DIOS++: A Framework for Rule-Based Autonomic Management of Distributed Scientific Applications," Proceedings of the 9th International Euro-Par Conference (Euro-Par 2003), Lecture Notes in Computer Science. August 2003.

[6] Li, D. and Muntz, R, Runtime Dynamics in Collaborative Systems. Department of Computer Science University of California, Los Angeles. 1999.

[7] D. Hirsch, P. Inverardi, and U. Montanari. Graph grammars and constraint solving for software architecture styles. In Proc. of the Int. Software Architecture Workshop, pages 69–72, Nov. 1998.

[8] D. L. M´etayer. Software architecture styles as graph grammars. In Proc. of the ACM SIGSOFT Symp. On Foundations of Software Engineering, pages 15–23. ACM Press, 1996.

[9] M. Wermelinger, A. Lopes, and J. L. Fiadeiro. A graph based architectural (re)configuration language. In Proc. of the Joint European Software Engineering Conference and Symp. On the Foundations of Software Engineering, pages 21–32.

[10] J.S. Bradbury, J.R. Cordy, J. Dingel and M. Wermelinger, "Supporting Self-Management in Dynamic Software Architecture Specifications", submitted to WOSS'04 - ACM SIGSOFT 2004 Workshop on Self-Managed Systems, submitted August 2004, 10 pp.

[11] Low PA. Autonomic nervous system function. J Clin Neurophysiol. 1993;10:14-27.

[12] Schore, A. N. Attachment and the regulation of the right brain. Attachment and Human Development, 23-47. 2000.

[13] Harris, Kelly F Matthews, Karen A. Interactions Between Autonomic Nervous System Activity and Endothelial Function: A Model for the Development of Cardiovascular Disease. *Psychosomatic Medicine* 66(2):153-164, April 2004.

[14] Janig W and McLachlan EM (1992a) Characteristics of function-specific pathways in the sympathetic nervous system. *Trends Neuroscience* 15: 475-481.

[15] Barron, K. D., & Chokroverty, S. (1993). Anatomy of the autonomic nervous system: brain and brainstem. In P. A. Low (Ed.), Clinical Autonomic Disorders, Evaluation and Management (pp. 3-15). Boston: Little, Brown and Co.

[16] M. E. Bratman 'Intentions, plans and practical reason'. Harvard University Press, Cambridge, Massachusetts. 1987

[17] J. Broersen, M. Dastani, J. Hulstijn, Z. Huang and L. van der Torre.'The Boid Architecture' Agents'01, Montreal, Quebec, Canada. 2001

[18] J. Filipe, 'A normative and intentional agent model for organisation modelling' Third International Workshop "Engineering Societies in the Agents World" Madrid, Spain. 2002.

[19] Elenkov IJ, Wilder RL, Chrousos GP, Vizi ES. The sympathetic nerve: an integrative interference between two supersystems: the brain and the immune system. *Pharmacol Rev* 2000;52:595–638

[20] Goslar, K., Buchholz, S., Schill, A., and Vogler, H. A multidimensional approach to context-awareness. In Proceedings of the 7th World Multiconference on Systemics, Cybernetics and Informatics (SCI2003), 2003

[21] Tierney, B., R. Aydt, D. Gunter, W. Smith, V. Taylor, R. Wolski, M. Swany. A Grid Monitoring Service Architecture. Global Grid Forum White Paper. 2001.

[22] H. J. Levesque, F. Pirri and R. Reiter (1998) 'Foundations for the situation calculus'. Linköping Electronic Articles in Computer and Information Science, Vol. 3(1998): nr 18. http://www.ep.liu.se/ea/cis/1998/018/
[23] Allen J. F. 'Towards a general theory of action and time'. Artificial Intelligence, 23, pp 123-154. 1984.
[24] Thielscher M. 'Introduction to the fluent calculus' Linköping Electronic Articles in Computer and Information Science, Vol. 3(1998): nr 14. http://www.ep.liu.se/ea/cis/1998/014/. 1998
[25] G. Frick and K. D. M¨uller-Glaser, "Generative development of embedded real-time systems," in ECOOP 2002 Workshop on Generative Programming, Malaga, Spain, June 2002.
[26] J.S. Bradbury, J.R. Cordy, J. Dingel and M. Wermelinger, "Supporting Self-Management in Dynamic Software Architecture Specifications", submitted to WOSS'04 - ACM SIGSOFT 2004 Workshop on Self-Managed Systems, submitted August 2004, 10 pp.
[27] M. Wermelinger. Towards a chemical model for software architecture reconfiguration. IEE Proceedings - Software, 145(5):130–136, October 1998.
[28] Miseldine P., Taleb-Bendiab A. Delivery of Autonomic Distribution: The Cloud Framework. Techincal Report. Liverpool John Moores University. February 2005.
[29] Miseldine P., Taleb-Bendiab A. Rainbow: An Approach to Facilitate Restorative Functionality Within Distributed Autonomic Systems. in the proceedings of the 6th Annual PGNet Symposium (PGNet 2005). Liverpool, June 2005.

Self-Organization and Autonomic Informatics (I)
H. Czap et al. (Eds.)
IOS Press, 2005

Collaboration Patterns for Adaptive Software Engineering Processes

Andrea Freßmann [a], Thomas Sauer [b], Ralph Bergmann [a]
[a] *University of Trier*
Department of Business Information Systems II
54286 Trier, Germany
{andrea.fressmann, bergmann}@wi2.uni-trier.de
[b] *rjm business solutions GmbH*
Sperlingweg 3
68623 Lampertheim, Germany
t.sauer@rjm.de

Abstract.
In the software engineering domain, workflow systems are an established technology for team coordination and cooperation. However, software development processes tend to be unknown, unrepeatable, and knowledge intensive, with changes occurring frequently. Thus, strategies to distribute best practices and lessons learned on how to collaborate effectively within this ever-changing environment are required, especially when project participants may work autonomously. In this paper, *collaboration patterns* are introduced to describe collaboration strategies explicitly. The presented approach is based on adaptive workflows, allowing to tailor the collaboration patterns as changes occur. An example of software project planning and enactment shows how collaboration patterns can be put into practice.

1. Introduction

In the software engineering domain, workflow systems are an established technology for team coordination and cooperation. That is, workflow support covers overall structuring of activities (e.g. by providing project plans, or a to-do list), reuse of task descriptions to efficiently handle situations encountered before, or effective scheduling to ensure that users always work on the most important tasks. However, software development processes tend to be unknown, unrepeatable and knowledge intensive, with changes occurring frequently. Otherwise, in organizations norms [10] exist and shape the way of working within organizations. Hence, effective collaboration strategies following such norms are necessary to ensure that the efforts of individual team participants are all aiming in the same direction.

In this paper, *collaboration patterns* are introduced to describe these strategies explicitly. Collaboration patterns express best practices, lessons learned, and norms representing contracts between agents, which provide an abstract representation for arbitrary project participants. Agents represent human team members as well as machines used, and collaboration among agents may constitute either a loose or strict contract. The type of contract depends on how close these agents work in sync. For example, a strict contract collaboration pattern might describe a search strategy by joining several different retrieval services offered by machine agents, while a procedure how two human agents may work together most effectively is a collaboration pattern denoting a loose contract.

The approach presented in this paper uses adaptive workflows to represent collaboration patterns, with adaptive workflow management provided by the *Collaborative Agent-based Knowledge Engine* (*CAKE*) [6]. CAKE has been devised for knowledge distribution throughout an organization to support planning and enactment of empirical processes [4], including, but not limited to, those found in the software engineering domain [5].

In the following section, the CAKE overall architecture is sketched, as CAKE provides the foundations collaboration patterns are built on. In Section 3, collaboration patterns are explained in detail, followed by an illustrative example in Section 4 showing how to put the presented approach into practice. Related and adjacent research is discussed in Section 5. Finally, a short conclusion and outlook to future work close this paper.

2. The CAKE System

The idea of CAKE as supporting system for processes results from investigations in different application domains in which similar user requirements are identified. Hence, CAKE is developed as general domain-independent system that can be tailored for application domains like fire services and roadside assistance[1], medical knowledge management, geographical information systems, and software engineering. The presented paper focuses on the software engineering domain. Its main characteristic is that people involved in software projects require support for planning and executing highly flexible processes while taking advantage of reusing various kinds of knowledge gained in former processes. Project managers and software developers can be envisaged as users collaboratively working on the same software development project.

In CAKE, the representation of processes is regarded on two different levels. First, *workflow definitions* describe processes in general, building a repository for possible processes. Second, *workflow instances* (or simply *workflows*) are concrete instances of workflow definitions that emerge when a software project is executed. Hence, workflow definitions are set up by the workflow modeler during build-time and are stable afterwards, while workflow instances can be subject of change and refinement during project execution in terms of adaptive workflows.

Planning support is demanded by the workflow modeler for project planning and by software engineers for workflow modifications during execution. *Enactment support* comprises support during workflow execution including support for accessing required information and best practices. The support of adaptive work-

[1] Application domain of the AMIRA (Advanced Multimodal Intelligence for Remote Assistance) Project funded by the European Commission.

flows leads to modifications during run-time and probably to the creation of new workflow instances that have no corresponding descriptions as workflow definition.

The approach that CAKE follows to address the above mentioned support types in an integrated fashion is the combination of the core concepts of workflow technology, agent technology, and case-based reasoning (CBR) technology in a new architecture. While the workflow technology is used for representing workflows and workflow related data as well as for controlling their execution, the agent approach is used to mediate and control access and communication to different information services and users. The CBR technology can be seen as third key component of CAKE; it supports access to applicable best practice processes as well as the selection of appropriate agents in the context of a specific workflow. The three technologies are integrated by a central data and process model, which can be expanded by domain-specific data classes and processes, leading to domain-specific ontologies.

Workflow Representation and Execution. Albeit CAKE distinguishes workflow definitions and workflow instances, both are represented in a similar way as a set of task descriptions together with control flow relationships among them. A *task* is an arbitrary activity that is executed by a user (e.g. a software developer) or by an autonomous agent. A *task description* contains an informal or formal specification of the activity to be done together with a characterization of necessary competencies an agent must have in order to be able to enact that task. For example a task can be the writing of a report or the automated search in some database. Further, a task can also be only a placeholder for an activity not yet defined in detail. When such a task is executed, further planning is necessary in order to determine the detailed activities to be done. Finally, a task can also be used to initiate a new sub-workflow. Through this means the hierarchical decomposition of a complex activities can be achieved.

The control flow relationships describe the temporal dependencies among the execution of the tasks a workflow consists of. Tasks can be arranged in sequence, in parallel, and by using splits and branches. Hereby the flow of control is defined, but not the flow of data.

Workflows are executed by a *workflow engine*. To execute a particular workflow the workflow engine creates a workflow instance from the respective workflow definition. Further, each workflow instance gets an associated *local context* assigned that facilitates capturing, storing, and changing of workflow-specific data. The context comprises of administrative data, workflow control data, workflow relevant data, and application data [11]. Thereby, the context is an information container for any kind of data used by the workflow.

Agent Technology. CAKE includes an agent framework for integrating interfaces to various data sources, external services, and humans. Two kinds of agents are distinguished: *information agents* and *user agents*. Both agents can act as interface to technology components and humans as well. Information agents provide various kinds of services such as information provision by search engines or human experts. Such agents are invoked by CAKE but do not act autonomously. Contrary, user agents are either interfaces for end users or for any kind of software. Such agents may work autonomously, e.g. by communicating with an information agent via CAKE.

All agents are registered in a central *agent pool* with their respective competence profile. The competence profiles of information agents are important in order to know which agents are available when an agent for a particular task has to be assigned. Usually, task descriptions only contain agent roles that act as indicator for suitable competence profiles.

User agents can work collaboratively on the same workflow but on different tasks. This collaboration can be defined in workflow definitions and may include tasks running in parallel or in sequence each of which refers to a different user agents or a distinct user role. This is called *agent collaboration.* For instance, two software developers can be working on the same software project. First, one of them is playing the management role and plans the project by assigning both developers to the tasks at issue. After having instantiated the corresponding workflow definition CAKE manages the coordination of these tasks while considering the dependencies among them. Further, agent collaboration is also used when information agents work together for achieving the common goal of fulfilling a certain information need by combining different information sources. The *collaboration patterns* are described in more detail in Section 3.

CBR Technology. Planning support as discussed above requires a search facility that allows searching for workflow definitions. Therefore, CAKE integrates a search engine supported by CBR [9,1]. CBR provides an approach to problem solving following the principle that similar problems have similar solutions. Therefore, CBR technology provides CAKE in expressing highly structured *workflow characterizations* with respect to domain-specific ontologies. Workflow characterizations are attribute-value representations specifying the goals and properties of workflow definitions. A workflow definition and its characterization are stored as *case*, leading to the *case base* that comprises available cases.

By searching on workflow characterizations the search engine determines whether they are reusable in a specific situation. If this is the case, the corresponding workflow definition is proposed as solution. For example, workflow definitions that represent the way to plan a project can be characterized with the goal "project planning". By comparing the specified goal with domain-specific ontologies the search engine performs a retrieval on workflow characterizations and presents a result set of such characterizations ordered by means of similarity and utility defined as similarity model [2] in CAKE. Then the corresponding workflow definitions can be used for further proceeding.

The CBR-based search engine can also be used for searching suitable agents. Similar to the workflow definition search, a CBR retrieval is implemented on agent characterizations allowing agent retrieval.

Combining workflow, agent, and CBR technology leads to scenarios whereas CAKE differs between user support during run and build-time. When registering into agent pool a user agent gets an appropriate workflow assigned for run-time support. The user agent can look for information provided by registered information agents. Otherwise, the support during build-time allows the user agent modeling a new workflow by reusing former workflows or sub-workflows.

3. Collaboration Patterns

For taking advantage of collaboration among various agents CAKE incorporates an approach that delegates and coordinates these agents, e.g. for realizing a meta search over more than one search engine. In terms of search strategies, simply providing access to each information source or service available is not sufficient because in this case it would be up to the end users to delegate their own requests to the various information agents. Specifying the details to query numerous information sources can be cumbersome if not impossible for the individual user, e.g. if he or she is not familiar with a particular type of information source. Moreover, evaluating and combining the result sets efficiently can be a challenging task, which might even require specialist knowledge.

In order to assist arbitrary agents in accessing data provided by information agent or to assist them in collaborating with other user agents, CAKE incorporates the concept of *collaboration patterns*. Collaboration patterns are workflow definitions that capture agent collaboration strategies based on best practices. Acting as contracts among agents collaboration patterns are used for managing, organizing, delegating, and coordinating agents. Patterns can occur as more strict or as more loose contracts depending on how strict or how loose the agents are in sync. For further distinguishing collaboration patterns from actual workflow definitions, which are formal representations of business processes, collaboration patterns produce an arbitrary artifact, just a by-product of a final business process product. For example, collaboration patterns can be used for realizing meta search facilities that have been developed in the area of information retrieval ; collaboration patterns can be used for human to human collaboration by describing a procedure how human agents are working together. In both examples the collaboration patterns originate on domain knowledge, best practices, and experiences made by domain experts.

While various approaches tend to avoid central-driven agent coordination strategies [3], this paper presents collaboration patterns as an approach that focuses on a mixed initiative of integrating central coordination and preserving flexibility of a highly dynamic agent pool. Central coordination is important for regarding dependencies between agents' actions, for meeting global constraints, and for exploiting multiple agents' competences [7]. Additionally, the presented approach covers integration of domain-specific knowledge and consideration of user requirements. First, using domain-specific knowledge for agent coordination leverages explicit specification which kind of agent or agent service is useful within which situation. Second, because of being highly flexible and configurable collaboration patterns can realize user requirements, e.g. for optimization purposes.

Representation of Collaboration Patterns. Collaboration patterns are workflow definitions as defined in Section 2. Thus, CAKE enables a variable arrangement of machine-executable tasks for realizing different agent collaboration strategies. An example for collaboration patterns is shown in Figure 1, which will be explained in detail in Section 4. Because of being represented as workflow definitions, collaboration patterns dispose the same technical possibilities as the actual workflow definitions. That is, the tasks and the control flow relationship are the building blocks of collaboration patterns. Therewith branches and the merging different branches are feasible.

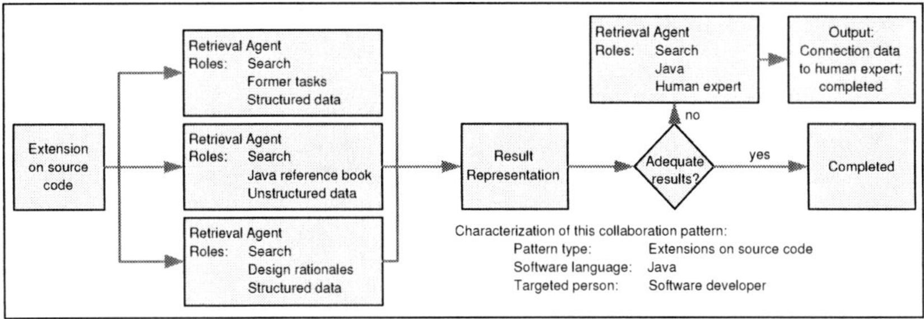

Figure 1. Example for Collaboration Patterns

For tasks that frequently occur the concept of *task templates* is used for specifying a common structure for such tasks. Task templates are predefined tasks offering a small set of parameters for tailoring, i.e. a domain modeler may conveniently incorporate standard activities into a collaboration pattern. For example, the task template "requesting an agent" requires the following parameters for being executed: the agent name or agent role has to be specified; the query, the predecessor task, and the successor task have to be assigned. Task templates already contain methods to find the respective agent, to get in contact to this agent, to perform a request, and to receive response.

Retrieval of Collaboration Patterns. Collaboration patterns base on best practices and are represented as workflow definitions, with their characterizations being stored in the respective characterization case base. Based on these characterization collaboration patterns can be retrieved during the build-time or during the execution of a workflow. As nesting of workflows is supported by CAKE, collaboration patterns can be applied at any time and on various levels. The underlying domain-specific ontology and the similarity model allow searching for collaboration patterns in consideration of semantic classifications. Consequently, retrieval of collaboration patterns is similar to retrieval of workflow definitions during build-time and workflow instances during run-time.

Characterizations of collaboration patterns consist of information about these patterns like their purposes and their contexts. Particularly, assigned *agent roles* stored in the context provide information about classification of the corresponding collaboration pattern because of describing what agent should provide, e.g., descriptions of services or descriptions of database contents.

Agent Allocation. CAKE integrates collaboration patterns that can be specified independently from specific agents. Consequently, collaboration patterns can consist of task descriptions in which only agent roles are specified, being independent from concrete agents. This leads to the need of retrieval facilities based on agent roles as requests for agent competence profiles.

When specifying a task description where a service is demanded the workflow modeler can utilize domain-specific ontologies for assigning agent roles like "Java expert" or "expert for design rationales". These roles describe which kind of agent is able to carry out the demanded service. The CBR-based search engine utilizes the ontologies for searching the most competent agent. CAKE allocates this agent to the task description before the task can be executed.

Figure 2. Example for a Loose Contract Collaboration Pattern

4. Example

A fleet management project has proceeded to the first iteration, and an intermediate system has been deployed at the customer site. Negotiations with the customers throughout the first iteration have resulted in the customer preference that a online booking system should be added to the system. Now, the developers playing management roles have to settle on the most basic workflows, which will control the very next steps. So, a developer specifies the artifact goal "booking system" and retrieves a workflow outlining best practice from a previous project. This workflow was previously performed within another application and experiences leads to the conclusion that a collaboration between a domain expert and a software developer was a benefit, which is represented through a loose contract in Figure 2: this collaboration pattern consist of a requirement analysis to be done by a domain expert with respect to what the customer really needs, the technical implementation to be done by a software developer, and a customer feedback process to ensure customer's satisfaction.

Next, the team starts to instantiate the very first workflow definition and suitable user agents playing the role of domain expert and software developer are assigned. The latter is supported by the CAKE retrieval facility on agent's characterizations, in combination with a collaboration pattern, which takes organization-specific constraints into account (by querying a human resources management information agent providing availability data, and a Wiki system containing tentatively planned leaves). These agents can now autonomously decide how to proceed for fulfilling the tasks. For example, the software developer starts work on the assigned task by adapting the actual loose workflow. Because of not being familiar to the actual project the software developer starts a CAKE retrieval process on comprehensive information about this project with respect to similar tasks, to relevant Java references, and to relevant design rationales. As consequence, the task 'implementation' is adapted with the collaboration pattern presented in Figure 1. That is, relevant information is presented to the developer that he can be familiarized to the work at issue. Additionally, the developer gets a link for searching a Java expert for booking systems if more explicit information is necessary. Primarily the idea is to support the developer through computerized agents but if this is not enough there is an option to ask human experts. Therefore, the collaboration pattern contains a task that organizes a retrieval on human Java experts. The result is to provide the developer with information about how to get in contact to such an expert. Surely, the developer can proceed further in adapting the assigned task until the task of the initial loose contract collaboration pattern is completed. Either he uses a collaboration pattern which already includes useful domain knowledge or he carries out his own adaptations on the task for achieving the envisaged task artifact.

5. Related Work

For realizing collaboration patterns CAKE makes use of agent enhanced workflows [12]. Such approaches focus agent collaboration without assigning workflow control to agents as the concept of agent-based workflows does. The management of business processes realized through agent enhanced workflows is still under control of workflow management systems. Agent enhanced workflows imply an agent level on top of the workflow management system in order to allow automatic changes to working environments through negotiations [8]. Here, agent collaboration is realized by usage of standard contract net protocol instead of workflows based on best practices like collaboration patterns.

Focusing traditional workflow management systems Workbrain merges Organizational Memory and workflow technology [13]. For structural planning a CBR application is integrated and can be utilized by a human workflow planner before the respective workflow will be enacted. By providing retrieval on workflow characterizations the CBR application allows users to describe a problem and retrieves similar solutions that can be used to construct new workflows.

Similar to CAKE, CBRFlow [14] combines workflow technology with conversational CBR for coping with changing and unpredictable environments. For realizing the concept of workflow modification during run-time workflow instances can be annotated with cases that consist of a set of question-answer pairs and one action. After the annotation these cases can be directly used as assistance for decision-making processes because of representing additional knowledge, maybe in form of instructions, which can be used during workflow execution. Modifications on the workflow definitions is done by workflow modelers and no search facility on workflow definitions is supported.

However, both Workbrain and CBRFlow approaches lack techniques for accessing external information sources. Particularly, agent technology is not supported that facilitates methods like collaboration patterns for addressing agents and organizing collaboration among them.

6. Conclusion

In this paper, collaboration patterns have been presented as an approach for describing collaboration strategies between arbitrary agents contributing to a software development joint effort. Since software development processes tend to be unknown, unrepeatable, and ever-changing, collaboration patterns have to be adaptive so they can be tailored in order to reflect new situations. This is achieved by representing collaboration patterns using the adaptive workflow technology provided by CAKE.

Collaboration patterns are integrated in CAKE that has been implemented domain-independently and that has been proven to be useful in a variety of application domains, including fire services and roadside assistance in the AMIRA project, medical services, as well as for workflow support in geographical information systems. Tailored to these different application domains CAKE is implemented and takes advantages of separating domain-specific knowledge and the underlying technologies, e.g. a linear CBR retrieval is utilized for retrieving matching similar workflow definitions and agent competence profiles.

Acknowledgements. The authors acknowledge the European Commission for funding AMIRA under grant number FP6, project IST-2003-511740.

References

[1] A. Aamodt and E. Plaza. Case-based reasoning: Foundational issues, methodological variations and system approaches. *AI Communications*, 7:39–59, 1994.

[2] R. Bergmann. *Experience Management.* LNAI 2432. Springer, 2002. ISBN 3540441913.

[3] E. H. Durfee. Scaling up agent coordination strategies. *IEEE Computer*, 34(7), 2001.

[4] A. Freßmann, K. Maximini, R. Maximini, and T. Sauer. Cbr-based execution and planning support for collaborative workflows. In *Workshop "Similarities - Processes - Workflows" on the Sixth International Conference on Case-Based Reasoning (ICCBR 2005)*, pages 271–280, Chicago, Illinois (USA), August 2005.

[5] A. Freßmann, K. Maximini, R. Maximini, and T. Sauer. Collaborative agent-based knowledge support for empirical and knowledge-intense processes. In *MATES 2005/CIA 2005*, volume 3550 of *LNAI*, Koblenz, Germany, September 2005. Springer-Verlag.

[6] A. Freßmann, R. Maximini, and T. Sauer. Towards collaborative agent-based knowledge support for time-critical and business-critical processes. In K.-D. A. e. al., editor, *WM 2005, LNAI 3782*, pages 420–430. Springer-Verlag, 2005.

[7] N. Jennings. Coordination techniques for distributed artificial intelligence. *Foundations of Distributed Artificial Intelligence*, pages 187–210, 1996.

[8] D. Judge, B. Odgers, J. Shepherdson, and Z. Cui. Agent enhanced workflows. *BT Technology Jounal*, 16:79–85, 1998.

[9] J. L. Kolodner. *Case-Based Reasoning.* Morgan Kaufmann, 1993.

[10] K. Liu, L. Sun, A. Dix, and M. Narasipuram. Norm-based agency for designing collaborative information systems. *Information Systems Journal*, 11:229–247, 2001.

[11] H. Maus. Workflow context as a means for intelligent information support. In *Proceedings of CONTEXT-01 Third International Conference on Modeling and Using Context*, volume 2116 of *LCNS*, pages 261–274, Dundee, Scotland, 2001. Springer.

[12] J. Shepherdson, S. Thompson, and B. Odgers. Cross organisational workflow co-ordinated by software agents. In *Proceedings of the workshop on Cross-organisational Workflow Management and Co-ordination*, 1999.

[13] C. Wargitsch, T. Wewers, and F. Theisinger. Workbrain: merging organizational memory and workflow management systems. In *Proceedings of KI'97 Workshop on Knowledge-Based Systems for Knowledge Management in Enterprises*, Freiburg, Germany, 1997. 21st Annual German Conference on AI'97.

[14] B. Weber and W. Wild. Towards the agile management of business processes. In K.-D. Althoff, A. Dengel, R. Bergmann, and T. Roth-Berghofer, editors, *WM2005: Professional Knowledge Management Experiences and Visions*, pages 375–382, Kaiserslautern, Germany, 2005. DFKI GmbH.

Self-Organization and Autonomic Informatics (I)
H. Czap et al. (Eds.)
IOS Press, 2005

Service Agents Based Collaborative Workflow Management Implementation

Leo PUDHOTA and Elizabeth CHANG
School of Information Systems, Curtin University of Technology
PO Box U1987, Perth WA 6845, Australia
PudhotaL@cbs.curtin.edu.au

Abstract: *Workflow management systems in collaborative logistic companies require strong information systems to support in a distributed environment. These IT integration requirements have expanded considerably with the advent of collaborative e-business; utilizing B2B (Business to Business) and P2P (Partner to Partner) e-commerce. This paper deals with adaptation management of collaborative workflow changes in such consortia and proposes architecture for implementation of these changes through the process of component integration and agent based workflow management system where by existing workflow systems adapt to the changes. This paper describes conceptual framework required for prototype implementation resulting in new collaborative workflow adaptation.*

1. Introduction

In this paper we aim to prototype adaptation management for dynamic business processes of large logistic consortia, often we see that the business processes are composed of several parts, a structured operational part and an unstructured operational part, or they could be composed of semi-structured parts with some given and some unknown details. Unpredictable situations may occur as a result of changes in decisions made by the management. The inability to deal with various changes greatly limits the applicability of workflow systems in real industrial and commercial operations. This situation raises problems in workflow design and workflow systems development. We propose workflow implementation methodology through the process of component integration techniques for development of new workflow using existing workflow components.

2. Collaborative Workflow for an Extended Enterprise

Collaborative workflow is a new type of workflow that allows an organization or enterprise to be added to the existing workflow model and be used in the extended organization or enterprise. The advent of the internet has provided mechanisms for binding organizations to work together as they need collaborative workflow management for carrying out sales over great distances and at any time. Collaborative workflow is important for marketing and enabled partnerships, previously inconceivable within a wide array of business, as well as other human activities. Management of collaborative workflow helps the connectivity and information richness that one is faced with in an increasingly dynamic business environment and workflow. Collaborative workflow management helps the shift from old business paradigms to new business paradigms. New collaborative organization workflow systems that *transcend* the *previous static, closed, competitive models* and *move* to *dynamic open re-configurable, often collaborative models* that are able to respond to the business environment dynamics inherent within the networked economy [13]. Several factors characterize collaborative workflow management for extended enterprise; namely:

1) A strong information infrastructure that extends beyond the original closed walls of the *individual enterprise*.
2) High connectivity and electronic handling of information, of all sorts including data and documents.
3) An increasingly *collaborative approach* between what were more traditional *individual enterprises*.
4) Utilization of new forms of electronic interaction, provision of services and utilization of services.
5) Ability to self organize and reconfigure the business of the organization, perhaps even the organization as a whole.
6) Use of multiple channels for sales and marketing.
7) Capture and utilization of business intelligence from data and *smart information use*.

These features are increasingly exhibited by successful modern business organizations, for instance, in collaborative supply chains, collaborative consortia for marketing, strategic partnerships, alliances and selling services, utilization of web sales, marketing and customer service and creation of multiple modes of user interaction with the business.

A key factor in the success of such collaborative workflow management for Extended Enterprise is the creation of the underpinning *information infrastructure* to carry out the required services and development to enable and support the creation and the strengthening of small-medium enterprises (SMEs) to achieve some of the characteristics of collaboration. However current workflow techniques do not address the collaborative workflow issues and management.

3. Challenges of Collaborative Workflow

Activities and artefacts do not quite constitute a process. We need a way to describe meaningful sequences of activities that produce some valuable result, and to show interactions between processes. Changes in collaborative workflow have to be incorporated into the integrated enterprise system; we have proposed a prototype of its working in our previous papers [8, 9, 10, 11]. In this paper we are concentrating on:

1) Implementation aspects of integrating and adaptation of changes in the new workflow into an already existing workflow.
2) Information systems can change at run time so that new workflow can synchronize with existing workflow to adapt quickly.

3) Other issues like
4) Management of data scattered over multiple origin systems/legacy systems, for example, a company will have consolidate data in one logical view with a unified architecture, thereby enabling data-source independence. Because application data continues to live and change in the origin systems, the new software layer must be able to retrieve origin data on the fly and also propagate changes back to the origin systems.
5) Provide support for transactions/interaction across multiple back-end systems. "The hard part is getting a transaction model wrapped around those back-end systems; so if it didn't work in the third system, it was able to roll back in those first two systems," *[16]*.

To keep pace with changing business needs, these systems must be periodically updated. However, modifying such *legacy systems* is in general very difficult: the system's structure and internal operation will become corrupted with the passage of time, designs and documentation are lost, and individuals with an understanding of the software move on. Completely rewriting such software tends to be prohibitively expensive, and is often simply impossible. Therefore, in the long term, the only way to keep such legacy systems useful is to incorporate them into a wider cooperating community, in which they can be exploited by other pieces of software. This can be done, for example, by building an 'agent wrapper' around the software to enable it to interoperate with other systems [21,19]. The process of transforming a software application such as a database into an agent is sometimes referred to as *agentification* [22].

These challenges will help in having a uniform data processing environment for the whole enterprise, which would lead to changes and improvements in customer services, control of receivables and increase efficiency in communication, sales, marketing as well as minimization of warehouse stocks, streamlining inventory and logistics flows.

Provide control to Consortium management to monitor the collaborative enterprise's condition, its stock, order and its general financial condition on a routine basis, this is indispensable to the management processes and enhances decision-making and changes which need to be taken on the short term and long term bases for the consortium to compete in the global market.

4. Conceptual Model of Service Oriented Framework

Conceptual Model provides an architectural separation of business functionality from technology implementation. This separation allowed designers to use business rules defined in a UML model to drive four distinct steps in implementing such systems.

Step1. Create two platform independent models in UML. The first model is a generic domain model, used to build a common understanding and vocabulary among warehouse Logistics domain experts.

Step2. The domain model is then mapped into a second platform independent model (PMI) representing warehouse logistic business. Each of the models includes a detailed set of UML Class Diagrams, Use Cases and associated Activity Diagrams describing the system [12].

Step3. Using this business model, we can create one or more subsystems to represent the logical functions of each of the enterprise systems. This business model contains both the details of the business logic, as well as the mapping of the logic into the major subsystems. The business model forms the basis for managing all changes to the current systems.

Step4. System Integration using Conceptual Model of Platform Specific Models (PSM's), for each of individual systems to form enterprise system. These models were each derived from one or more subsystems in the business model. The relationships are shown in Figure 4.

System construction consists of customizing each of the enterprise systems, and creating the business logic. Business logic that spanned systems is constructed using components technology and deployed in the application server [12].

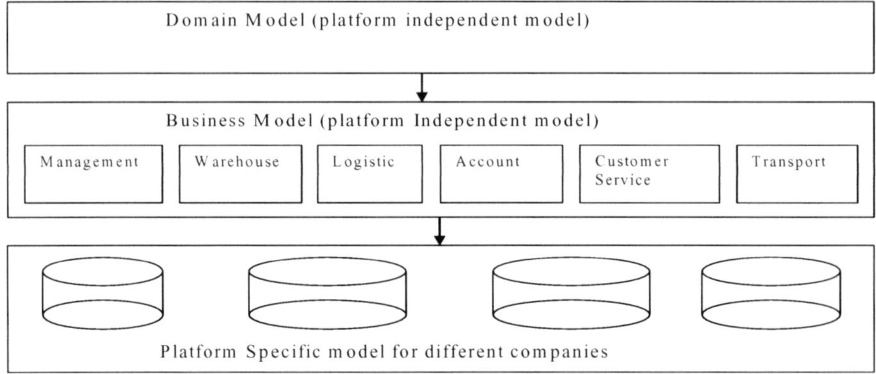

Figure 1. Service Oriented for system integration

Enterprise monitor repository helps by monitoring the front end as well as the back end of the automated enterprise system as shown in figure 3, this frame work helps to balance across one or more application server processes (also called instances) running on one or more machines. Once connected and running, Enterprise component service instances help in monitoring the whole workflow of the logistics enterprise by the scanning process of the goods which takes place at each and every entry and exit process in the industry and they are stored in the database to be reviewed as per user requests; they maintain themselves, their session's state for users, and their database connections. Automated values are derived from each node/sensor at entry and delivery point to know the exact state and they are stored once in 6 hrs any errors or inappropriate data matching is automatically notified by the monitor repository. Notify derivatives process helps the logistic industry to keep track of all the items in the enterprise, this also helps handling and transportation of material efficient, fast, and by definition redundant.

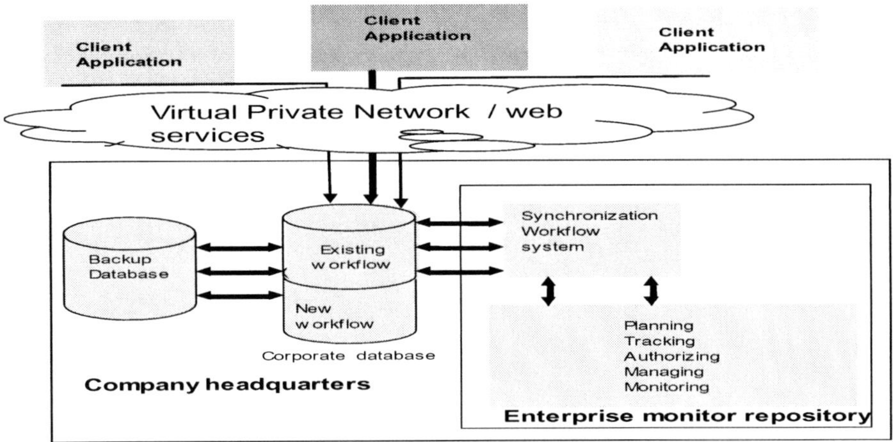

Figure 2. Enterprise monitor repository

5. Service Components Oriented Framework

We represent each agent as a distinct role or department in the enterprise and is capable of providing one or more services. For example, a customer service department provides the services to the customers who are making a booking for the logistic services, a transportation department offers the service of picking up the goods once the payment is made, and the accounting department provides the service of costing and delivering the docket. Agents who require a service from another agent enter into a negotiation for that service to obtain a mutually acceptable time, resources and degree of adjustment for quality implementation of activities. Successful negotiations result in binding agreements between agents. This agent-based approach offers a number of advantages over more typical workflow solutions to this problem. The proactive nature of the agents means services can be scheduled in a just-in-time fashion (rather than pre-specified from the beginning), and the responsive nature of the agents means that service exceptions can be detected and handled in a flexible manner. Project ADEPT [18] tackles similar problem by viewing a business process as a community of negotiating, service providing agents.

5.1. Service agent's internal structure.

Manager Module Controls all the lower layer modules along with other functions like adaptations to change situation, decision on other services requirements and when required initiating negotiations with other gents and merging with other services agents. Networking Module deals with connects to processing automation, connection to information system, communication between agents, reasoning and calculations if and when required.

Intelligence module deals with reasoning like how to combine symbolic and numerical processes, assisting other agents proactively providing information they require, and shielding them from information that is not of interest thus acts both as an information provider and filterer and an information gatherer. Terminology is based on genuine automation ontology and process related variables having references, reasoning based on the input from monitor and sensor devices and taking a decision on the changes in the workflow

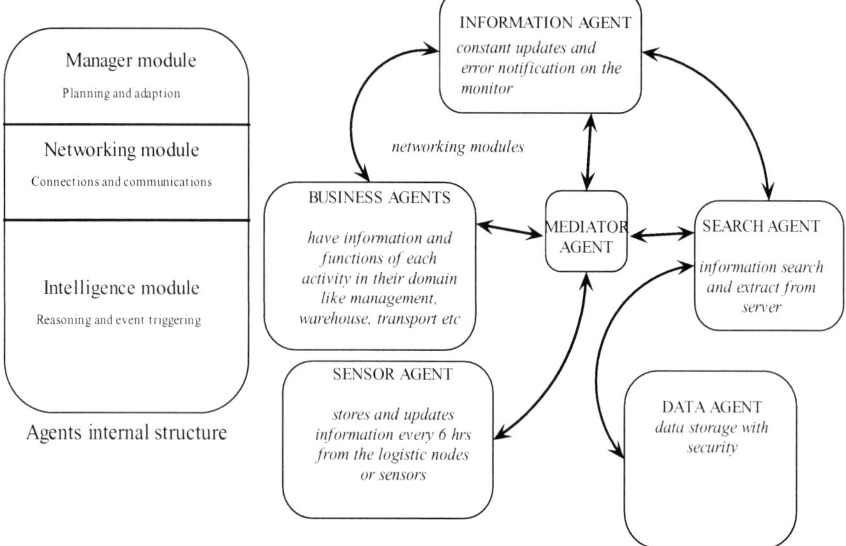

Figure 3. Agents internal structure and their functions

6. Virtual Collaboration via Service Communication

The method we propose is Dynamic Virtual agent Clustering approach, the dynamic virtual agent clustering pattern plays a crucial role in that it embeds the self-organizing properties of components. The main responsibility of this pattern is to configure the enterprise to minimize cost enabling for flexible, re-configurable agent structures. At all levels of the architecture, task propagation occurs by a process of virtual cluster formation. This implementation (which is suitable for manufacturing and logistic industries) places the control into a mediator agent and by this reduces the degree of autonomy of the individual components.

Agents provide a better means of conceptualizing and/or implementing a given application. Here, three important domain characteristics are often cited as a rationale for adopting agent technology [17]:

- data, control, expertise, or resources are inherently distributed,
- the system is naturally regarded as a society of autonomous cooperating components, or
- the system contains legacy components, which must be made to interact with other, possibly new software components.

Dynamic virtual clustering of synergetic partnerships of collaborative organizations aiming to achieve a common goal, finding and accessing an unknown service that is available on the Web, matching of different templates from different sources to better services of their customers, and time management. Such a dynamic Web synergistically glues different dispersed organizations/resources into an added value enterprise capable to accomplish more then the sum of each individual organization/resource could. The basic architecture is based on different component type's behavior as describes in holonic enterprise model [13, 14] like Product Component (PC), Product Model Component (PMC), Resource Component (RC) business component (BC), transport component (TC) and Mediator Component (MC). Product Component will hold information about process status of various goods during logistics operations, time constraint variables, goods status and decision knowledge relating to the order request; it basically acts as a dual for physical component and information component. Here the physical component develops from its state of order forms to an intermediate form like storage in warehouse to delivery of goods as its final stage, its hold all information relating to logistics of the products life cycle. A resource component holds physical and information components, physical component like transportation mode, path of transport, transport preferences (example cold storage) etc and information component like planning and scheduling. For adaptive architecture to work we need a Mediator Component as an intelligent logical mediator to link the orders, goods data and specific organizational resources dynamically. This mediator component can create a dynamic mediator component if a new task needs to be implemented, such as changes in the workflow. When the task is completed this dynamic component is stored in a work-log for further reuse. This PMC identifies the order related resource cluster, (a cluster can be considered as component grouping) and manages task decomposition associated with their clusters. The life cycle of dynamic virtual cluster has four stages 1.Resource grouping 2.Control component creation 3.Execution processing and 4.Termination/worklog [15, 16]. In resource grouping control, we have schedule controller cluster shown in figure 3, a cluster can be considered as component grouping. The controller cluster is associated with one or more physical control like real time operation system and its hardware devices; it functions as a distributed node transparent resource platform for execution of cluster control tasks at the resource level. The dynamic mediator component records and traces local dynamic information of individual components in its association with virtual cluster community, this mediator component can also pass instance information of the partial resource components to some new virtual cluster communities while assigned tasks are being executed. This is a mathematical model approach which will be dealt in detail in following papers.

7. System Implementation

An agent system design will both describe the various different roles of each agent that exist within the system and characterize the relationships that exist between these roles. The building process is broadly classified as 1. Workflow logistics automation Ontology 2. Logistics and General monitoring services (Enterprise services) 3. Ontology of reasoning rules (Enterprise rules), having identified the various agent roles in a system, our next step is to determine how each of these roles can be best realized. Agent architecture needs to be devised or adopted for each role, which will deliver the required functional and non-functional characteristics of the agent's role. Many agent architectures have been developed by the intelligent agent's community, with many different properties [19]. At one extreme, there are 'strong AI' systems, which allow users to build agents as knowledge-based systems, or even as logic theorem proves. In order to build agents using such systems, one goes through the standard knowledge-based system process of knowledge elicitation and representation, coding an agent's behavior in terms of rules, frames, or semantic nets. At the other extreme, there are many agent frameworks that simply offer enhanced versions of (for example) the Java Applications of Intelligent Agents programming Selanguage; they include no AI techniques at all. Neither of these extremes is strictly right or wrong: they both have merits and drawbacks [20]. In general, of course, the simplest solution that will effectively solve a problem is often the best. There is little point in using complex reasoning systems where simple Java-like agents will do. Obviously, more detailed guidelines to assist with this decision making process are desirable.

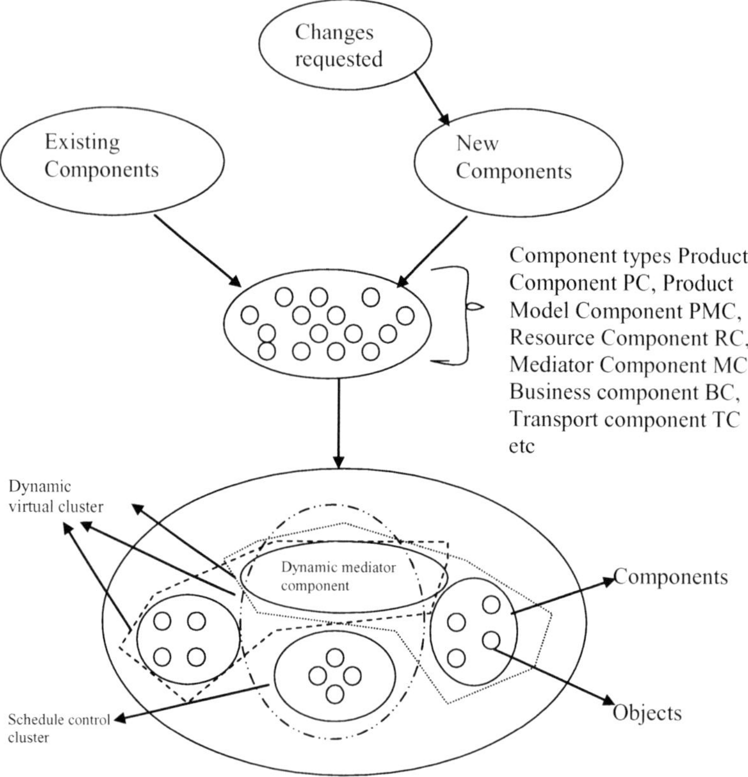

Figure 4. Dynamic virtual agent cluster component framework

8. Conclusion

In this paper, we have discussed logistics e-business development of adaptive workflow systems based on changes made at managerial levels. We have also discussed various approaches for dynamic adaptation of the management workflow system and the frame work for implementation. We propose new models of adaptive systems by the process of dynamic virtual agent clustering; our future research will be to develop a prototype based on this methodology of a working adaptive system.

Reference:

[1] Marshak, R.T.: "Falling in Love with Distinctions", In "New Tools for New Times: The Workflow Paradigm", Future Strategies Inc., 1994

[2] Miers, D: "The Workware Evaluation Framework", Enix Limited, 1996

[3] Chang, E: Requirement Specification of Logistic Manager for Software Engineering Project, Department of Computer Science and Software Engineering, The University of Newcastle, 2000.

[4] Haake, J.M., Wang, W.: "Flexible Support for Business Processes: Extending Cooperative Hypermedia with Process Support".

[5] Denning, P.J.: "The fifteen level", In Proceedings of ACM SIGMETRIC Conference on Measurement & Modeling of Computer Systems, May 1994.

[6] Sheth A.: "State-of-the-art and future directions", In Proceedings of the NSF Workshop on Workflow and Process Automation in Information Systems, May 1996.

[7] David Neumann, An Introduction to WebObjects. Retrieved: July 30, 2004, from http://.mactech.com/articles/mactech/Vol.13/13.05/WebObjectsOverview.

[8] Pudhota L, Chang E. et al. International Journal, Computer Science, System and Engineering, "Extension of Activity Diagrams for Flexible Business Workflow Modeling "volume 18 no3 May 2003, UK.

[9] Pudhota L, Chang E "collaborative workflow management for logistics consortium" ICEIS April 2004 Porto, Portugal.

[10] Pudhota L, Chang E "Modelling the Dynamic Relationships between Workflow Components" ICEIS April 2004 Porto, Portugal.

[11] Pudhota L, Chang E, Venable J "E- Business technology adaptation through workflow mining" MSV June 2004 Las Vegas, Nevada, USA.

[12] http://www.omg.org/mda/mda_files/UNextMDA4.pdf

[13] lieru. M, Robert W. Brennan Scott S. "The holonic enterprise: a model for Internet-enabled global manufacturing supply chain and workflow management", Canada 2000.

[14] Ulieru. M, Stefanoiu. D, et al. "Holonic metamorphic architecture for manufacturing" University of Calgary, Calgary, Canada 2000.

[15] Christensen, J.H. (1994), "Holonic Manufacturing Systems: Initial Architecture Standards Directions", Proceedings of the First European conferenceManufacturing systems, European HMS Consortium, Hanover, Germany.

[16] Available at http://www.journee.com/n_hl_020703b.html, retrieved on 10 Jan 04

[17] Hala, S.M., workflow concepts and techniques, Available: Http://www.loria.fr/~skaf/cours/workflow/workflow1/sld001.htm], retrieved on 21 Feb. 04

[18] N. R. Jennings, P. Faratin, M. J. Johnson, T. J. Norman, P. O'Brien, and M. E. Wiegand. *Agent-based business process management*. International Journal of Cooperative Information Systems, 5(2&3):105--130, 1996.

[19] Wooldridge, M. J., and Jennings, N. R. 1995. Intelligent agents: Theory and practice. Knowledge Engineering Review 10(2). http://citeseer.ist.psu.edu/article/wooldridge95intelligent.html

[20] N. R. Jennings and M. Wooldridge available at http://agents.umbc.edu/introduction/jennings98.pdf Applications of Intelligent Agents retrieved on 12 Sept 05

[21] Genesereth, M. R. and Ketchpel, S. P. (1994). Software agents. *Communications of the ACM*, 37(7):48-53.

[22] Shoham, Y. (1993). Agent-oriented programming. *Artificial Intelligence*, 60(1):51-92.

Self-Organization and Autonomic Informatics (I)
H. Czap et al. (Eds.)
IOS Press, 2005

An Agent-Based P2P System to Distributed Knowledge Management

Jiuyang TANG, Weiming ZHANG, Weidong XIAO and Daquan TANG
School of Information System & Management, National University of Defense Technology, Changsha, 410073, Hunan, China
jiuyang_tang@hotmail.com

Abstract. The problem of distributed knowledge management (DKM) is eminent in the industry enterprises area and is receiving a lot of attention in the research community. Peer-to-Peer (P2P) has shown to be the suitable distributed computing paradigm to the high dynamism knowledge management environment. However, there are several technical challenges in building scalable P2P based DKM systems. One of them is the intelligent coordination, which is complicated by the deficiency of P2P in exchanging complex information and coping with semantic heterogeneity. To address this problem, an agent-based P2P architecture is proposed. Each agent-based peer (agent peer) creates and manages its own knowledge. Moreover, agent peers intelligently collaborate with each other to deal with knowledge management actions over the P2P network.

Keywords. agent, P2P, distributed knowledge management

1. Introduction

In current knowledge-based economy, the competitiveness of enterprises lies in the effective management and sharing of knowledge both within and across several cooperating enterprises. Knowledge Management (KM) refers to the process of creating, codifying and disseminating knowledge within complex organizations [1,2]. KM has been a predominant trend in enterprises and attracted a lot of attention and investments during several years.

Traditional KM systems use a centralized approach, which leads to the construction of one or a few knowledge repositories, organized around a global ontology or other meta-structures [3]. However, it is impossible to centralize knowledge at a global scale, which needs a long development phase and is difficult to maintain. Recent applications are seeing the emerging need to support KM in heterogeneous, distributed and dynamic environments. To tackle this issue, some researchers turn to distributed knowledge management (DKM): the exchange and integration of knowledge across local, autonomous KM systems or communities, without the requirement of neither centralized repositories nor shared ontologies and other categorization structures [4].

To support the DKM approach, we develop an agent-based P2P architecture. The architecture integrates and extends concepts taken from the P2P and agent communities, where each agent peer takes in charge of its own knowledge and interacts with others to deal with knowledge management actions over the P2P network.

The rest of this paper is organized as follows: first, we briefly examine technical background in DKM and argue why agent-based P2P system (APS) provides a suitable infrastructure for DKM; second, we describe the main components of APS and elaborate the knowledge discovery procedure; then we give the related work; finally, we draw some conclusions.

2. Background

Recently, P2P computing [5] has emerged as a new distributed computing paradigm of sharing resources available on the Internet. A P2P system is a fully distributed cooperative network, which offers extensive resource sharing, scalability, self-organization and fault-tolerance. Emerging P2P concepts prove to be quite appropriate and suitable to the increasingly decentralized nature of modern enterprises and their dynamic networks. The application of the P2P technology within the DKM context makes it possible for different organizations and individuals to maintain different views of the world while exchanging information.

Nevertheless, today's P2P solutions rely on keyword search and rather simple knowledge representation techniques, which make it unable to exchange complex information and to cope with information heterogeneity problems.

Agent approaches have already been successfully employed in AI area. Agents are flexible entities and are capable to adapt themselves to new environments. Agents also can discover needed resources and negotiate with other agents. Furthermore, agents work together, use the distributed resources and work as a team to solve problems. Because of such characteristics, some researchers claim that the possible solution to the current drawbacks of the P2P approach is to use agent technology [6,7].

It's reasonable to use agents to represent peers and implement different P2P services, so that the peers can behave intelligently. Agents are better suited to capture reasoning and collaboration among distributed components. The combination of agent and P2P is highly innovative solution to DKM with prospective benefits to maintain its own knowledge structure while intelligently exchanging complex information in a distributed network.

3. Agent-Based P2P Architecture

The APS consists of an open-ended network of distributed agent peers. Every participating peer is required to make its knowledge available to other peers, and to search relevant knowledge from other agent peers.

Figure 1 shows the general architecture of a peer. Descriptions of the individual components are as follows.

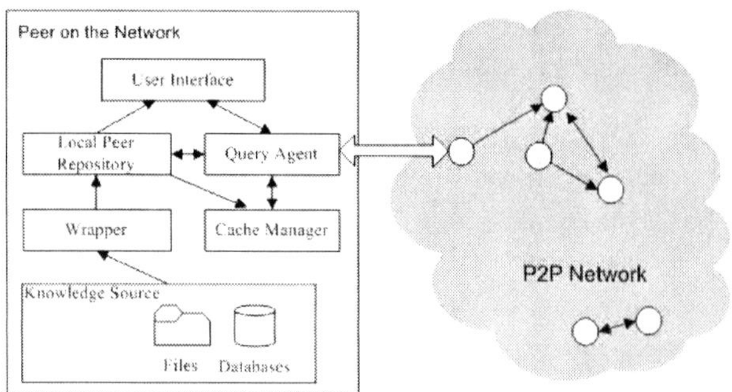

Figure 1. The Architecture of An Agent Peer

• Knowledge Source: the place where the information of a peer is stored. In general, it can be files, or relational knowledge.

• Wrapper: which takes the responsibility of wrapping the underlying knowledge. It is a component that extracts ontology-like structures from local knowledge source according to user's requirement.

• Local Peer Repository (LPR): which is materialized through wrapper by semantics-based information to be shared with other peers. It is the real resource for the external information searching and internal user browsing and navigating. The information in a LPR are semantically annotated.

• User Interface: which provides individual views on the information available in LPR. The views can be implemented using different visualization techniques (topic hierarchies, thematic maps, etc). Through the user interface, users can construct and submit queries by clicking in the views graph or entering manually as text.

• Query Agent: which is responsible for query processing and answering. When receiving requests, it passes them onto the peers it knows about, while at the same time performs the search on its own LPR. The query agent communicate both with other peers on the network and other applications running on the local peer. Moreover, they are designed and built to match any search query to the resources. The results are passed back to the originator's query agent.

• Cache Management: which supports caching of answers returned from agent peers, which helps to reduce the response time for subsequent answers, forward the query to other peers, or reuse corresponding stored relationships.

To discover resources in the APS, user issues a query on a peer. A query agent will be created with the following operations performed:

(1) This peer is set as the "creator peer" for that particular query agent. The query agent first parses the query, and updates itself with information from LPR. By the way, it takes the responsibility of wrapping the semantics-based knowledge, in other words, makes possible the communication between agents and non-agent applications. Agent Communication Languages (ACLs) are used as wrapper languages because ACLs implement a knowledge-level communication protocol that is unaware of the choice of content language and ontology specification mechanism.

(2) Then the address of creator peer is set as the query agent's home address and TTL (Time-to-live) for the query agent is set.

(3) Initially the query agent is given the addresses of some other peers participating in the network, either from cache or from learning.
(4) The query agent clones itself enough times to allow a mobile agent to be sent to each of these peers.
(5) After arriving at each peer the mobile agent decrements its TTL, and tries to translate query in the schema of creator peer into a corresponding query in the schema of local peer, which depends on the capabilities of query agent to do ontology-based schema mapping. If matching successes, the results and corresponding mapping returns to the creator peer, and are stored in Cache Management.
(6) If the mobile agent's TTL has expired, then it returns to the creator peer. Otherwise the mobile agent clones itself enough times to allow a clone of itself to be sent to each peer that was known to the current peer before the mobile agent arrived.

4. Related work

There has been a lot of research work done in the area of DKM.

The InfoQuilt [8] system provides a platform for creating, exchanging and sharing of knowledge. Under the new concept a P2P Semantic Web (PSW), it supports knowledge discovery and learning from a multitude of diverse autonomous distributed resources. Like Napster [9], the P2P network in InfoQuilt requires every peer to register itself to a directory or directories, where provide the necessary contact information to connect to a specific peer. To get semantically relevant information and services, users perform queries by way of searching the directory.

Edamok [10] is a research project focused on P2P infrastructures to implement DKM applications. It aims to realize knowledge sharing among peer communities of interest. In its P2P architecture, each peer (K-peer) provides all the services needed to create and organize local knowledge. In order to cope with knowledge discovery and search, Ctx-Match algorithm is used to establish semantic mapping between concepts among autonomous K-peers to deal with semantic coordination.

SWAP [11] combines Semantic Web and P2P to facilitate efficient knowledge sharing and finding. The knowledge of a particular peer is extracted from several knowledge sources and stored in the Local Node Repository. Knowledge discovery is achieved by rewriting user queries and selecting the peers that are likely to know the answer based on semantic topology.

SEWASIE [12] is a mediator-based P2P system that supports search for information through a flexible integration of semantic information. Each peer creates and maintains its own personalized knowledge base and no centralized global information system is required. It introduces query agents and brokering agents, while the former take charge of query processing and answering, and the later maintain a view of knowledge handled by the network. Each brokering agent directly controls at least one peer and handles the weaker semantics mappings among distinct peers at a wider level. To some certain, SEWASIE adopts a hybrid P2P architecture while our proposed APS is a pure P2P system.

5. Conclusion

In this paper, we focused on DKM in agent-based P2P infrastructure. The proposed APS is a loosely coupled network of agent peers that cooperate to facilitate effective knowledge management and sharing that may be beyond the individual capacities or knowledge of each particular agent peer. APS combines two highly successful technologies, viz. agent and Peer-to-peer computing. Agent technology is employed to tackle complex knowledge representation and semantic reasoning while P2P brings to distributed computing paradigm.

While there are several areas of the work presented here that require further investigation, we believe that agent-based P2P approach do provide a viable means to DKM.

Acknowledgement

This work is supported by the National Natural Science Foundation of China (Grant No.60172012); the Hunan Provincial Natural Science Foundation of China (Grant No. 03JJY3110).

References

[1] W. Clyde, Holsapple, editor, Handbook on Knowledge Management, Springer, international handbooks on information systems edition, 2003.
[2] M. Bonifacio, P Bouquet, and R. Cuel, Knowledge Nodes: the Building Blocks of a Distributed Approach to Knowledge Management, Journal of Universal Computer Science 8(6):652–661, 2002.
[3] D. Bertolini, P. Busetta, A. Molani, M. Nori, and A Perini, Peer-to-Peer, Agents, and Knowledge Management: a Design Framework, Technical Report tr0207, Istituto Trentino di Cultura, 2002.
[4] M. Bonifacio, P. Bouquet, and P. Traverso, Enabling Distributed Knowledge Management: Managerial and Technological Implications, Novatica and Informatik/Informatique, III(1), 2002.
[5] D. S. Milojicic, V. Kalogeraki, R. Lukose, K. Nagaraja, J. Pruyne, B. Richard, S. Rollins, and Z. Xu. Peer-to-Peer Computing, Technical Report HPL2002-57, HP Technical Report, 2002.
[6] H. Homayounfar, F. Wang, and S. Areibi, Advanced P2P Architecture Using Autonomous Agents, IN: 15th International Conference on Computer Applications in Industry and Engineering, San Diego, California, 2002, 115-118.
[7] G. Moro, A. M. Ouksel, and Sartori C, Agents and Peer-to-Peer Computing: A Promising Combination of Paradigms, IN: AP2PC 2002, 1-14.
[8] M. Arumugam, A. Sheth, and I. B. Arpinar, The Peer-to-Peer Semantic Web: A Distribuited Environment for Sharing Semantic Knowledge on the Web, Workshop on Real World RDF and Semantic Web Applications (WWW'2002), Honolulu, Hawaii (USA), 2002.
[9] http://www.napster.com
[10] M. Bonifacio, P. Bouquet, G. Mameli, and M. Nori, Peer-Mediated Distributed Knowledge Management, AMKM 2003, 31-47.
[11] M. Ehrig, C. Tempich, J. Broekstra, F. V. Harmelen, M. Sabou, R. Siebes, S. Staab, and H. Stuckenschmidt, SWAP: Ontology-based Knowledge Management with Peer-to-Peer Technology, WOW 2003, 17-20.
[12] D. Beneventano, S. Bergamaschi, A. Fergnani, F. Guerra, M. Vincini, and D. Montanari, A Peer-To-Peer Agent-Based Semantic Search Engine, SEBD 2003, 367-378.

Part V

Autonomic Communications

Self-Organization and Autonomic Informatics (I)
H. Czap et al. (Eds.)
IOS Press, 2005

Intelligent Traffic Control on Internet-Like Topologies

DATRA - Dynamically Adjusted Traffic Rate Alterations

Antonia Katzouraki [a,1], Philippe De Wilde [a,2] and Robert Ghanea Hercock [b,3]
[a] *Imperial College London and Heriot Watt University*
[b] *BTexact Future Technologies*

Abstract. *The growing scale and importance of technological networks throughout the world has highlighted the devastating consequences of catastrophic network failures. In this paper, we address this crucial issue through a detailed analysis of network traffic distribution across network nodes, with the aim of developing an intelligent traffic control model. Specifically, we develop and demonstrate the Dynamically Adjusted Traffic Rates model, which aims to fairly distribute traffic amongst network nodes according to their network characteristics. Our model is independent of topology and is based on dynamically adjusted traffic rates and properties similar to those observed in real Internet traffic. In this paper our model is numerically analyzed over a variety of network topologies to display its chaotic features leading to self-similarity. The model is inspired by biological evidence.*

Keywords. Dynamic Adjustments, Networks, ODEs, Intelligent Traffic Control, Equilibrium

1. Introduction

Catastrophic network failures are one of the major problems arising from resource overloading and the potential propagation of network deficiencies. In this work, we attempt to prevent such catastrophic failures by strategically controlling the distribution of traffic to the available network resources according to their observed characteristics.

Cascading failures of networks have proved to result in devastating consequences on everyday human life. A characteristic example is that of the 10th August 1996, when

[1]Department of Electrical and Electronic Engineering, Imperial College London, London, SW7 2AZ, UK
E-mail: antonia.katzouraki@imperial.ac.uk
[2]Department of Computer Science, Heriot Watt University, Edinburgh, EH14 4AS, UK
E-mail: pdw@macs.hw.ac.uk
[3]BTexact Future Technologies Group Adastral Park, Martlesham Heath, Ipswich, IP5 3RE, UK
E-mail: robert.ghanea-hercock@bt.com

a voltage-based failure occurred in two power lines. Major power disturbances were propagated through a series of *cascading failures*, ultimately leading to blackouts in 11 US states and 2 Canadian provinces, leaving 7 million people without power for up to 16 hours [1]. Similarly, on the 4th May 2000, the VBS/Lovelet-A virus (also known as the Love Bug or ILOVEYOU worm) caused considerable damage as it infected computers world-wide. The virus fooled computer users into thinking they had received a love letter in their email. However, when the attached file was opened, the virus would forward itself to other email addresses found on the infected computer [2]. Such events demonstrate the importance and necessity of intelligent traffic control, in mitigating the effects of these failures.

In this work, we develop a biologically inspired deterministic model, through which we aim at controlling the distribution of traffic to the available network resources with the goal of fairly re-distributing traffic and preventing node overloading. In our model traffic rates are dynamically adjusted according not only to the link costs but also to each node's buffer capacity, so that all network nodes will behave in a synchronized manner to achieve *equilibrium* (provided that the exogenous parameters allow equilibrium to be reached). This strategy not only distributes the traffic load amongst network nodes, but also synchronizes the traffic and link cost changes with buffer and topology variations.

In contrast to previous approaches (e.g. [3,4,5,6]) our model is *independent* of topology, and focuses on controlling traffic and cost rate dynamics rather than improving the network's topology. The majority of the previous research studies the influence of network structure on the dynamics of the system, whereas through our model we aim to combine *numerical analysis* with observed *empirical studies* on real networks applied on diverse Internet like topologies.

In an approach different to that commonly used in research studies on multi-agent systems, we provide cost strategies designed to achieve competitiveness by utilizing non-linear functions (e.g. [7,8,9,10]). In this way our model controls traffic and cost rate dynamics to avoid overloading of nodes and deliver global load balancing.

2. Previous Work

Recently there has been much interest in studying real-world networks and using them to investigate various phenomena, such as network resilience in random or deliberate attacks and various spreading. Real-world networks can be characterized accurately by Poisson-distributed, Scale-Free, Watts-Strogatz graphs as well as other more recently invented graphs [11,12,13,14,15]. E–R graphs are the oldest type of graphs used to represent real-world networks. Whilst E–R graphs do not possess important properties of real-world networks [16,17], they can be used in order to investigate various phenomena, such as network resilience in random failures, deliberate attacks and virus spreading.

Recently scientists have investigated the network resource allocation problem to deliver more reliable networks by applying market-based control mechanisms on multi-agent systems. According to their architectures agents represent various network resources to coordinate their resource allocation decisions without assuming *a priori* cooperation. The design of the agents' trading behavior utilizes either simple linear functions or concepts from game theory [7,8,9,10].

In other research, multi-scale networks were designed to construct a framework for exploring organizational robustness with respect to a range of environmental stresses.

As a result, they produced a class of networks to reduce the likelihood that individual nodes will suffer congestion-related failure and ensure that the network as a whole will disintegrate when failures do occur [6].

Another hierarchical design of interest is the Hypergrid topology, which also seems to offer considerable advantages in respect of robustness. It appears to have the best connectivity robustness of all recent topologies. The absence of just a few highly connected nodes (homogeneous degree) makes the topology resilient enough to intentional attacks without propagating propagating deficiencies of individual nodes [3].

It is only recently that scientists have started to study the behavior of network resilience using models defined on random graphs and ideas from percolation theory. To address the resilience of the network, scientists have studied the consistency of the *giant component* [12,15] on different types of topologies.

One of the first approaches to study node congestion is Benohamed's control algorithm, which utilizes a non linear system, whilst applying a saturation function to impose bounds to traffic rate. It describes network load that consists of traffic corresponding to all source-destination pair – connection – flows. They study a single congested node, where the congestion is due to saturation of one of its outgoing links [18].

An alternative approach of studying node congestion control utilizes deterministic modeling which is based on the analytical study of Ordinary Differential Equations (ODE). Frank Kelly provided a flow control model and described how the flows operate as a result of the fine operation of the TCP-software at the packet-level [19] and how his model is extended utilizing concepts from game theory, in order to provide Quality of Service [20,21].

Additionally, TCP-Vegas is congestion control algorithm, based on a dual method, utilizing a discrete-time non linear system with a continuous update function. The main concept of this method is to treat source rates as primal variables and congestion measures as dual variables looking for the global optimization [22].

In this paper, we propose a source control algorithm to adjust the sources' rates in order to fairly re-distribute traffic and prevent node overloading and thereby nodes' congestion. Various network topologies have been simulated by a numeric solution of the coupled ODEs to prove that the system always reaches *equilibrium* and even when the total demand is larger than the network capacity the input source's rates continue to adjust to avoid the overloading whereas the packets continue circulating across the network reaching *stable limit cycle*. Therefore, our source algorithm dynamically adjusts source rates bases on both topology's and sources' characteristics. It monitors the nodes/links capacities via the link cost and traffic equations, which adjust their rates automatically when the buffer tends to being full.

3. Model Formulation

We implement dynamic traffic distribution by assigning costs to the links connecting network nodes. The link cost will depend on the amount of packets inside the node. The load of each link depends on the cost. That is, *the higher the cost, the fewer packets are sent through the link*. Our goal is that all packets will receive the least possible congested service through the *DATRA* model depending on network conditions. On the contrary, packets willing to use an almost congested link would face a *high* cost. Thus, assuming that packets choose the least congested path (denoted by the lower cost of the specific link), we can prevent pathological conditions, such as congested links or failing nodes by intelligent, adaptive cost management and load balancing on the links.

Besides the obvious benefits (e.g. avoiding overloaded nodes) of smart cost management, our model can offer other less obvious advantages such as Quality of Service (QoS). According to the model the packets are selected to be sent to an almost congested node only when the packet is willing to pay the high link-cost. Our model identifies where there is link capacity, but does not specify which individual packets can use the available link capacity. Each packet's header has a flag where the type of service it needs to receive is indicated. In other words, this flag indicates the packet cost/ quality it requires. Enabling packets to choose their link according to their flag indication results in QoS.

Here we outline our model, the parameters that it is dependent on, the concept it is based on and the scope of its applications.

Consider a directed graph with bidirectional links and a set of nodes. A graph with *n* nodes is completely specified by an $n \times n$ *matrix*, which is called *adjacency matrix* $A = [a_{ij}]$ consisting of zeros and ones. Each entry $a_{ij} = 1$ represents a connection between the nodes i and j, whereas an entry such as $a_{ij} = 0$ implies that there is no connection between the nodes i and j.

Consider single nodes *i=0,1,2,3,...,n* whose number of packets at time *t* is denoted by $x_i(t)$. To keep the notation simple we denote $x_i(t)$ by x_i where $x_i \geq 0$, since the number of queued packets at any given node in the network is at least zero. The amount of packets at each node is specified by an $1 \times n$ matrix, $X = [x_{1j}]$. Finally, the initial number of packets at each node must be larger than its service rate ($x_i[0] > s_i$).

Link cost from node i to node j is specified by an $n \times n$ matrix, $C = [c_{ji}]$. It partly controls the rate and the way in which packets flow through the network.

The arrival rates are specified by an $n \times n$ matrix, $\Lambda = [\lambda_{ji}]$, which represents the fraction of traffic relayed to each node. It is the routing probability that the traffic flows from the same node through different links. We assume that the input of each node is described in terms of the outputs from the nodes that are connected to it. The aforementioned routing probability not only depends on the link cost, but also on the input and output of this link. In order to calculate this routing probability, it is first assigned values which are produced by the formula:

$$\Lambda = \frac{1}{C} \tag{1}$$

Then the routing probabilities are normalized and thereby their sum becomes equal with unity. Finally, those routing probabilities are multiplied with the amount of traffic that leaves the node producing the fraction of traffic that traverses each link, respectively.

The service rates are specified by an $1 \times n$ matrix, $S = [s_{1j}]$ and depend on the nodes' processing power. The change of the service rate depends on both the routing probability (λ_{ij}) and the link cost (c_{ij}).

The buffer spaces for each node *i* are represented by an $1 \times n$ matrix, $B = [B_{1i}]$. B_{1i} controls how many packets are injected at each node in the network and at which point Hysteresis (immediate change) [23] takes place.

At node i, scheduling and queue management mechanisms are responsible for the packet processing and forwarding. Therefore, the loss caused can be avoided by preventing packets from entering that node. In the event that a network node overloads or fails, the network paths can be changed to compensate utilizing the DATRA model. As a result there are no fixed paths through the network.

3.1. Dynamic Traffic Rate Alterations

A major problem is the competitive demands that applications make for network resources, such as link bandwidth and buffer spaces in routers or switches [9,7,8,10]. Related competitive interactions have also been studied in the context of *biological* population models – A. Lotka (1925) and Voltera (1926), the dimensionless system form of which we partially utilize in this research.

We assume that once a node's almost full buffer has emptied, the traffic rate then increases exponentially. This assumption plays a vital role in the *burstiness* of the network traffic. For instance, packets from NNTP, FTPDATA and WWW connections are clustered in bursts. These large bursts are likely to completely dominate traffic dynamics whilst their inter-arrival times are regulated by network factors, such as shortage of essential networking supplies [24]. We express resource competitiveness utilizing non-linear functions. These incoming traffic alterations are illustrated by the first part of Eq.(2), which is called the Logistic Differential Equation and which is commonly used in population models, where the population growth rate decreases with increasing population due to various factors such as limited food supply, overcrowding and disease [25,26]. Similarly, users and applications compete for limited network resources.

Conversely, when the buffer approaches being full, the traffic rate is taken to be zero otherwise the node will become congested. The node has to employ internal mechanisms to avoid congestion, otherwise the node will start dropping packets and the buffer will overflow. In order to avoid the frequent use of those mechanisms, the arrival rate needs to be reduced to zero value at the appropriate times.

Traffic alterations at each node are modeled by nonlinear dynamic equations:

$$x_i' = \sum_j \lambda_{ij} \cdot x_j \cdot s_j \cdot \left(1 - \frac{x_j}{B_j}\right) - s_i \cdot \left(1 - \frac{x_i}{B_i}\right) \cdot x_i \qquad (2)$$

Particularly, the terms inside the brackets in the above equation decrease the traffic rate for large values of traffic, whilst the summed term represents the arrival rate of traffic from node j to the node i.

3.2. Dynamic Link Cost Alterations

Each link cost alteration is modeled by nonlinear dynamic equations, in order to deliver the least congested service possible. The link cost strategy we apply provides a robust way to distribute the traffic amongst the nodes. In this way we prevent node overloading by dynamically adjusting the traffic and link cost rates. The link-cost to each node depends on its traffic.

Specifically, we assign a cost, c_{ij} to each link between the nodes i and j, which depends on the queue length x_i at the node j. The link cost increases rapidly with the output queue length and decreases rapidly with the input queue length, so that the packets are held back when faced with a high link cost. Through the use of these *dynamically adjusted traffic rate alterations* we ensure that all packets travel through the least congested route and the packets should be forced to pay a very high cost if they want to use an almost congested link. Consequently, the link-cost function not only avoids an overload in congested nodes but it also reduces the load of the relatively overloaded nodes:

$$c'_{ij} = \lambda_{ij} \cdot s_j \cdot \left[\frac{x_i^2}{1+x_i^2}\right] - \lambda_{ij} \cdot s_i \cdot \left[\frac{x_j^2}{1+x_j^2}\right] \tag{3}$$

The term inside the squared brackets controls the flow of traffic. During busy periods it allows the node to relay as many packets as possible, whereas during non-busy periods it allows the node to relay the number of packets that are currently available.

3.3. DATRA Model on Multiple Nodes

By combining the non linear Ordinary Differential Equations (ODEs) of the traffic alterations coupled with the ODEs of the link cost alterations we achieve the following set of ODEs:

$$\begin{aligned} c'_{ij} &= \lambda_{ij} \cdot s_j \cdot \frac{x_i^2}{1+x_i^2} - \lambda_{ij} \cdot s_i \cdot \frac{x_j^2}{1+x_j^2} \\ x'_i &= \sum_j \lambda_{ij} \cdot x_j \cdot s_j \cdot \left(1 - \frac{x_j}{B_j}\right) - s_i \cdot x_i \left(1 - \frac{x_i}{B_i}\right) \end{aligned} \tag{4}$$

Over the next sections we observe, using two case studies, that our model's results include extreme sensitivity to the initial conditions of the model, creating chaos, which generates self-similarity along with odd periodicity.

4. Case Studies on Multiple Nodes

In this section we study the performance of our model through a series of simulations on multiple nodes forming diverse topologies. Solving the resulting nonlinear system of ODEs can be addressed through the use of numerical procedures especially developed for solving ODEs. In the next section we utilize the classic R-K method of order four applied on a 64-bit computer to produce more accurate results. Even though our model has been tested for over 1000 nodes, for simplicity here we will illustrate our results through two case studies: the first based on a 3-node with 2-degree network; and the second based on a 20-node with 3-degree network.

4.1. Case study 1: Directed Graph of 3 Nodes With Degree 2

Consider a network of 3 nodes with degree 2

X = (250, 50, 100)
S = (10, 10, 10)
B = (450, 450, 450)

We assume that maximum number of packets to each node equal with their buffer sizes, so that nodes do not overflow. If a node overflows, the program deactivates the node and its links. The deactivation of a node withholds the amount of packets it contains just before its deactivation. This phenomenon can be interpreted as a *dropping mechanism*, in order to reduce the traffic in the network.

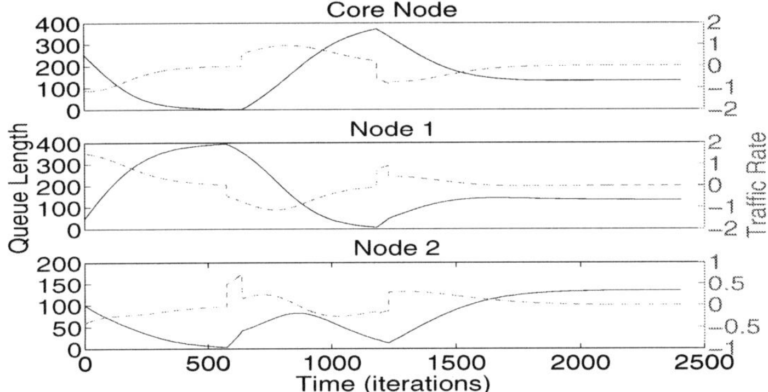

Figure 1. Solid line: Traffic alterations, Dashed line: Traffic rate alterations.
The traffic rates reversals occur when the buffer approaches towards being full or idle, at which points it jumps to a lower or higher value respectively. This phenomenon is well known in chaos theory as *Hysteresis*

Figure 1. highlights the relationship between the traffic levels in each node and their rate alterations. Each subgraph within Figure 1. consists of two phases –that of the *topping rate* and that of the *sinking rate*, reflecting a two reversal pattern. Additionally, both the total traffic and the rates oscillate rhythmically between upper and lower limits, which are determined by the buffer capacity of the node. The traffic rates reversals occur when the buffer approaches towards being full or idle, at which point it jumps to a lower or higher value respectively. This phenomenon is well known in chaos theory as *Hysteresis*, which is the immediate jump that reduces the load of an overloaded node. After or before Hysteresis both the traffic and the rate vary exponentially/logarithmically.

Figures 2. and 3. plot the phase portrait for traffic and its rates over time (iterations). Think of the *trapezoid-shaped spiral* as having four distinct phases: (i) a slow topping up traffic phase; (ii) an immediate change of the traffic rate; (iii) a slow sinking traffic phase; and (iv) an immediate change of traffic rate.

Note that as time passes the trapezoid-shaped spirals are repeated in identical fashion other than the fact that their size becomes slightly smaller each time. This *self-similarity* is a reflection of the fact that our model is attempting to achieve dynamical balance between queue lengths and traffic rates among nodes and hence the trapezoid-shaped spirals tend towards the non-zero equilibrium. The self-similarity property applies across all nodes simultaneously so that they behave in a synchronized manner to achieve the equilibrium. Self-similarity is a well-known and established property of Internet traffic in the networking and modeling community [27,28].

In Figure 4., it appears that the nodes have reached equilibrium after roughly 2,000 iterations. However, after closer investigation we observe that the link cost function continues to vary even after 2,000 iterations. Note that the magnitude of the oscillations steadily decreases as we approach the steady-state equilibrium.

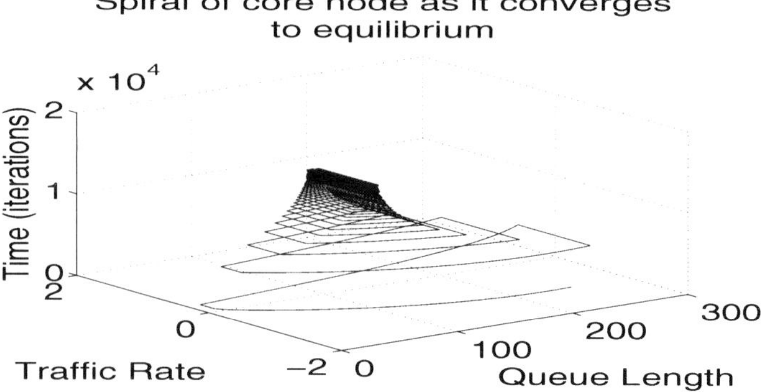

Figure 2. The *trapezoid-shaped spiral* has four distinct phases: (i) a slow topping up traffic phase; (ii) an immediate change (Hysteresis) of the traffic rate; (iii) a slow sinking traffic phase; and (iv) an immediate change (Hysteresis) of traffic rate.

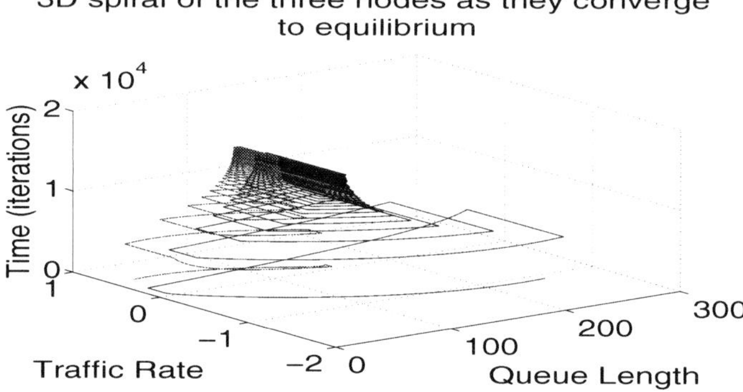

Figure 3. Note that as time passes the trapezoid-shaped spirals are repeated in identical fashion, other than the fact that their size becomes slightly smaller each time. This self-similarity of traffic applies across all nodes simultaneously so that they behave in a synchronized manner to achieve the equilibrium. Self-similarity is a well-known and established property of Internet traffic in the networking and modeling community.

Recall that the number of packets that each node sends to other nodes affects the queue length which influences the cost function. The cost function affects the traffic rate which describes the number of packets that each node receives. Given that the *traffic rate* is inversely proportional to the link cost, the *change in the traffic rate* is negatively proportional to the link cost and hence the graph of Sparky DCost (change of the link cost) in Figure 4. is a mirror image of the graph of the change in the traffic rate. Such behavior as the one described by this mirror image graph of the change in the traffic rate has been observed in modeling studies of Internet backbone traffic where the cumulative number of packet arrivals appears to closely follow a piecewise-linear function [29]. The spikes in Figure 4. highlight the points at which the rate immediately changes (Hysteresis), whilst the height and direction of each spike represents the relative magnitude of change in the slope. Finally, we have identified the Sparky DCost pattern through graph-

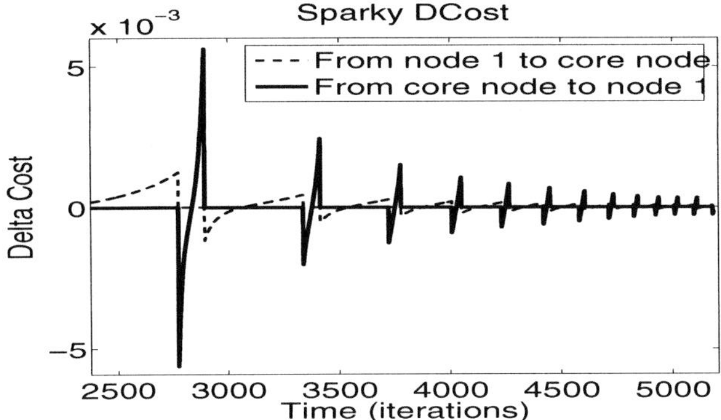

Figure 4. Sparky cost change over time. This mirror image graph of the change in the traffic rate has been observed in modeling studies of Internet backbone traffic where the cumulative number of packet arrivals appears to closely follow a piecewise-linear function [29]. The spikes highlight the points at which the rate immediately changes (Hysteresis), whilst the height and direction of each spike represents the relative magnitude of change in the slope.

ing the change in cost, since the spikes are not distinguishable from simply graphing the cost (due to the large scale involved).

4.2. Case study 2: Directed Graph of 20 Nodes With Degree 3

The conclusions from our first case study hold for large number of nodes and large degree. Through both case studies we aim to show that the DATRA model is independent of topology. Consider a homogeneous tree topology of 20 nodes with degree 3.

$X = (250, 30, ..., 50)$
$S = (10, 10, ..., 10)$
$B= (1450, 1450, ..., 1450)$

Unlike the first case study where there were enough links to ensure that all nodes could potentially be allocated with packets at the first iteration, in this case study there are only enough links to allow three first neighbor nodes to be allocated with packets at the first iterations. Consequently, the traffic at the three first neighbor nodes increase exponentially over time, whereas the traffic at the more distant nodes from the core node oscillate at lower magnitude since much of the distribution of packets has already taken place at earlier node levels. This is illustrated in Figure 5.

The first graph in Figure 5. illustrates the traffic and its rate alterations at the core node, whereas the second graph illustrates the traffic and its rate alterations at the 17th node. The curves in this case study exhibit similar phenomenon as the previous case study. Any differences are simply attributable to the different input, such as network dynamics.

Figures 6 and 7 plot the phase portrait for traffic and its rates over time (iterations). Think of a spiral consisting of trapeziums that are repeated in an identical fashion other than the fact their size becomes slightly smaller or larger each time step. The differences in their sizes are attributable to the different input, such as network dynamics. This self-similarity is a reflection of the fact that our model takes into consideration network fac-

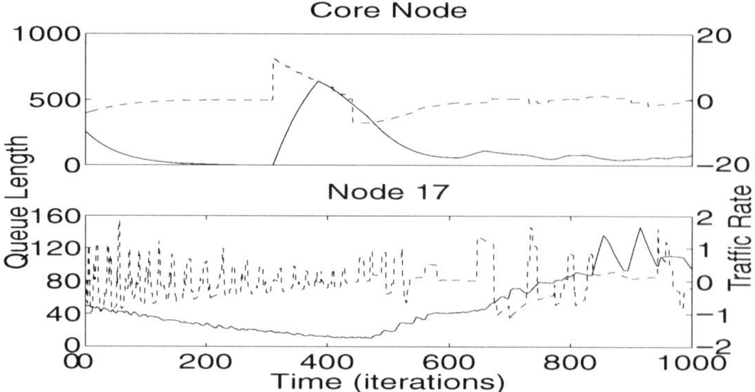

Figure 5. Solid line: Traffic alterations Dashed: Traffic rate alterations

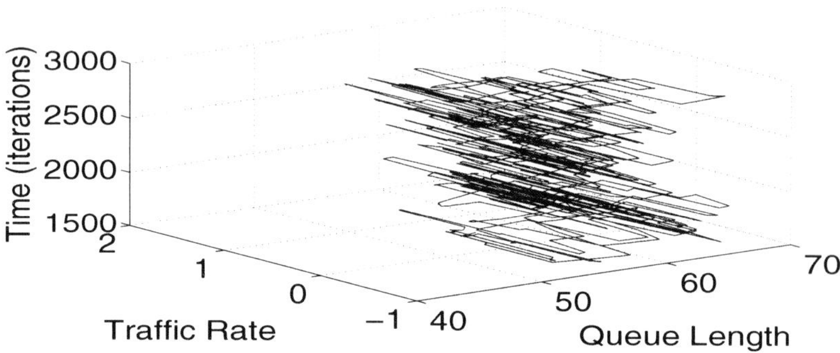

Figure 6. The trapezoid-shaped spirals are repeated in identical fashion other than the fact that their size becomes slightly smaller or larger each time step

tors (e.g. available bandwidth) which leads to a process quite different from memoryless packet arrivals. Moreover, in Figure 7. nodes behave in a synchronized manner, similar to that observed in modeling studies of TCP/IP [30].

In Figure 8., it appears that the nodes have reached equilibrium after roughly 5,000 iterations. The spikes in Figure 8. highlight the points at which the rate immediately changes, whilst the height and direction of each spike represents the relative magnitude of change in the slope.

5. Conclusions and Future Work

'No man is an island' (John Donne). We make our way in the world through human networks, and thereby networks play a vital part in our lives. Failing networks can cause severe adverse consequences.

Our overarching goal is to develop a traffic-control model. According to our model's strategy, we control the distribution of traffic to available network resources and thereby

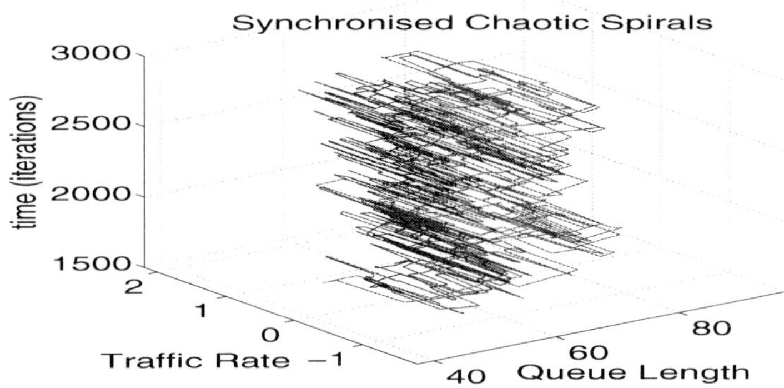

Figure 7. Nodes behave in a synchronized manner, similar to the one observed in modeling studies of TCP/IP [30]

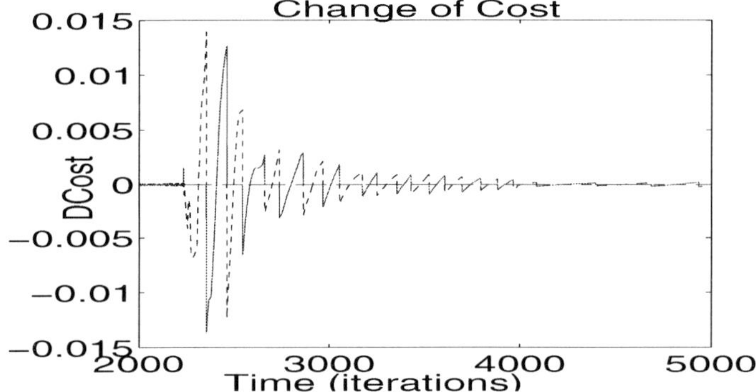

Figure 8. Sparky Cost change over time (last 3000 iterations). The spikes highlight the points at which the rate immediately changes (Hysteresis), whilst the height and direction of each spike represents the relative magnitude of change in the slope

minimize network failures by disallowing resource overloading. Traffic rates are dynamically adjusted based on link costs so that all network nodes behave in a synchronized manner. Synchronized reconfiguration ensures that deviations from equilibrium are prevented (provided that the exogenous parameters allow equilibrium to be reached). This strategy not only distributes the traffic load amongst network nodes, but also synchronizes the traffic and link-cost changes with buffer and topology variations.

Our model currently takes into consideration only the node's buffer capacity. In the future we intend to extend our analysis to consider other factors such as link capacity. Additionally, throughout our model's description, we use the concept of packets having neither a destination nor a source, and all having equal size. In particular, these packets circulate around the network until the network reaches equilibrium. If there is not enough buffer space for them, then the packets will continue circulating for ever reaching a limit cycle, instead of equilibrium. In future work, the model will be extended to allow us to use a variety of routing strategies, where the packets not only will have a appropriate size, but also they will be generated by a source node sending them to a destination node. Taking into consideration that networks change in time and thereby are products of evolution, we intend to incorporate this evolving feature of real-world networks into our model.

Achieving intelligent traffic control in today's networks depends on our ability to correctly model the mechanisms involved in the dynamics of complex networks. As numerous studies over the last decades have shown this is an overwhelming intellectual challenge. We believe, our approach is a first step towards this direction.

6. Acknowledgments

The research was funded by General Dynamics and BTexact. The authors would like to thank Prof. Patrick Purcell, Matthew Johnson and Thomas Karagiannis for constructive technical discussions and comments.

References

[1] Western Systems Coordinating Council (WSCC): Disturbance report for the Power System Outage that occured on Western Interconnection at 1548 PAST, *http://www.wscc.com*, (1996).

[2] *http://www.pewinternet.org/pdfs/LoveBug.pdf.*

[3] Fabrice Saffre: Adaptive Security and Robust Networks, *Information Security Bulletin*, **7, 11**, (2002).

[4] Yiong Chen, Zhongshi He: Bounds on the Reliability of Distributed Systems with Unreliable Nodes and Links, *IEEE Transactions on Realibility*, **53, 2**, (2004).

[5] Karp, Koutsoupias, Papadimitriou, Shenker: Optimisation problems in congestion control, *41st IEEE Symposium on Foundations of Computer Science*, (2000), 66–74.

[6] S. H. Strogatz: Information exchange and the robustness of organizational networks, *Nature*, **100, 21**, (2003), 12516–12521.

[7] N.Haque, N.R. Jennings, L. Moreau: Resource allocation in communication networks using market-based agents, *Knowledge-Based Systems*, **18**, (2005).

[8] M.A. Gibney, N. R. Jennings: Dynamic Resource Allocation by Market-Based Routing in Telecommunications Networks, *2nd Int. Workshop on Multi-Agent Systems and Telecommunication*, (1998), 102–117.

[9] E. Altman, H. Kameda: Equilibria for Multiclass Routing in Multi-Agent Networks, *Advances in Dynamic Games*, **7**, (2005), 343–368.

[10] M.A. Gibney, N.R. Jennings, N.J. Vriend, J.M. Griffiths: Market-Based Call Routing in Telecommunications Networks using Adaptive Pricing and Real Bidding, *Third International Workshop on Intelligent Agents for Telecommunication (IATA)*, (1999).

[11] D. J. Watts and S. H. Strogatz: Collective dynamics of "small-world" networks, *Nature*, **393** ,(1998), 440–442.

[12] M. E. J. Newman: The structure and function of complex networks, *SIAM Review*,**45, 2**, (2003), 167–256.

[13] Albert-Laszlo Barabasi and Reka Albert: Emergence of Scaling in Random Networks, *Science*, **286**", (1999), 509–512.

[14] M. E. J. Newman,S. H. Strogatz, D. J. Watts: Random graphs with arbitrary degree distribution and their application, *Santa Fe Institute, Working Papers*, (2000)

[15] S.N Dorogovtsev, J. F. F. Mendes: Evolution of Networks, *Advances in Physics*, **51, 4**, (2002),1079–1187.

[16] B. Bollobas: Random Graphs, 2nd Edition, Cambridge University Press (2001).
[17] Michalis Faloutsos and Petros Faloutsos and Christos Faloutsos: On Power-law Relationships of the Internet Topology, *SIGCOMM '99*, (1999) 251-262.
[18] Lotfi Benmohamed, Semyon M. Meerkov: Feedback Control of Congestion in Packet Switching Networks: The Case of a Single Congested Node, *IEEE/ACM Transcactions of Netaworking*, **1, 6**, (1993), 693–708.
[19] Jacobson: Congestion avoidance and control, *ACM SIGCOMM*, **88**, (1988), 314–329.
[20] Peter B Key and Derek R McAuley: Differential QoS and Pricing in Networks : where flow-control meets game theory, *IEEE Software*, (1999)
[21] Frank Kelly: Mathematical Modelling of the Internet, *Springer-Verlag*, (2001), 685–702.
[22] Steven H. Low, Larry L. Peterson, Limin Wang: Understanding TCP Vegas: A duality Model, *Journal of ACM*, **49, 2**, (2002), 207–235.
[23] Steven. H. Strogatz: Nonlinear Dynamics and Chaos, *WestView press*, (2000).
[24] Vern Paxson, Sally Floyd: Wide-Area Traffic: The Failure of Poisson Modeling, *IEEE/ACM Transactions on Networking*, (1995).
[25] B. Drossel, P. G. Higgs and A.J. McKane: The influence of Predator – Prey population dynamics on long–term evolution of food web structure, *Theoretical Biology*, **208**, (2001), 91–107.
[26] V. Krivan, J. Eirner: Optimal foraging and Predator-Prey Dynamics III, *Theoretical Biology*, **63**, (2003), 269–279.
[27] W. E. Leland, M. S. Taqqu, W. Willinger, and D. V. Wilson: On the self-similar nature of Ethernet traffic (extended version), *IEEE/ACM Trans. Networking*, **2**, (1994), 1–15.
[28] Mark E. Crovella, Azer Bestavros: Self-similarity in World Wide Web traffic: evidence and possible causes, *IEEE/ACM Transactions on Networking*, **5(6)**, (1997), 835-846.
[29] Thomas Karagiannis, Mart Molle, Michalis Faloutsos, Andre Broido: A Nonstationary Poisson View of Internet Traffic, *IEEE INFOCOM*, (2004).
[30] Andras Veres, Miklos Boda: The Chaotic Nature of TCP Congestion Control, *INFOCOM*, **3**, (2000).

Self-Organization and Autonomic Informatics (I)
H. Czap et al. (Eds.)
IOS Press, 2005

MA-DIDS: A Multi-Agent Based Distributed Intrusion Detection System

Huihua YANG [a,b], Yong WANG [a,b], Hongmei ZHANG [a,b] and Xingyu WANG [b,1]
[a] *Guilin University of Electronic Technology, P R China*
[b] *East China University of Science and Technology, P R China*

Abstract. In this paper, a novel architecture of multi-agent based distributed intrusion detection system is presented. MA-DIDS developed the frameworks of Common Intrusion Detection Framework (CIDF) and Autonomous Agent for Intrusion Detection (AAFID), and actualized distributed data collection, detection and response. MA-DIDS consists of 7 kinds of agents, namely, data collection agent (DCA), data preprocessing agent (DPA), intrusion detection agent (IDA), event analyzing agent (EAA), management agent (MA), intrusion responding agent (IRA), and communication agent (CA). MA-DIDS is platform independent, dynamically structure scalable and function shrinkable, it endows network security manager more power and flexibility to configure a DIDS. In this paper, the network-based DCA and DPA, and Linux host-based and Windows host based DCA and DPA are briefly illustrated. In MA-DIDS, all the anomaly detection IDAs are implemented using support vector machines; and CAs are specifically designed to enhance the communication security and response speed, with them, the agents can communicate safely and cooperate harmony. The architecture of MA-DIDS can efficiently cut off the network traffic added up by IDS, and promote the detected performance.

Keywords. distributed intrusion detection system, multi-agent, mobile agent, anomaly detection

1. Introduction

Intrusion detection system (IDS) plays an increasing important role in our information world. According the principles of detection, IDSs are usually classified into misuse detection and anomaly detection. Misuse detection works fast, but can't detect attacks unknown to it. Anomaly detection is a model based intrusion detection approach, it first learns a profile of the normal and abnormal user behaviors, and then generalizes this learnt model to unseen samples. On the other hand, people want to build the IDS that can adapt itself to the scalable network range and changing environment. As a countermeasure, people try to present some novel detection system based on a distributed architecture.

Researches have already presented some useful distributed IDS architectures, among them, CIDF (Common Intrusion Detection Framework) [1] and AAFID (Autonomous Agent for Intrusion Detection) [2] are most familiar to the practitioners.

[1] Corresponding author: Xingyu WANG, College of Information Science and Engineering, East China University of Science and Technology, PO Box 579, Shanghai 200237, P R China; email: xywang@ecust.edu.cn

As a generic model for IDS, CIDF divides IDS into following five components: event generator, event analyzer, response unit, communication unit, and event database. The COAST group in Purdue University presented the well-renowned AAFID scheme, which is the first IDS based on autonomous agents. An AAFID system consists of agents, transceivers and monitors. The AAFID architecture solves the load balance problem very well, and the level division makes it work efficiently. But it's difficult to configure, the failure of a single monitor agent may cause failure of the whole system, and the collaboration and control mechanism need further improvement.

Based on multi-agent framework, this paper presented a much more scalable and re-configurable IDS architecture – MA-DIDS, it features distributed data collection, and distributed detection and response. MA-DIDS is composed of 7 kinds of agents, and an event database. As for detection principle, this paper mainly focuses on anomaly detection; however, misuse detection can also be applied in. In order to promote the performance of detection agent, the up-to-date computational intelligence algorithm - support vector machines (SVMs) are used as detect engine. In MADIDS, D-S evidence theory can also be used to fuse information from different detection agents so as to boost the prediction performance.

The remainder of the paper is organized as following: section 2 illustrates the framework of MADIDS, and section 3 elaborates their implementation and the experiments, finally, section 4 concludes the paper.

2 The Framework of MADIDS

As illustrated in Fig 1, MA-DIDS consists of a event database (ED) and 7 kinds of agents, namely, data collection agent (DCA), data preprocessing agent (DPA), intrusion detection agent (IDA), event analyzing agent (EAA), management agent (MA), intrusion responding agent (IRA), and communication agent (CA). All these 7 kinds of agents work harmony in detecting intrusions under the control and coordination of the MA. The main functions of these agents are brief listed below.

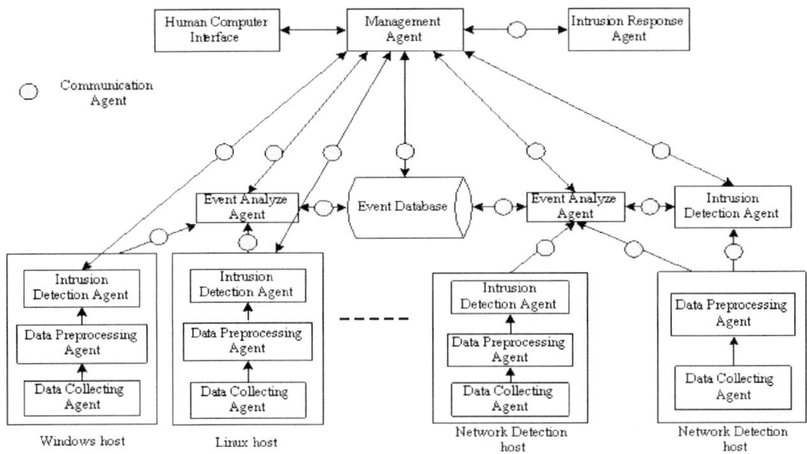

Fig 1 the layout of Multi-Agent based Distributed Intrusion Detection System

There are two kinds of data collection agent (DCA) in MADIDS. The first kind DCA collects information from packet grabbed from the network, is called network-based DCA, or NDCA. Another kind of DCA is host-based data collection agent, or HDCA. Because the HDCA must be designed conforming to the platform and platform's secure characteristics, so one should design HDCA for each platform.

Data preprocessing agent (DPA) transforms the information received from DCA into format suitable for the processing of intrusion detect agent (IDA) and the event analyzing agent (EAA). Generally speaking, DPA's processing procedure heavily relies on the input of DCA. Therefore, DPA also differs with OS. However, some procedures are common in DPA, such as data regularization and normalization. Some sophisticated DPA even include a powerful feature selection module.

Intrusion detection agent (IDA) is intended for discriminating the misuses or attacks from those normal connections, and this procedure is actualized based on either rules in misuse detection or pre-trained models in anomaly detection.

The main functions of event analyzing agent (EAA) is to analyze data transported from the DPAs and the decision result of IDAs, transform them into events, and then transfer them into the event database for the enquiries of other IDAs, MAs, and other kind of agents.

Management agent (MA) manages the entire MA-DIDS, it provides a human computer interface for the security manager to custom secure strategy, configure parameters, check intrusion message, deal with the alert messages coming from EAAs, and judge the secure status in a global view.

Intrusion response agent (IRA) responds to the alert of IDA by moving to the source of the alert directly or according the intelligent strategy mentioned in [3], and then takes possible action such as releasing the connection, killing the process, etc.

Communication agent (CA) is to provide a communication way for different agents in MA-DIDA. It is designed to be mobile agent and deployed where it is needed.

Event database (ED) stores all the midterm and final data, events, it uses large scale relational database management system (DBMS), such as Oracle or SQL server, so as to provide efficient data query and processing capability.

Finally, human-computer interface (HCI) provides a convenient approach for engineers to design and deploy the distributed IDS, get running information, and make configuration and regulation.

Compare with the CIDF and AAFID architecture, MA-DIDS is more flexible and powerful. Except the features of the platform independence and multi-agent cooperation, there are still many advantages to use multi-agents and mobile agents in MA-DIDS such as dynamically movable, distributed computation, dynamically scaleable, intelligence detection.

3 The Implementation and Experiments

In this section, we give a brief description of the design of agents of MADIDS.

3.1 Data Collection Agent

DCA includes NDCA and HDCA. NDCA gets information from the packets taking from the network, and it is usually deployed at the key point of the network or on the border of the network. Though, there are many tools to get network packets, such as

Libpcap, Jpcap, Tcpdump, and Snort, but they work not as fast as we need. We designed a tool using the packet sockets to capture primitive datagram directly form the data link layer [4]. Compare with Libpcap, our tool works more efficiently, and can be easily embedded into other applications.

Because the diversity of host operation system, services and logs, we have to design HDCA for each platform. We introduced the mobile agent technique to design HDCA agents, and implement them on the IBM Aglet developing toolkit, which is based on platform-independent language - JAVA.

Generally speaking, HDCA collects information from system log, system status, and application status, user input information, user instruction list, system log information, system call sequence, process information, CPU and memory use information, etc. In general, HDCA is more complicated compared with NDCA.

Since the most popular platform are Windows and Linux, the case study of the paper will based on them. For anomaly detection, to select and obtain the features which can accurately reflect the status of host security is of great importance.

3.2 Data Preprocessing Agent

The main work of network DPA is to analyze the datagram, checks some key fields, or reassemble the network connections, and make some statistics, such as the length of packets, the network delay etc. In MADIDS, the network connection constructed in real-time is shown in Fig 2 and the features designed for NDPA is listed in Table 1. Because the features from source to destination were the same as the ones form destination to source, for simplicity, we just list the features of one direction.

Host DPA handles more complex of raw data than network DPA, the information mainly includes log files, running information of the system, changes of some key files. For detail information of the features designed for Linux DPA and Windows DPA, please reference to [4].

However, some procedures are common in DPA, such as data regularization and normalization. In MADIDS, a feature selection module is optional in DPA, the feature selection algorithm could be the simple Pearson correlation criterion, Fisher discriminative criterion, and the complex support vector machines based recursive feature elimination [5]. The feature selection module finds out important variables, and so makes detection shrinkable when data collection time is to be reduced. In this case, the DPA is to be designed as a mobile agent.

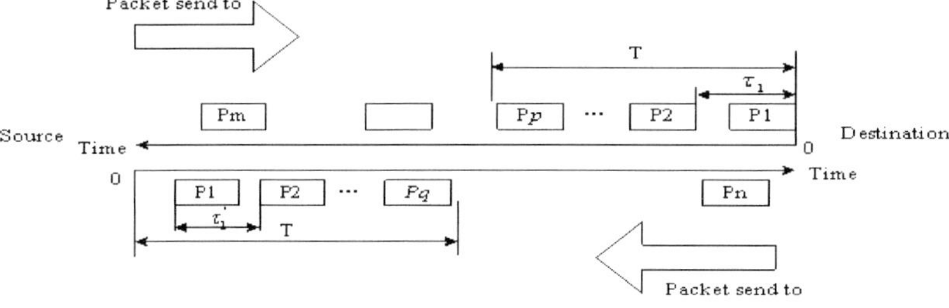

Fig 2 illustration for the real-time intrusion detection features extracted from network links (Pm: the m-th packet)

Table 1 the real-time intrusion detection features extracted from source to destination

Snum	The number of packets sent from source to destination within the sampling window T
Sbyt	The bytes sent from source to destination within the sampling window T
Sfrag	The percentage of fragmented packets
Surg	The percentage of URGENT packets
Ssyn	The percentage of SYNC packets
Sack	The percentage of ACK packets
Sfin	The percentage of FIN packets
Srst	The percentage of RESET packets
Spsh	The percentage of PUSH packets
Slen_mu	The averaged length of packets
Slen_sig	The standard derivation of the length of packets
Stime_mu	The averaged time delay of packets
Stime_sig	The standard derivation of the time delay of packets

3.3 Intrusion Detection Agent

Intrusion detection agent plays a key role in MADIDS. It roams through the system to tell attacks from normal uses and then reports suspicious activity by diffusing message. As we main focus on anomaly detection, the state-of-art classifiers - support vector machines (SVMs), are used as detector. In MADIDS, based on SVMs, we designed IDA for network-based data, for Linux and Windows host based data respectively. The basic principles and implementation of SVMs please refer to [6].

3.4 Event Analyzing Agent

EAA can also be divided into two categories, which are network EAA and host EAA, in terms of the events they handle with. In MADIDS, for misuse detection, the pattern matching is developed to combine with protocol analysis; it works more efficient, and gets more accurate result. In a 100 Mbps Ethernet, we tested EAA on a PC with a Pentium IV 2.5GHz CPU, it can reach the 60% load without loss any packet.

Host EAA analyzes the information according to domain knowledge, it is more sophisticated than the network EAA, because it is platform dependent, and it can't be transplant from one platform to another. The detail is omitted here.

3.5 Management Agent

Management agent (MA) provides a human computer interface, through this interface, security manager can custom secure strategy, initialize the MA-DIDS system, configure parameters, check alert and intrusion messages, deal with the alert messages coming from EAAs and consult the current agents' activity. Another main function of MA is to initialize, dispatch, reconfigure and recall all the mobile agents in MA-DIDS. MA could include an information fusion module, which makes global decision by taking into account all the information collected from agents scattered in the network.

Consequently, it will provide a more accurate result than a single intrusion detection agent.

3.6 Intrusion Response Agent

IRA is an indispensable component to IDS, and is generally implemented with software. The general response to an intrusion includes giving alert, killing the corresponding process, changing file properties, forcing user to log out, and even sending EMAIL or short message to network security manager.

However, security is a very complex problem; one can't count on to solve the entire problem by using a kind of secure approach, but should try to integrate multiple facets and diverse approaches. In order to build a solid safety-secure system, the following intrusion response cooperation mechanism can help: 1) IDS cooperates with firewall; 2) IDS cooperates with routers and switches; 3) IDS cooperates with anti-viruses systems; 4) IDS cooperates with honey-pots.

3.7 Communication Agent

Referring to the work of IDWG, we designed the intrusion alert protocol and event presentation protocol with XML. In our protocol, the secure communication is based on SSL protocol rather than the encryption and identity verification used in IDWG. Referring to the IDMEF protocol, the alert message is described with DTD of XML.

CAs provided a convenient and secure channel for agents to exchange information and to collaborate. With the help of CA, multiple agents could build a function scalable network, could cut off some unimportant functions to ensure the real-time response for time-critical applications, or add up some computational resources to ensure detection accuracy.

3.8 Simulation Experiment

In order to test our data collection and feature selection approach of those DCA and DPA, and detection method of IDA, we established a simulation environment. For detail please reference to [4]. We deployed the DCA and DPA, IDA in Linux host, Windows host, and network key node, and then used a series of intrusion tools to launch attacks. So far as the network IDA is concerned, we got a train set of 3371 samples, in which 2701 are attack samples, the rest are normal samples, and a test set of 308042 samples, in which 308234 are attack links. As listed in Table 2, by changing the length of sampling-time window T, we got a best comprehensive detect performance when T is 500 ms, with detect rate 98.02%, false alarm rate 0.6% and the undetected rate 1.98%. As for the Windows host IDA, the detect rate is 89.25%; and for the Linux host, the detect rate is 90.16%. Moreover, all the information of these distributed IDAs and EAAs could be further reused by the information fusion component [4] in the management agent so as to get a better prediction.

4 Conclusions

In this paper, a novel IDS architecture based on multi-agent is presented. With the help of communication agents, the other 6 kinds of agents in MA-DIDS can communicate safely, and cooperate well. The novel architecture can efficiently cut off the network traffic, and promoted the detected rate. To give some examples, the network, Linux, and Windows based DCA and DPA are designed and implemented, and they all use SVM as IDA's detector. In a word, the MA-DIDS architecture developed the AAFID and CIDF framework, and gave more flexibility to configure a DIDS, moreover, it is range scalable and function shrinkable.

Table 2 The performance of network SVM IDA on the test set with different sampling window

Length of Sampling Window	5ms	10ms	50ms	100ms	250ms	500ms	1sec	2sec	10sec	1min	5min
Train accuracy(%)	99.73	99.73	100	100	100	100	100	100	99.94	99.53	99.53
Test accuracy(%)	95.27	95.34	97.11	97.31	97.95	98.02	98.31	98.4	98.58	98.52	98.53
False alarm rate (%)	4.76	4.76	1.19	0.6	0.6	0.6	1.79	1.79	1.79	0.6	0.6
Undetected rate(%)	4.73	4.66	2.89	2.69	2.05	1.98	1.69	1.6	1.41	1.48	1.47

Acknowledgements

The support from the National Research Foundation for the Doctoral Program of Higher Education of China under grant 20040251010 is acknowledged here.

References

[1] S. Chen, B. Tung, and D. Schnackenberg. The common intrusion detection framework-data formats. Internet draf-draft-cidf-data-formats-00.txt, 1998.

[2] J.S. Balasubramaniyan, F.O. Garcia, and D. Lsacoff. Architecture for intrusion detection using autonomous agent[R]. COAST Technical Report, COAST Laboratory, Purdue University, 1998.

[3] N. Foukia. IDReAM: Intrusion Detection and Response executed with Agent Mobility. The International Conference on Autonomous Agents and Multi-Agent Systems (AAMAS'05), pp 264-270, Utrecht, The Netherlands, 2005.

[4] Y.Wang. Distributed Intrusion Detection Based on Computational Intelligence Methods. PhD Dissertation, East China university of Science of Technology, 2005

[5] A.H. Sung, S. Mukkamala. Identify important features for intrusion detection using support vector machines and neural networks. In: IEEE Proceedings of the 2003 Symposium on Application and the Internet, pp 209-216, 2003.

[6] V.N. Vapnik. The Nature of Statistical Learning Theory. Springer, New York, 1995

Self-Organization and Autonomic Informatics (I)
H. Czap et al. (Eds.)
IOS Press, 2005

A New Access Network Architecture for the Next Generation Wireless Networks

Jin TIAN [a], Guangguo BI [b] and H. TIANFIELD [c]

[a, b)] *State Key Laboratory of Mobile Communications, Southeast University*
2 Sipailou, Nanjing 210096, China
[a)] *Tel: +86 25 5241 8518 ext. 2277, E-mail: tj.tech@postel.com.cn*
[b)] *Tel: +86 25 8379 3267, E-mail: bigg@seu.edu.cn*

[c)] *School of Computing and Mathematical Sciences*
Glasgow Caledonian University
70 Cowcaddens Road, Glasgow, G4 0BA, UK
E-mail: h.tianfield@gcal.ac.uk

Abstract. The rapid development of mobile and wireless communication technologies requires that wireless devices should support voice, video, and data communications and should flexibly access IP based core networks. This paper proposes a new wireless module architecture for the next generation wireless networks and discusses its access methods based on access bridge point. In addition, the protocols for accessing IP based core networks are discussed for the proposed access network architecture.

Key words. wireless module, access bridge point (ABP), access network architecture

1. Introduction

The wireless communication is a fast growing area, where the number of users is increasingly expanding and the user demands for better services are becoming increasingly higher and higher. these have posed numerous challenges for the next generation wireless networks. Most of the current research works are in the area of distributed computing, mobile agents, multimedia support, etc. Some other research works are to improve the Quality of Service from both the user's and service provider's points of view. The next generation wireless network infrastructures are expected to be deployed in an environment where many other types of wireless and wired communication systems are already in existence. The 4th Generation (4G) wireless communications are envisaged as hybrid broadband networks that integrate different types of network topologies and platforms. The overlapping of different network boundaries represents the integration of different types of networks in 4G. There are two types of integration. The first is the integration among heterogeneous wireless networks of varying transmission characteristics such as Wireless LAN (Local Area Networks), WAN (Wide Area Networks), PAN (Personal Area Networks), as well as MANET (mobile ad hoc networks) [1]. The second is the integration of wireless networks with the fixed network backbone infrastructures, the Internet, and Public Switched Telephone Network (PSTN). Much work remains to be done to enable a seamless integration, which, for example, can extend IP to support mobile network devices.

Wireless networks markets may divide into two parts logically, i.e., voice oriented market and data market oriented. Wireless voice devices mainly deal with mobile handsets adapting the 2nd Generation (2G) mobile communication with Global System for Mobile Communications (GSM), the 3rd Generation (3G) with Code Division Multiple Access (CDMA), and the 3G beyond with Orthogonal Frequency Division Multiplexing (OFDM). At the same time, wireless data communications utilize wireless networks cards, such as those plugged in portable computer or personal computer, adapting the WPAN (Wireless PAN) with Bluetooth and the WLAN (Wireless LAN) with 802.11b/g/a. New wireless communication technologies, especially OFDM, have mainly supported voice communications in fast moving environments. However, mobile communications are challenged by hard high rate data and video communications. Apparently, time is critical for data and video communications in much fast moving environments because jounce makes it impossible to do anything. The R&D of telecommunication industries is working on future technologies that can meet these demands at even lower costs. So a great leap has to be taken towards the next generation wireless communications that promise to bring an end to most of the problems faced.

In the environment of next generation wireless network infrastructures, how should a wireless module be designed and how will it access IP based core networks? To address these challenges, this paper puts forward a new architecture for wireless modules for the next generation wireless networks and further analyzes the access methods by which wireless modules can connect the terminal devices to IP based core networks based on Access Bridge Point (ABP) via wired or wireless interfaces.

2. A New Architecture for Wireless Modules

In a pervasive environment almost every device in our everyday life will be network enabled. In a consumer-electronic terminal device the wireless module co-exists with other modules as exemplified in Table 1.

The wireless module in a consumer-electronic terminal basically needs to perform two functionalities, i.e., communication and the base band process potentially necessary in wireless communications. A wireless module can have three types of interfaces, i.e., data generic interface such as PCI (Periphery Component Interface) for connection with other devices, UWB radio interface, and 3G radio interface. UWB interface is needed for the delivery of large throughputs. Therefore, in order to support video as well as voice and data communications, a wireless module should comprise ASIC (Application Specific Integrated Circuit) modem for base band process, OFDM based digital frequency transformation, and UWB (Ultra Wide Band) antenna and 3G antenna. Specifically a new generic architecture can be put forward for wireless modules for the next generation wireless networks, as shown in Figure 1.

Table 1. Wireless module along with other modules

<table>
<tr><th rowspan="2">Consumer-electronic terminal</th><th colspan="6">Other modules in addition to wireless module</th></tr>
<tr><th>Display screen</th><th>Microphone</th><th>Speaker</th><th>User Interface</th><th>Power</th><th>CPU</th></tr>
<tr><td>TV</td><td>√</td><td></td><td>√</td><td>√</td><td rowspan="10">√</td><td rowspan="10">√</td></tr>
<tr><td>Voice handset</td><td>√</td><td>√</td><td>√</td><td>√</td></tr>
<tr><td>Printer</td><td>√</td><td></td><td></td><td>√</td></tr>
<tr><td>Laptop</td><td>√</td><td></td><td></td><td>√</td></tr>
<tr><td>PDA</td><td>√</td><td></td><td></td><td>√</td></tr>
<tr><td>Portable computer</td><td>√</td><td></td><td>√</td><td>√</td></tr>
<tr><td>Personal computer</td><td>√</td><td></td><td>√</td><td>√</td></tr>
<tr><td>Home Appliance</td><td></td><td></td><td></td><td></td></tr>
<tr><td>Digital Camera/ Camcorder</td><td>√</td><td></td><td></td><td>√</td></tr>
<tr><td>Digital Projector</td><td></td><td></td><td></td><td>√</td></tr>
</table>

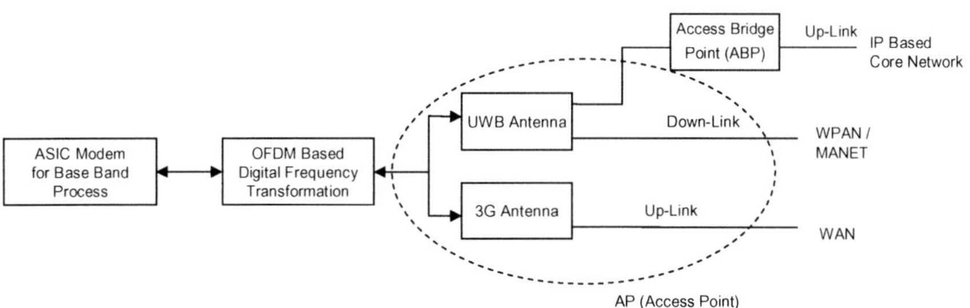

Figure 1. A new generic architecture for wireless modules for the next generation wireless networks

A wireless module only using UWB radio interface can enable up-link into network infrastructures based on ABP (Access Bridge Point) and support voice services and high rate data services. A wireless module only using 3G radio interface can support voice services but is very limited in supporting data services to communicate anywhere. Moreover, a wireless module equipped with both UWB radio interface and 3G radio interface can also relay as AP (Access Point) by which the UWB radio interface acts as the down-link into WPAN or MANET and the 3G radio interface as the up-link into WAN. This characteristic is very important, which means that in practical situations a terminal device can first join in a WPAN or MANET and if one of the nodes in the WPAN or MANET has the wireless module of the generic architecture as shown in Figure 1, then the terminal device can take this node as an AP to establish further connection to IP based core networks.

Terminal devices with wireless modules embedded, and ABP as well, can readily form WPAN or MANET. Examples of WPAN are the so-called Piconet and Scattered Net that are formed at home or in office by Printer, PDA, Laptop, TV, Notebook PC, etc. Normally a MANET connects a number of devices via serial nets. However, in rural places where there are few devices to form MANET and thus 3G radio interface will be used for communications as the only approach available.

MAC (Media Access Controller) and route protocols between terminal devices with wireless modules embedded, and ABP as well, are ad hoc network protocols and the UWB interface, including Cognitive Radio (CR) interface, serves the physical layer technology.

MANET are peer-to-peer communication with no fixed infrastructure. In these networks, each node functions as a router that maintains routes toward other nodes in the network, and each node relies on intermediate nodes to relay its packets to the destination [2]. The core functionalities of the network layer are routing and packet forwarding. Routes from the source to the destination are established and maintained by the routing protocols, while data packets are forwarded by intermediate nodes along the established route to the destination.

Mobile devices communicate with ABP through UWB interface. The PHY (physical layer) entity for a UWB interface was specified that utilizes the unlicensed 3.1 - 10.6 GHz UWB band, as regulated in the United States by the Code of Federal Regulations. Everything started out as impulse radio but after the FCC (The Federal Communications Commission) published the regulations for commercial UWB devices, the field has split in two: an impulse radio UWB approach backed by Motorola, and a multi-band OFDM solution backed by a 90-company industry alliance, MBOA (Multi-Band OFDM Alliance). The two opposing proposals for standard have been presented (IEEE, 2004), and now both parties continue to develop their own products, as well as participate in the formal standardizing process. Multi-Band OFDM (MB-OFDM) with OFDM used in mobile communications beyond 3G, is an alterative method for UWB network. The support of transmitting and receiving at data rates of 53.3, 110, and 200 Mb/s is mandatory [3].

DS-UWB (direct sequence UWB) has definitely selected IEEE 802.15.3 standard adapted to high rate (20Mb/s) physical layer transmission as its MAC protocol. MBOA has forged a new MAC for UWB networks that will meet the myriad needs of the PC, consumer electronics and mobile markets. The new MBOA MAC, while deviating substantially from the 802.15.3 version, was deemed necessary by the alliance. The new MAC will solve all the mobility and consumer electronics requirements and issues as well as the needs of all the isochronous and asynchronous applications and related issues for UWB systems [4]. No company has designed a commercial chip with MBOA MAC because MBOA have not decided the final MAC solution. IEEE 802.22 Working Group was set up to develop a standard for a cognitive radio-based PHY/MAC/air-interface for use by license-exempt devices on a non-interfering basis in spectrum that is allocated to the TV Broadcast Service [5].

3. Access Methods Based on Access Bridge Point

Figure 2 and Table 2 show the access methods for wireless modules based on ABP. Five types of ABPs support a wireless module to access IP core networks through wired or wireless technology.

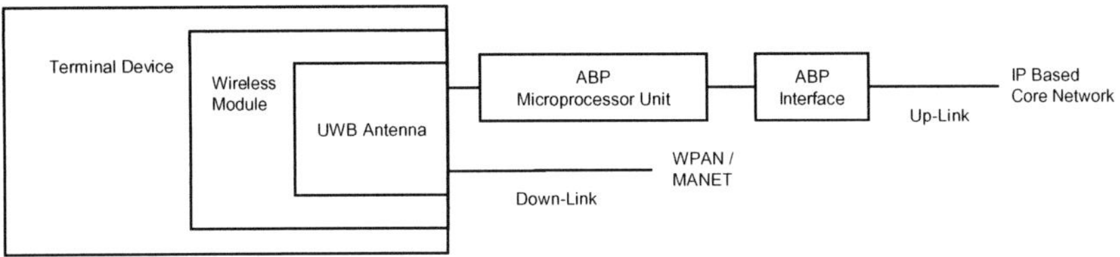

Figure 2. Architecture of Access Bridge Point

Table 2. Physical layer technologies for ABP interfaces

ABP Interface	Physical Layer
Type I	UTP or Fiber
Type II	Fix wireless
Type III	WiMAX
Type IV	WLAN
Type V	Satellite

Type I can directly access IP based core networks via router, Type II to V require a variety of wireless technologies to access IP based core networks. Each ABP includes one UWB based interface, at least, for connection to WPAN and MANET, and an additional interface based on wireless technology with one of the interfaces, Fix wireless, WLAN, WMAN or Satellite, or wired technology compatible with IEEE 802.3 via unshielded twist paired (UTP) cables catalog 5 or 6. When ABPs access IP based core networks, they are either through one of the wireless technology via their corresponding physical layers or through wired interface. A very important point is that ABP device acts as a bridge between mobile devices and IP based core networks.

Operation protocols between ABPs and IP based core networks adopt the Open Systems Interconnection (OSI) model protocols stack: physical layer, data link layer, network layer, etc.

The Data-Link Layer is the protocol layer in a program that handles the moving of data in and out across a physical link in a network. The Data-Link Layer is layer 2 in the Open Systems Interconnection (OSI) model for a set of telecommunication protocols. The Data-Link Layer contains two sub-layers [6] that are described in the IEEE-802 LAN standards, i.e., Media Access Control (MAC) and Logical Link Control (LLC).

The network layer defines the network address [7], which differs from the MAC address. Some network layer implementations, such as the Internet Protocol (IP), define network addresses in a way that route selection can be determined systematically by comparing the source network address with the destination network address and applying the subnet mask. Because this layer defines the logical network layout, routers can use this layer to determine how to forward packets. Because of this, much of the design and configuration work for internetworking happens at Layer 3, the network layer.

4. Summary

2G or 3G based handsets mainly support voice services. Traditional data service modules are separate from traditional handsets for voices. Data are delivered via a wireless network card plugged into portable computer or PC. So it is inconvenient to use wireless devices anywhere since the user has to make ready both types of separate terminal devices, i.e., one for voice services and the other for data services. The diversity of terminal devices from 3G, UWB, Bluetooth, IEEE 802.11g/b/a, etc., and the terminal dependent nature of services, have largely restricted the development of wireless communications. Wireless modules of the proposed generic architecture are versatile in supporting video, voice and data services, capable of being flexibly plugged in to or out of consumer-electronic terminals to support moving and wireless requirements, and thus services are provided independent of wireless terminals.

Five types of ABPs upwards interfaces, i.e., UTP or Fiber, Fix wireless, WiMAX, WLAN and Satellite technology, can connect mobile devices to IP based core networks. Although one ABP only has one interface for up-link, multiple ABPs existing around can provide a variety of interfaces. Wherever a mobile device is, an ABP can easily connect it to IP based core networks because many possibilities are provided by the multiple ABPs through wired and wireless interfaces. Therefore, the proposed access network architecture efficiently meets the requirements posed by the next generation wireless communications.

On-site temporary network formation can easily be built up by ad hoc network protocols without requiring fixed infrastructure, which effectively helps overcome the hardness of building network physical infrastructures and the occurrence of network dynamism and breakdowns. Terminal devices at home or in office can form MANET based on UWB physical layer.

Legacy AP devices with catalog 5 or 6 cable up-link interface to access IP based core network can control operation instructions and distribute radio resources in networks topology architecture. New ABPs are similar to legacy AP. 3G based handsets are now increasingly available in markets. Transition from 3G to the 3G combined with UWB is made much simple.

The proposed generic architecture of wireless modules with largely relies upon the most promising technologies including cognitive wireless technology, such as UWB physical layer technology, ad hoc MAC, and network layer technology. UWB physical layer supports high rate up to 500 Mb/s when slowly moving in 10 meter short range to enable DVD, media streaming, home cinema and games.

Acknowledgement

The first and the second authors are supported by Intel's University Research Program under Grant No. 4507177326.

References

[1] Imrich Chlamtac, Marco Conti, Jennifer J.-N. Liu, Mobile ad hoc networking: imperatives and challenges, Ad Hoc Networks 1 (2003), 13-64.

[2] Hao Yang, Xiaoqiao Meng, Songwu Lu. Self-Organized Network-Layer Security in Mobile Ad Hoc Networks, Proceedings of the 3rd ACM workshop on Wireless security, Atlanta, GA, USA, September 28 - 28, 2002, pp. 11 - 20

[3] MultiBand OFDM Alliance (MBOA) SIG (2004). MultiBand OFDM Physical Layer Proposal for IEEE 802.15 Task Group 3a. September 14, 2004 (http://www.multibandofdm.org).

[4] Patrick Mannion (2004). Alliance defines new MAC for UWB networks. EE Times – Global news for the creators of technology (online). 16 March 2004 (http://www.eetimes.com/).

[5] Jim Lansford, UWB Coexistence and Cognitive Radio: How Multiband-OFDM Leads the Way, White Papers, 2004 (http://www.alereon.com/technology/white-papers/).

[6] Data Link Layer. Whatis?com ---- The leading IT encyclopedia and learning center (http://whatis.techtarget.com/)

[7] Internetworking Technology Handbook, Documentation of Cisco Systems (http://www.cisco.com/univercd/cc/td/doc/cisintwk/ito_doc/index.htm)

Self-Organization and Autonomic Informatics (I)
H. Czap et al. (Eds.)
IOS Press, 2005

Identity and Trust Management Directions for User Controlled Light-Path Establishment

R.J. HULSEBOSCH[1], M.S. BARGH, J.H. van der SPEK,
P.H. FENNEMA and J.F. ZANDBELT
Telematica Instituut, The Netherlands

Abstract. This contribution investigates the role of identity management and trust in user-controlled provisioning of end-to-end light-paths across multiple optical networks managed by different (optical) network service providers. Lack of trust between the user's identity provider and the optical network service providers hampers effective exchange of user identity information needed for authorizing the reservation and utilization of network resources in such user-controlled light-path establishment. The paper proposes a Virtual Organization as a trusted third party to overcome this lack of trust and analyzes the possible roles it can play to facilitate trusted and secure exchange of identity credentials.

Keywords. Identity management, light-path, trust, virtual organization, secure DNS

1. Introduction

Grid computing witnesses a paradigm shift that extends the supercomputing parallelism, as started a decade ago, to the super-networking parallelism [1]. According to this new vision, also optical networks are considered as schedulable resources for Grid applications. This super-networking vision turns the problem of bandwidth-guaranteed end-to-end path provisioning into a typical resource allocation problem.

Nowadays many research institutions, schools, and large enterprises purchase optical wavelengths or entire strands of optical fiber in order to use their own switching equipment for connecting to each other or to Internet service providers [2]. This trend promotes the development of novel high-performance Grid applications that empower the end-points (i.e., users or applications) to establish deterministic (i.e., circuit switched) paths on demand in a condominium fashion to reduce infrastructure costs. Such user/application-empowered networks ensure ultimate QoS by dedicating applications their own guaranteed end-to-end light-paths/wavelengths on demand. This so-called *user controlled light-path provisioning* also puts forward the possibility of provisioning light-paths across multiple management domains, through collaborative trading of light-paths and shared control over user-owned switching equipment [3],

[1] Corresponding Author: R.J. Hulsebosch, Telematica Instituut, P.O. Box 589, 7500AN Enschede, The Netherlands, tel. no. +31 53 485 0498, fax no. +31 53 485 0400, email Bob.Hulsebosch@telin.nl.

without relying on Optical Network Service Providers (ONSPs), e.g. hybrid ISPs or optical exchange operators, to configure core network equipment.

To make the access control process efficient over a heterogeneous set of ONSPs, fragments of user profiles have to be collected, stored and interpreted at different places and by different players. These user profiles, moreover, need to be communicated and matched against each other. Thus, identity and its management are required to effectively deliver access control services. As identity usage extends beyond the confines of a single business or institution, federation of identity need also be taken into account.

Generally most ONSPs are not aware of the identity of the user that requests access to their network resources. The user identities and related attributes are managed by and stored at so-called Identity Providers (IdPs), belonging to the institutions of users. Usually there exists no trust relationship a priori established between an ONSP and a user's IdP. The main focus of this contribution is to investigate the ways to (dynamically) establish trust between a user's IdP and ONSPs in order to securely exchange identity information such as authentication status and attributes required for authorizing light-path establishment. The trust establishment approaches investigated here rely on a trusted third party that is called a Virtual Organization (VO) in Grid communities [4]. While such VOs are typically centralistic entities with a lot of management overhead, the paper also enumerates a number of ways to lighten the load imposed upon VOs.

The paper is structured as follows. Section 0 provides an overview of several fundamental developments in the area of user controlled light-path establishment. Section 0 outlines solution directions for trust establishment between users and ONSPs. Section 0 discusses, evaluates and compares the most promising solution directions. Finally, the concluding Section 0 outlines challenges and future work.

2. Fundamentals and Developments

This section describes three key issues in user controlled light-path establishment.

2.1 Network Management

A light-path usually spans over multiple ONSPs. Each of these ONSPs deploys network management technologies to control their optical network resources. A number of emerging technologies are available today to facilitate end-users to construct light-paths on demand without relying on the ONSPs to configure core network equipment. The most advanced network management solutions that have emerged recently are DRAC [5] and UCLP [6].

It is unrealistic to expect that applications for light-path establishment are capable of communicating with this variety of network management solutions. Therefore, some kind of adapter component is needed that offers light-path applications a standardized and uniform interface towards these solutions. For this purpose we envision Network Management Controller (NMC) components that, as shown in Figure 1, intervene the control signaling between the application and the heterogeneous management components of network resources. The NMC also glues the identity and network management control planes together by having interfaces to the local AAA-servers, functioning as a policy decision point to enforce authorization policies, or streamlining the process of resource reservation.

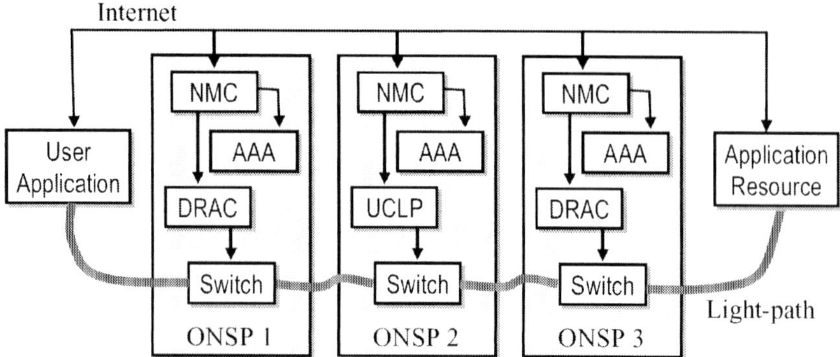

Figure 1. Managed light-path provisioning over multiple ONSPs.

2.2 Identity Management

An identity management framework is required to control the pieces and types of information pertaining to a party, i.e. attributes and identifiers, made available to other parties. Identity management solutions take care of effectively managing definitive identifiers of users and ensure that users have fast, reliable and secure access to resources, services and data of other parties [7]. Identity information about a party that is stored and secured in one domain should effectively be shared or made portable across contexts and organizational boundaries. Such portability is usually realized by means of federation.

There are a number of industrial standards that deal with simplifying federated identity management and that also facilitate Single Sign On (SSO). Examples are Liberty Alliance [8], SAML by OASIS [9], and WS-Federation as part of the WS-* Roadmap by Microsoft, IBM, and partners [10]. Shibboleth and A-Select are two other identity management initiatives that may also be relevant for this work. Shibboleth is an attribute authority service developed by the Internet2 community to support inter-institutional authentication and authorization for access to websites [11]. A-Select is a framework where users can be authenticated by several means with Authentication Service Providers [12]. Recently, OASIS has standardized SAML 2.0 that combines the best of SAML 1.1, Liberty ID-FF 1.2 and Shibboleth [13].

Federated identity management involves a central authority, i.e. IdP, to manage the authentication, provisioning, lifecycle, linking, privacy, attribute discovery and sharing of a party's digital identity. The IdP links all user identifiers and discovers attribute information with the goal of SSO and personalized service offering in a privacy-preserving way. In particular SSO is a desirable feature for establishing light-paths across multiple ONSPs, enabling users to use multiple digital identifiers for path establishment. Each user has her own preferred IdP, which is usually the IdP of the company or institution he/she is working at. This IdP is often federated with an attribute service provider that manages the attributes belonging to a user identity. The user trusts both of them, but can the ONSP do the same?

2.3 Trust Establishment

Efficient and seamless provisioning of light-paths requires establishment of trust relationships in order for parties to share identity information securely. Trust management for end-to-end light-path provisioning can be implemented in two ways: by delegating trust from one party to the other (hop-by-hop trust), or via a trusted third party (single-hop trust).

Delegation of trust for setting up an optical path seems the most natural way as it can be based upon the physical existence of a dark fiber between any neighboring ONSPs on the path. Both neighboring parties know each other after all and thus each one can rely on the other's trust for an incoming request. User controlled light-path provisioning based upon delegated hop-by-hop trust principle has been demonstrated by Oudenaarde et al. [14]. In large systems, however, there are a number of drawbacks of such delegated trust. Firstly, it results in potentially long and fragile chains of trust with limited accountability. Moreover, it will be difficult to trace the identity of the party that has been delegated trust. Consequently, this lack of identity hampers the exchange of additional attribute information, making fine-grained access control impossible. Secondly, certain types of trust relationships are difficult to express and validate with delegated trust management. Suppose P allows Q to perform actions a_1 and a_2 on its resources. If Q subsequently wants to delegate R to only perform a_1, expressing and validating this type of partial delegation can be difficult.

Given that the participating administrative parties reside in different domains and may have different security infrastructures (e.g. PKI, Kerberos), it is necessary to realize the required trust relationships through some form of federation among these parties. *A trusted third party* is required to create the necessary trust. In Grid communities such a party is represented in the form of a Virtual Organization (VO) [4]. VOs could serve specific communities like the high-energy physics, astronomers or bio-informatics. The GLIF community is a good example of such a VO [1]. Users' institutions or companies are member of a VO, whereas ONSPs typically are partners of the VO, i.e. ONSPs have a trust relation with the VO. Usually, this trust is based on contractual agreements. As VOs increasingly turn into distributed multi-institutional collaborations, secure authentication and authorization becomes a growing challenge. VOs are typically used in the Grid computing environments to authenticate and authorize users, based on the identities of the interacting entities [15]. While this approach is straightforward and intuitive, it becomes impractical to administer as soon as VOs expand, VO membership changes dynamically, and user rights vary on a periodic basis or according to a user's role in an organization. It is impractical and unscalable for the VO itself to store and manage all the identities and attributes. This paper envisions that this information could better be stored locally at the individual IdP of the registered members. Identity management technology can then be used to communicate the attributes, for doing which the VO somehow must guarantee their authenticity and integrity.

3. Roles of Virtual Organization

The task of the VO is to offer the necessary trust infrastructure for the ONSPs to rely on the identities and attributes provided by IdPs and federated attribute service providers. The key principle here is to separate between two issues: (1) establishing trust between an arbitrary ONSP and the IdP of the user, i.e., verifying the federation memberships of the parties involved, and (2) verifying the user's credentials or attributes by the IdP upon request from the ONSP. The VO can fulfill its trusted role in the exchange of identity and attribute information in a number of ways described below.

VO as Attribute Signer and Issuer. The VO can sign attributes like for example "VO Gold membership". The signed attribute is either stored at the VO or transferred and stored at the home IdP of the user. The former option is not preferable as it creates a VO that has to manage all the individual users of its members and their corresponding attributes. The latter option provides for distributed management of VO user memberships and local responsibility of the attributes. ONSPs who request for attributes can verify the validity of attributes by checking the VO signature. Revocation of the signed attributes will be difficult as they are not under direct control of the VO. Setting signed expiration dates on the attributes may be a solution for this issue.

VO as Attribute Verifier. The VO just verifies the attributes provisioned by the member organizations themselves. The VO trusts its members to do this properly and consciously. In our light-path scenario, the ONSPs cannot directly approach the IdP of the user for user authentication and attributes, as there is no trust relationship between the ONSP and IdP. Therefore, the ONSPs reach the VO which on its turn contacts the IdP to check the true identity of the user and to provide the required attributes of the user. As a mediator, the VO may first check the VO-related attributes like Gold or Silver membership, and then sign and return the result to the ONSP. In this case the VO operates like a relaying IdP for both identification and attribute information. This approach results in a more controllable situation regarding the control of the release of the user attributes by the VO. The attribute verification implies that the VO either (a) keeps its own database of user associations and attributes or (b) keeps just its own database of member organizations and delegates the verification of a user's association with a member organization and his attributes to the local IdPs again. The latter option is preferable, as it prevents the VO from becoming too centralistic and overloaded with identity management functionality.

VO as Certificate Authority. The VO behaves like a certificate authority and issues certificates for its member organizations. The IdP of the user can use the membership certificates to sign the attribute assertions. The ONSPs can use this signature to determine the VO, check if the VO trusts this IdP, and verify the attributes issued by the IdP. The main difference between this approach and the second approach is that here the VO is not involved operationally.

VO as Discovery Provider. The VO fulfils the role of a discovery service provider similar to the Liberty ID-WSF [16]. The VO tells the ONSP where to find the required attributes related to the user. Since the VO and the ONSP have a trust relationship, the ONSP trusts the validity of the pointer to the attributes and also it trusts the attributes themselves. Note that in this scenario, the VO verifies the authenticity of the user's light-path request and checks if the user is a member of the VO. Moreover, the VO must generate credentials that authorize the ONSP to fetch the attributes of the user. These features imply that the VO must administrate the individual users of its members. It is questionable if this is a desired situation and fits in the organizational role the VO fulfils.

VO as DNS Tree Administrator. Network providers may use DNS as a key component in finding people's identity and attributes in order to create a "persistent identity". The security extensions of DNS (DNSsec, [17, 18, 19]) can be used to allow the ONSP to verify whether the user belongs to a VO that is trusted by him. This is achieved by establishing a trust relationship between the IdP and the ONSP that implies both IdP and ONSP belong to the same VO federation and that the ONSP may rely on the IdP for user authentication and attributes. To this end, DNSsec facilitates establishing a secure channel between the IdP and the ONSP. DNSsec requires registration of a domain name for each VO (like, e.g., 'vo-astroph.org'). The VO should manage this domain. The entries registered on this domain represent the user institutions and ONSPs that are associated with the VO. The owner of 'vo-astroph.org' should ensure that it only lists the parties that adhere to the policies of the VO that have been agreed upon. The DNS-entries should list a public key to enable establishing a secure connection between the user's IdP and the ONSPs. Hereby, moreover, DNSsec guarantees that both the ONSP and the user's institute are members of the VO because otherwise they would not appear in the entries of the VO's DNSsec. User attributes are then communicated over the established secure channel since the ONSP has becomes assured of the user's IdP in verifying user identity and providing trustful user attributes.

4. Discussion

History has shown that for a VO to be successful, it has to be lean and mean, and above all, it must have a clear functional role. In the context of user controlled light-path provisioning, this implies that the VO should not be responsible for identity related services such as storage or authentication. These services should be delegated to the user IdPs. In addition this excludes the VO as attribute signer or discovery service provider. Turning a VO into a signing or certificate authority isn't the right way as it is not one of the core-functionalities of a VO. This leaves the relaying and DNSsec variants as potential viable solutions. In the first solution, the VO merely acts as a relaying IdP between the ONSP and the IdP of the user. In the DNSsec solution, the VO isn't involved in the identity communication path at all and it just induces trust between the user's IdP and ONSPs. These two solution directions regarding an identity management framework for user-driven light-path provisioning are discussed in the following sections in more detail.

4.1 VO as Attribute Verifier (Relaying VO)

In this solution direction, the VO takes a relatively central position. Note that multiple VOs exist and that ONSPs are associated, depending on their policies, with (almost) all of them. For scalability purposes, one could imagine a decentralized and distributed configuration of the VO over all parties. By keeping a list of its member organizations, the VO relays all identity and attribute communication between the IdPs of its members and the ONSPs.

Looking at the available technologies for identity management it appears that, at least theoretically, all of them are suitable for building our solution. In particular, the Cross A-Select implementation already offers all the functionality to realize the relaying IdP VO scenario [12]. Also Liberty Alliance / SAML 2.0 allows for such a relaying VO by means of the specified IdP proxying mechanism [13]. For Shibboleth the same holds because it is also based on SAML 2.0. The WS-Federation specification describes an example of indirect trust between IdPs (Figure 3 of [10]). This means that the building blocks of the WS-* roadmap can provide a solution for our scenario based on the WS-Federation specification.

4.2 VO as DNS Tree Administrator

In the DNSsec scenario, the home IdP of the user is directly communicating with the ONSP via a secure channel in between. The VO doesn't play an active role in the communication of identity and attribute information thereby increasing the efficiency of identity communication. DNS allows the ONSP to discover the proper IdP. The fact that a user is associated with a certain VO cannot be hidden by the DNSsec solution and may harm the user privacy. DNSsec on its turn guarantees correctness of DNS records and can therefore be used as a safe certificate store. Such stored certificates can be self-signed and need not be part of a PKI. The certificates are obtained from the DNS system and used for establishing a secure channel for the exchange of identity management messages between the IdP and the ONSP. Such a message exchange can be based on SAML 2.0, A-Select, etc.

In the DNSsec solution, trust is established via the DNS system. However, this solution doesn't address the attribute authenticity issue, unless one relies on the trust that has been established via DNSsec in the form of the secure channel. We believe it is safe to assume that, as soon as the user or its institution trusts the VO, and the ONSP also trusts the VO, the ONSP may trust the credentials that it obtains from the user or the IdP of his/her institution as well. If the VO finds out that a member is using invalid attributes the trust relationship is damaged resulting in an exclusion of the member.

The use of DNSsec to dynamically establish trust relationships between IdPs and ONSPs to our knowledge has been barely addressed in the literature. There is however strong overlap with the proposed solution described in [20]. Here DNSsec is combined with RADIUS authentication server for dynamic trust establishment between domains.

4.3 Requirements

The two proposed solution directions fulfill a number of high-level requirements. First of all, both solutions preserve the distributed identity character. Secure authentication and attribute exchange is facilitated over multiple domains. Authentication is performed in the user's home domain, where the user is best known. Attributes can be collected from multiple, distributed attribute authorities. User authorization can, irrespective of the true identity of the user, be granted based only on trustworthy attributes provided by the user's IdP. As a consequence, the ONSP does not have to deploy user authentication solutions. Furthermore, the IdP guarantees the privacy of the user towards the ONSPs. Access management to user attributes remains an issue in the DNSsec case. The user must specify which ONSP has access to what attributes. This is complex as the user often does not know on beforehand which ONSPs will ask for attributes. Moreover, a potentially large number of ONSPs may need to be taken into account in the policy. In the relaying model, attribute access can solely be granted to the VO. Another important criterion that is met by the solutions is their economical viability due to the relatively lightweight role of the VO in the light-path establishment process and its simple administrative tasks. Finally, standard technologies like SAML, TLS and DNS can be used to successfully implement the solution.

5. Conclusions and Future Work

To allow users to setup light-paths across multiple ONSPs in a user-friendly and easy way requires identity management functionality. An identity management framework facilitates identity and attribute information exchange and thereby offers SSO functionality for authorized reservation of optical bandwidth.

For establishing the required trust between the involved parties a trusted third party, called Virtual Organization (VO), is required. The role of the VO in the enforcement of trust between the involved parties can be implemented in multiple ways. Two preferred ways have been identified and described: the VO as relaying IdP or as DNS domain administrator. Figure 2 shows the high-level interaction diagrams of the solution directions. Both solutions tackle issues as how to eventually establish trust between users, their IdP's and ONSPs, and how to provide the necessary credentials in a secure and trustworthy manner to enforce authorization. Moreover, both solutions seem economically viable and can be implemented with existing identity standards.

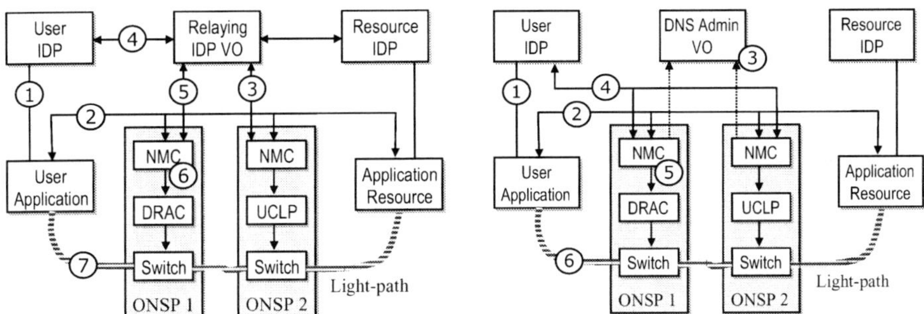

Figure 2. The two solution directions: The VO as relaying IdP (left) or as DNS administrator (right). The numbers represent the order of the interactions between the components.

A critical component in the identity frameworks that has been identified is the NMC. This component glues the identity and network management planes together by providing the proper interfaces. Lack of standard technologies for such NMC components makes them a challenging topic for further research. Other future work will encompass making a detailed description and analysis of the solution directions, followed by an implementation of the most feasible and challenging solution. The implementation of the core identity framework will preferably be based on the SAML 2.0 protocol. It will be challenging to introduce the VO as a relaying IdP or discovery provider in SAML 2.0. Also, the potential of DNSsec in the identity framework seems worthwhile to investigate in more detail, as it has been barely associated with identity management.

Acknowledgements

The work described in this paper was conducted in the GigaPort Next Generation program, sponsored by the Dutch government and coordinated by SURFnet. We thank Roy Arends, Remco Poortinga, and Henk Eertink for their useful and inspiring input.

References

[1] Global Lambda Integrated Facility (GLIF) website: www.glif.is.
[2] W.Golab and R.Boutaba. Resource Allocation in User-Controlled Circuit-Switched Optical Networks, Proceedings of the 7th IEEE International Conference on High Speed Networks and Multimedia Communications, June 2004.
[3] R.Boutaba, W.Golab, Y.Iraqi, and B.St.Arnaud. Lightpaths on Demand: A Web-Services-Based Management System. The IEEE Communications Magazine 42[7], 2-9. 2004.
[4] I. Foster, C. Kesselman, and S. Tuecke, The Anatomy of the Grid - Enabling Scalable Virtual Organizations. International .Journal of High Performance Computing Applications 15[3], pp. 200-220. 2001.
[5] Nortel Networks, Application Brief: Dynamic Resource Allocation Controller (DRAC). 2004.
[6] D.L.Truong, O.Cherkaoui, H.Elbiaze, N.Rico, and M.Aboulhamid. A Policy-based approach for User controlled Lightpath Provisioning. 9, no 1, pp. 859-872. 2004. IEEE/IFIP Network Operations and Management Symposium (NOMS 2004), April 2004.
[7] A. Reed, The Definitive Guide To Identity Management, SafeNet: The Foundation of Internet Security, Realtimepublishers.com, Inc., 2004.
[8] Liberty Alliance website: www.projectliberty.org.
[9] OASIS SAML website: www.oasis-open.org/committees/tc_home.php?wg_abbrev=security.
[10] S. Bajaj et al., Web Services Federation Language (WS-Federation), Version 1.0, July 8 2003.
[11] Shibboleth Project website: http://shibboleth.internet2.edu/.
[12] A-select website: http://a-select.surfnet.nl.
[13] OASIS Security Assertion Markup Language (SAML) 2.0 Technical Overview - Working Draft 04, 2005.
[14] B. van Oudenaarde, The Multi-Domain Control Challenge, Super Computing 2004 Demonstrator, Pittsburgh PA, US, see also www.science.uva.nl/research/air/projects/aaa/sc2004.
[15] I. Foster and C. Kesselman, The Grid 2: Blueprint for a New Computing Infrastructure, Morgan Kaufmann Publishers, November 18, 2003.
[16] J. Beatty et al., Liberty ID-WSF Discovery Service Specification, Version 2.0-02, 2004, see www.projectliberty.org/specs/draft-liberty-idwsf-disco-svc-v2.0-02.pdf.
[17] R. Arends, R. Austein, M. Larson, D. Massey, and S. Rose, RFC 4033, DNS Security Introduction and Requirements, IETF, 2005.
[18] R. Arends, R. Austein, M. Larson, D. Massey, and S. Rose, RFC 4034, Resource Records for the DNS Security Extensions, IETF, 2005.
[19] R. Arends, R. Austein, M. Larson, D. Massey, and S. Rose, Protocol Modifications for the DNS Security Extensions, RFC 4035, IETF, 2005.
[20] H. Eertink, A. Peddemors, R., Arends, and K. Wierenga, Combining RADIUS with Secure DNS for Dynamic Trust Establishment between Domains, TERENA Networking Conference (TNC'05), Poznan, Poland, 2005.

Self-Organization and Autonomic Informatics (I)
H. Czap et al. (Eds.)
IOS Press, 2005

Tag-Based Cooperation in Peer-to-Peer Networks with Newscast[1]

Andrea Marcozzi [a], David Hales [a], Gian Paolo Jesi [a], Stefano Arteconi [a], Ozalp Babaoglu [a]

[a] *University of Bologna Dept. of Computer Science*
Via Mura Anteo Zamboni, 7
40127 Bologna, Italy
{marcozzi, hales, jesi, arteconi, babaoglu}@cs.unibo.it

Abstract. Recent work has proposed how socially inspired mechanisms, based on "tags" and developed within social science simulations, might be applied in peer-to-peer overlay networks to maintain high cooperation between peers even when they act selfishly. The proposed mechanism involves a dynamic re-wiring algorithm called "SLAC". The algorithm assumes a random sampling service over the entire population of nodes but does not implement this itself. In this paper we re-implement SLAC on an open source peer-to-peer simulation testbed called "PEERSIM". For the random sampling service we utilize an existing protocol called "NEWSCAST". We present the results of some experiments we performed in which peers play the Prisoner's Dilemma game with their neighbours. Our results demonstrate that SLAC augmented with NEWSCAST produces high levels of cooperation. This increases our confidence that previous results from SLAC are generally applicable and valid and also that SLAC could have applications in real implemented systems. Finally we discuss the open issues that need to be addressed for SLAC to progress to a valuable deployable protocol.

Keywords. peer-to-peer, cooperation, social networks, adaptivity, tags

1. Introduction

In recent works a novel socially inspired algorithm based on the "tag" idea [7] has been applied to the problem of sustaining cooperation in peer-to-peer networks composed of nodes behaving selfishly [4].

In the earlier tag models, individual agents interact randomly (in the form of mean-field interaction) under the constraint that they are more likely to interact with agents sharing an identical or similar tag.

In the context of the social scientific interpretation tags represent arbitrary surface markings attached to agents that can be observed and copied by other agents. They have no direct behavioural significance [7]. In human society tags can be viewed as fashions:

[1]Based on the Laurea thesis of Andrea Marcozzi, March 2005. Partially supported by the EU within the 6th Framework Programme under contract 001907 "Dynamically Evolving, Large Scale Information Systems" (DELIS).

styles of dress, colors of hat, brand logos or cosmetic makeup. The key property of tags s that although they are distinctive and immediately observable they can be quickly changed and copied.

A number of simulation models have demonstrated that over a broad range of parameter values high levels of cooperation and altruism emerge when agents bias their interaction towards others sharing the same tag [5,11]. These previous models follow an evolutionary approach in which agents reproduce and mutate their behaviours and tags. In such models those gaining higher utility are more likely to reproduce their tags and behaviours with the addition of small amounts of mutation - noise or random variation. The assumptions underlying this approach is that such an evolutionary mechanism can capture the essential elements of cultural learning within a society - that agents are bounded optimizers, copying the tags and behaviours of those who are gaining higher utility than themselves.

We have previously described, in detail, the modification needed to transform (via a series of simulation models) the evolutionary models into an algorithm more applicable to a target application of peer-to-peer file sharing [4]. However, in that work the final algorithm (the SLAC algorithm) was tested on a non-publicly available and highly abstracted simulation test bed. Also, SLAC relies on a random sampling service over the entire P2P network and this was assumed rather than implemented.

In this paper we describe a re-implementation of, and experiments with, SLAC on the open source PEERSIM system [13]. PEERSIM offers a more realistic P2P simulation environment - protocols previously tested on it have been successfully implemented [17]. In addition the missing random sampling service was provided by the NEWSCAST protocol which is already implemented on PEERSIM. Both SLAC and NEWSCAST are highly scalable (up-to millions of nodes), robust (recovering from noise and the removal of nodes) and completely decentralized (requiring no centralized services).

In the following sections we describe the original tag-based evolutionary algorithm and the derived SLAC P2P algorithm. We then discuss briefly the application task we tested our system with - the Prisoner's Dilemma game followed by an overview of the PEERSIM P2P simulation environment and the NEWSCAST protocol. We then discuss the SLAC implementation details and present the results of experiments performed. Finally we conclude with a discussion of the results and review some of the open issues that need to be addressed in order to progress towards a valuable deployable implementation.

2. Previous Tags Models

The basic algorithm has been adapted from previous (quite different) simulation work using "tags". This work demonstrates a novel method of maintaining high levels of cooperation in environments composed of selfish, adaptive agents. The emphasis of the previous work has been towards understanding biological and social systems [5]. Tags are markings or social cues that are attached to individuals (agents) and are observable by others, often represented in models by a single number, they evolve like any other trait in a given evolutionary model. The key point is that the tags have no direct behavioural implication for the agents that carry them. But through indirect effects, such as the restriction of interaction to those with the same tag value, they can evolve from initially random values into complex ever changing patterns that serve to structure interactions.

The simulated environments in which tags have been applied have generally been very simple with interactions based on pair-wise games with immediate payoffs. Nevertheless, we have attempted to adapt the salient features of such tag systems for application in P2P networks. These features are that agents:

- Restrict interaction to those with whom they share a group defined by tag value
- Selfishly and greedily optimize by preferentially copying the behaviour and tag of others with higher utility
- Periodically mutate their tags and behaviours

By copying and mutating tags, agents effectively move between interaction groups. By restricting interaction within groups free riders tend to kill (reduce the membership) of their own group over time because exploited agents will tend to move elsewhere to get better payoffs, while cooperative groups tend to spread via mutation of the tag. Previous tag models have demonstrated high levels of cooperation in "commons tragedy" [6] scenarios (e.g. in the Prisoners Dilemma – see below). We will not cover the results of the previous tag models in detail here, since the emphasis is not relevant and space precludes detailed treatment, rather we will present our newly derived algorithm (based on the salient features outlined above) and the results we obtained when applying it to two different simulated P2P scenarios.

3. Cooperation and the Prisoner's Dilemma

Distributed P2P applications often require that nodes behave cooperatively or altruistically to help others in the network. For example, in a file-sharing system, nodes are required to host and upload files on demand to other nodes that require them. Also they need to reply to queries concerning what files they host. But why should nodes do this? In an open system there is an incentive for nodes the behave selfishly - saving their own storage and bandwidth but using other nodes. This problem is not limited to file-sharing because any application that requires other peers to perform actions on their behalf, in some sense, relies on a degree of cooperation from those others. Obviously cooperative behaviours can be built into the peer client software but in an open system how can we ensure that such software will not be changed?

The fundamental issue, then, is: *how can one maintain cooperative (socially beneficial) interactions within an open system under the assumption of high individual (peer) autonomy*. An archetype of this kind of social dilemma has been developed in the form of a minimal game called the Prisoner's Dilemma (PD) game. Economic and social scientists have often deployed this minimal game as a canonical form of the contradiction that can arise between individual and collective interests.

In the PD game two players each selected a move from two alternatives and then the game ends and each player receives a score (or pay-off). Figure 1 shows a so-called 'pay-off matrix' for the game. If both choose the 'cooperate' move then both get a 'reward' — the score R. If both select the 'defect' move they are 'punished' — they get the score P. If one player defects and the other cooperates then the defector gets T (the 'temptation' score), the other getting S (the 'sucker' score). When these pay-offs, which are numbers representing some kind of desirable utility (for example, money), obey the following constraints: $T > R > P > S$ and $2R > T + S$ then we say the game represents a

Table 1. A payoff matrix for the two-player single round Prisoner's Dilemma (PD) game. Given $T > R > P > S \wedge 2R > T + S$ the Nash equilibrium is for both players to select Defect but both selecting Cooperate would produce higher social and individual returns. However, if either player selects Cooperate they are exposed to Defection by their opponent — hence the dilemma

	Cooperate	Defect
Cooperate	R, R	S, T
Defect	T, S	P, P

Prisoner's Dilemma (PD). When both players cooperate this represents maximizing of the collective good but when one player defects and another cooperates this represents a form of free-riding. The defector gains a higher score (the temptation) at the expense of the co-operator (who then becomes the 'sucker').

A game theoretic analysis drawing on the Nash equilibrium solution concept (as defined by the now famous John Nash [10]) captures the intuition that a utility maximizing player would always defect in such games because whatever the other player does a higher score is never attained by choosing to cooperate.

In the context of a P2P system how do we solve this problem without going back to centralized control or closed systems? In the following section we describe the "tag" inspired SLAC algorithm.

4. The SLAC Algorithm

The SLAC algorithm [4] assumes that peer nodes have the freedom to change behaviour (i.e. the way they handle and dispatch requests to and from other nodes) and drop and make links to nodes they know about. In addition, it is assumed nodes have the ability to discover other nodes randomly from the network, compare their performance against other nodes and copy the links and (some of) the behaviours of other nodes.

As discussed previously we assume that nodes will tend to use their abilities to selfishly increase their own utility in a greedy and adaptive way, that is if changing some behaviour or link increases utility then nodes will tend to select it. The algorithm relies on Selfish Link and behaviour Adaptation to produce Cooperation (SLAC) - a task domain independent outline is given below.

Over time nodes engage in some activity and generate some measure of utility U. This might be number of files downloaded or jobs processed etc, depending on the domain. Periodically, each node (i) computes its performance and compares this against another node (j), randomly selected from the population. If $Ui < Uj$ node i drops all current links and copies all node j links and adds a link to j itself - see Figure 1.

Also, periodically, and with low probability, each node adapts its behaviour and links in some randomized way using a kind of "mutation" operation. Mutation of the links involves removing all existing links and replacing them with a single link to a node randomly drawn from the network. Mutation of the behaviour involves some form of randomized change - the specifics being dictated by the application domain. Previous tag models, on which SLAC is based have indicated that the rate of mutation applied to the links needs to be significantly higher than that applied to the behaviour - by about one order of magnitude [3].

When applied in a suitably large population, over time, the algorithm follows a kind of evolutionary process in which nodes with high utility tend to replace nodes with low

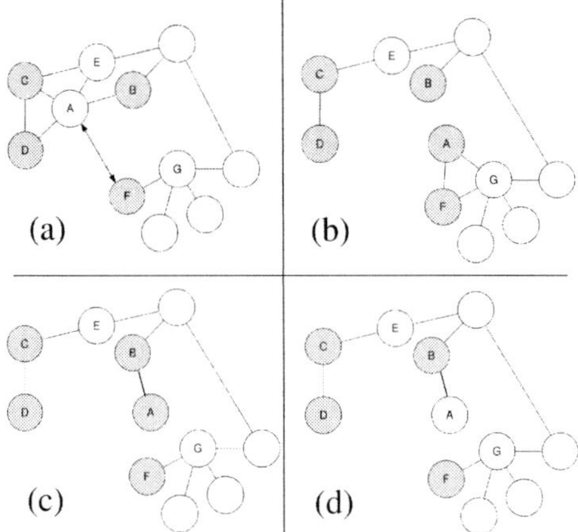

Figure 1. An illustration of 'replication' and 'mutation' as applied in the Selfish Link-based Adaptation for Cooperation (SLAC) algorithm from [4]. Shading of nodes represents strategy. In (a) the arrowed link represents a comparison of utility between A and F. Assuming F has higher utility then (b) shows the state of the network after A copies F's links and strategy and links to F. A possible result of applying mutation to A's links is shown in (c) and the strategy is mutated in (d).

utility (with nodes periodically changing behaviour and moving in the network). However, as will be seen, this does not lead to the dominance of selfish behaviour – as might be intuitively expected – since a form of incentive mechanism emerges via a kind of ostracism in the network.

5. The Peersim System

Evaluating the performance of P2P protocols is a complex task. One of the main problems for their evaluation, is the extremely large scale that they may reach. P2P networks involving hundred of thousands of peers (or more) are not uncommon (e.g., about 5 millions machines are reported to be connected to the Kazaa/Fasttrack [15] network). In addition P2P systems are highly dynamic environments; they are in a continuous state of flux, with new nodes joining and leaving (or crashing).

These properties are very challenging to deal with. Evaluating a new protocol in a real environment, especially in its early stages of development, is not feasible. Distributed planetary-scale open platforms (e.g., Planet-Lab [16]) to develop and deploy network services are available, but these solutions do not include more than about 500 (at the time of writing) nodes. Thus, for large-scale systems, a scalable simulation test bed is mandatory.

The Peersim P2P simulator [13] has been developed with the aim to deal with the previously stated issues. Its first goals are: extreme scalability and support for dynamism. It is a GPL open-source Java based software project. Peersim has proved to be a valuable tool and it is used as the default experimentation platform in the BISON project [14]. In the following, we provide a brief description of its characteristics.

5.1. Peersim Design Goals

The Peersim simulator is inspired by mainly two objectives:

- High scalability: P2P networks may be composed by millions of nodes. This result can be achieved only with a careful design of the data structures involved, trying to avoid (when possible) any overhead. But the memory footprint is not the only problem: the simulator engine must be also efficient.
- Support for dynamism: the simulator must manage nodes joining and leaving the network at any time; this feature has tightly relations with the engine memory management sub-system.

Another important requirement is the *modular* or *component* inspired architecture. Every entity in the simulation (such as protocols and the environment related objects) must be easily replaceable with similar type entities.

The Peersim extreme performances can be reached only accepting some relaxing assumptions about the simulation details. For example, the overhead introduced by the low level communication protocol stack (e.g., TCP or UDP) in not taken into account because of the huge additional memory and CPU time requirements needed to accomplish this task.

5.2. Peersim Architecture

As previously stated, Peersim is inspired by a modular and very configurable paradigm, trying to limit any unnecessary overhead. The simulator main component is the *Configurator* entity targeted to read configuration files. A configuration file is a plain ASCII text file, basically composed by key-value pairs. The Configurator is the only not interchangeable simulation component. All the other entities can be easily customized.

In a Peersim simulation, the following three distinct kind of elements can be present: protocols, dynamics and observers. Each of them is implemented by a Java class specified in the configuration file. The network in the simulation is represented by a collection of nodes and each node can hold one or more protocols. The communication between node protocols is based on method calls. To provide a specific kind of service, each component must implement a specific *interface*. For example a protocol has to implement at least the `Protocol` or `CDProtocol` interface to run on Peersim.

Peersim has an utility class package to perform statistic computations or to provide some starting topology configuration based on well know models (such as: random-graph, lattice, BA-Graph,...).

The *Simulator* engine is the component that performs the computation; it has to run the component execution according to the configuration file instructions. At the time of writing, Peersim can perform simulation according to the following execution models:

- Cycle based: at each step, all nodes are selected in a random fashion and each node protocol is invoked in turn;
- Event based: a support for concurrency is provided. A set of events (messages) are scheduled in time and node protocols are run according to the time message delivery order.

This paper work is based on the first simulation model.

```
while(TRUE) do
 wait(Δt);
 neighbour = SELECTPEER();
 SENDSTATE(neighbour);
 n_state = RECEIVESTATE();
 my_state.UPDATE(n_state);
```

(a) Active Thread

```
while(TRUE) do
 n_state = RECEIVESTATE();
 SENDSTATE(n_state.sender);
 my_state.UPDATE(n_state);
```

(b) Passive Thread

Figure 2. The gossip paradigm.

6. the Newscast protocol

Newscast [9] is a gossip-based topology manager protocol. Its aim is to continoulsy rewire the (logical) connections between hosts. The rewiring process is designed in such a way that the resulting overlay is very close to a random graph. The generated topology is thus very stable and provides robust connectivity. This protocol has been used successfully to implement several P2P protocols, including broadcast [9] and aggregation [8].

As in any large P2P system, a node only knows about a small fixed set of other nodes (due to scalability issues), called *neighbours*. In Newscast, the neighbourhood is represented by a partial, fixed c size view of node *descriptors* composed by a node address and a logical *time-stamp* (e.g., the descriptor creation time).

Referring to the usual gossip scheme (see Figure 2), the protocol behaviour performs the following actions: selects first a neighbour from the local view, exchanges the view with the neighbour, then both participants update their actual view according to the received view. The data actually sent over the network by any Newscast node is represented by the node's own descriptor plus its local view.

In Newscast, the neighbour selection process is performed in a random fashion by the SELECTPEER() method. The UPDATE() method is the Newscast core behaviour. It merges ($\bigcup$ operation) a received view (sent by a node using SENDSTATE()) with the current peer view in a temporary view list. Finally, Newscast trims this list to obtain the new c size view. The node descriptors discarded are chosen from the most "old" ones, according to the descriptor time-stamp. This approach changes continuously the node descriptors hold in each node view; this implies a continuous rewiring of the graph defined by the set of all node views. This behaviour is shown in Figure 3.

Even though the system is not synchronous, we find it convenient to describe the gossip-scheme execution as a sequence of consecutive real time intervals of length Δ (see the "wait" statement in pseudo-code in Figure 2), called *cycles* or *rounds*.

The protocol always tends to inject new informations in the system and allows an automatic elimination of old node descriptors using the aging approach. This feature is particularly desirable to remove crashed node descriptors and thus to repair the overlay with minor efforts. In addition, the protocol does not require any clock synchronization, but only that the timestamp of node descriptors in each view are mutually consistent.

The topology generated by Newscast has a low diameter and it is close to a random graph having out-degree c. Experimental results proved that a small 20 elements partial view is already sufficient for a very stable and robust connectivity, regardless of the network size.

Newscast is also cheap in terms of network communication. The traffic generated by the protocol involves the exchange a few hundred bytes per cycle for each peer and is estimated in [9].

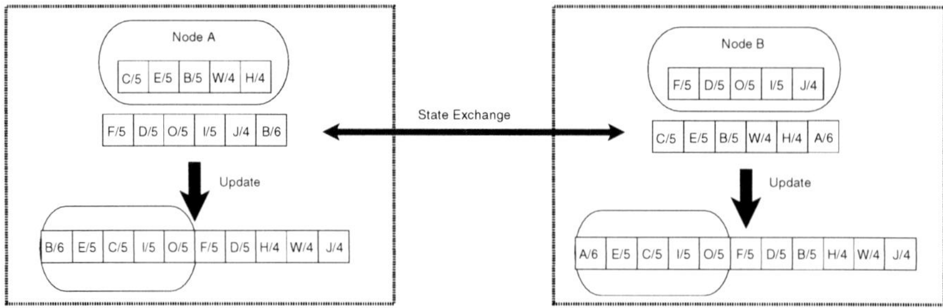

Figure 3. A Newscast exchange between node A (active) and B. Each node has its own 5 descriptor elements view depicted inside the ellipses. A descriptor is a *node-ID, timestamp* pair. After the state exchange node A has received the node B view and viceversa; then each partecipant merges the received view with its own. The result is depicted under the empty arrow: each node has selected the "freshest" descriptors at random and has discarded the others (those inside the ellipse) to obtain new 5 element view. Note that in this basic example, each node sends its entire view; however, the view can be purged by "old" descriptors before sending.

A protocol such as Newscast provides a service to pick random nodes from the whole network and we can call it *Randomizer Service*. The chance to extract a fresh new node, selected at random from the whole network, is a high desiderable source of information for P2P protocols. We can consider such a service as a *building block* for many P2P protocols. In this vision, the cost effectiveness of Newscast is very useful, because the *Randomizer Service* has to be always-on and run by all peers involved in the overlay.

Such a randomizer service can also be a key component during the initialization phase (*bootstrap*) for any higher level protocol in order to fill its view at the beginning. We use Newscast both as a randomizer service and as a bootstrap facility in the Peersim implementation of the SLAC algorithm. We describe this implementation in the next section.

7. Implementing SLAC in Peersim

Now we purpose to implement and test the solutions proposed by the SLAC algorithm into the Peersim platform, on top of Newscast. We have already seen that Newscast has lots of desirable properties such as scalability, robustness and the ability to maintain a random topology. This is why we based our experiments on top of such protocol. Newscast has been previously implemented within Peersim (as previously discussed) and therefore we do not discuss implementation details here.

For SLAC we implemented three classes in Peersim whose names are: **PdDistributionInitializer**, **PdObserver** and of course **PdProtocol**. The first one initializes the nodes of the network with a strategy (cooperate or defect), the second one calculates and prints the results from the simulations. **PdProtocol** is the core of our SLAC implementation and hence deserves a more complete discussion.

The PdProtocol (an overview)

As we have already seen, Peersim is highly modular. So it wasn't so hard to implement a new protocol based on the NetWorld [1] model specifications (which implements SLAC) compatible with its structure. Peersim simulations [13] are performed through a series of cycles and in each of these some operations are performed. During the simulation our protocol is involved in three phases:

- phase 0: initialization of an auxiliary array for neighbours list (nodes are provided by Newscast);
- phase 1: the PD game is played and the appropriate payoffs are distributed;
- phase 2: two nodes are randomly chosen from the network (these are provided by Newscast) and their payoffs are compared (this is the reproduction phase).

Phase 0 is performed at the first simulation cycle: each node just copies its neighbour list from its Newscast's instance.

In phase 1 the PD game is played between nodes. Each node is initialized with a random value chosen from a set of two (True = Cooperate, False = Defect) and at each cycle each node plays a round of the PD game with one of its randomly chosen neighbours. After this game interaction, payoffs are calculated and distributed.

With phase 2 is performed the reproduction task. It takes place every I cycles (in our experiments we have $I = 4$). In this phase we want to compare the average payoffs of two nodes chosen within the network. The first node we take into account is randomly chosen within the network; the second one, as in phase 0, is taken from Newscast: it is a randomly chosen neighbour of the Newscast instance of the first node. Once two nodes (i, j) are selected this way, reproduction phase can start. Let's consider the case in which node i has a higher average payoff than j, the following operations are performed (obviously if j's payoffs is greater than i's the symmetric rules apply):

- all j's neighbours delete their links to j itself; [1]
- j's neighbours list is cleared;
- j's neighbours list is filled in with new items: i's neighbours;
- the winner itself (i) is added to j's neighbours list;
- strategy is copied from i to j;
- some variables held by j are cleared (e.g. number of games played);
- j is added to the respective neighbours lists of the winners neighbours;
- j itself is added to i's neighbours list;
- mutation is applied with a certain probability.

Neighbours lists have a fixed size F, hence it is not permitted to add nodes when such a limit is reached. That is why this operation is performed by an appropriate function which ensures this limit is not exceed: before adding a new node, a check on the actual size of the list is performed; if it is equal to F, a randomly chosen node is deleted and then the new one is added, else, if the actual size is smaller than F, the new node is simply added.

This is what happens when i has an average payoff greater than j. Of course if j had an average payoff greater than i, the same algorithm will be performed but with i and j in inverted positions. When the two nodes hold the same value the winner is randomly chosen between the two nodes and then the same operations listed above are performed.

[1] Actually, in this implementation of SLAC, this first action was not implemented, interestingly, this change did not stop high cooperation from emerging.

Table 2. Simulation parameter.

Parameter	**Value**
Cycles	1000
Network size	4000,8000,12000,16000
Initial topology	Random
Maximum degree	20
Strategy mutation rate (MR)	0.001
Tag mutation rate (MRT)	0.001
Reproduction interval (I)	4

Table 3. PD payoffs adopted in the model.

1st player	**Strategies**	**2nd player**
1	DC	0
0.8	CC	0.8
0.1	DD	0.1
0	CD	1

At the end of each reproduction step *mutation* is performed. It is applied with low probability (*mutation rate, MR = 0.001*), to the node losing the comparison at the beginning of the reproduction phase. Mutation is applied to both strategy and tags (neighbor list). When applied to the strategy bit, it implies the flipping of such bit; when applied to the neighbor list, it involves the cancellation of all the links in the node's neighbors list and the substitution of such nodes with just one node that is randomly selected from the entire network.

8. Simulation Experiments and Results

A series of experiments were done with this new "tag-based" protocol on Newscast. Obtained results relate to experiments done varying several parameters: size of the network, simulation seed, strategy initialization.

The main parameters used are illustrated in Table 2 while the PD payoffs used are those in Table 3.

In the next subsections we will show some of these results and draw some conclusions.

8.1. Cooperation with different network sizes

Results shown in Figures 4 and 5 relates experiments done on networks having different sizes.

The diagram in Figure 4 represents the number of cycles needed to obtain high levels of cooperation (about 93%) over a series of 1000 cycles. The diagram compares the results obtained performing the experiments on four different network size. The results shown in the figure represents the average and standard deviation (out of N sims) of the

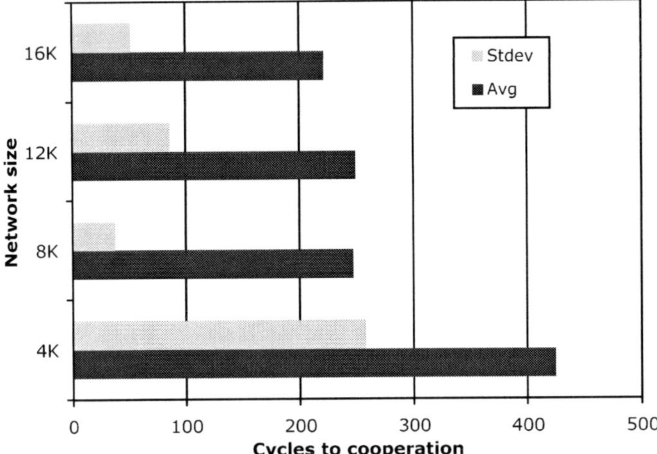

Figure 4. Average number of cycles needed to obtain high levels of cooperation (about 93%) and the relative standard deviation with four different seeds. MR = MRT = 0.001, I = 4, payoffs from table 3. Results from different network size are compared.

number of cycles needed to obtain cooperation. The network was started from complete defection, the mutation rate used was the same both for the strategy and the neighbour list ($MR = MRT = 0.001$) and the number of cycles occurring between a reproduction phase and the next one was $I = 4$. Let's note the average cycles number for a network of 4000 nodes: it is much higher than the average for the other network sizes and this is because with seed 1, the 93% of cooperation was obtained just after 772 cycles. The same experiment was also performed with a different seeds table but the results we obtained are nearly identical to those just given.

Figure 5 gives the percentage of cooperating nodes over a series of 4000 cycles. The figure just shows the first 150 cycles since after that point there are no relevant changes in the results. Even here the mutation rate was the same both for the strategy and the tag, experiments were performed over different network size and parameter I was set to 4. On the contrary of the previous experiment, here was used always the same seed, hence the percentages we give are not an average. Nodes were initialized at random: at the beginning of the simulation we had a population composed of about half cooperating nodes and half defecting nodes; after the first few cycle, the percentage of cooperating nodes decreased but soon after the 23th cycle it started increasing toward good levels of cooperation.

Observations

From both the diagrams it's easy to note the good level of scalability of the model we are testing: results of Figure 4 are an average of results obtained with different seeds and we succeeded in obtaining similar results for three different network size. The only difference is found with the network composed of 4000 nodes where we obtained a very

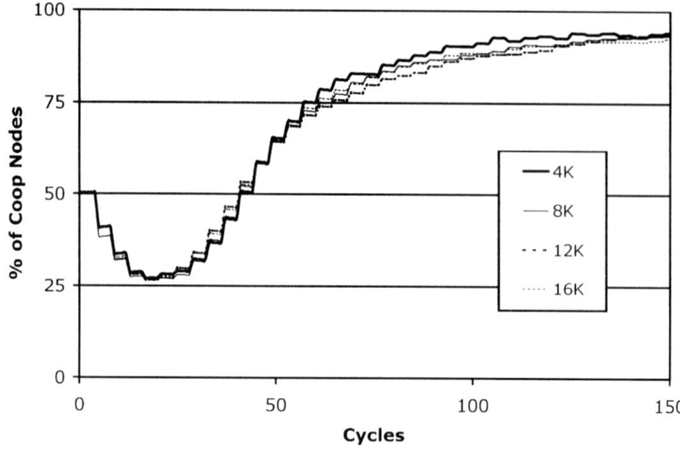

Figure 5. Cooperating nodes over a series of 4000 cycles (single run). MR=MRT=0.001, I=4, payoffs from table 3. Nodes are initialized with a random strategy. Results after cycle 150 do not increase significantly. Note: the "staircasing" effect is an artefact of the synchronous reproduction at every 4 cycles - with asynchronous reproduction the artefact is not visible.

high value with just one seed. It would be interesting to make further experiments with more seeds. Figure 5 gives the same important result: here using always the same seed, we obtained the same trend with all the sizes.

8.2. Cooperation with long runs

Some experiments were done with a big number of cycles. On a network composed of 4000 nodes were performed a series of experiments; for each of them we used a different random seed, $MR = MRT = 0.001$, $I = 4$ and performed 10000 cycles.

Results are very close for each seed used, hence in Figure 6 we just propose those obtained with one of them. The diagram shows that good levels of cooperation can be obtained from cycle 645 (95.5 %): from this cycle to cycle 10000 the average percentage of cooperating nodes is 95.40 and the standard deviation is 1.03.

Observations

The experiments just proposed have a great importance since they test the reliability of the model in the time. From all the tests we made (some even with 50000 generations), and also in those proposed in this section, we found that once cooperation has started, it never claps and can be sustained for long times; we also learned that cooperation can be sustained at good levels.

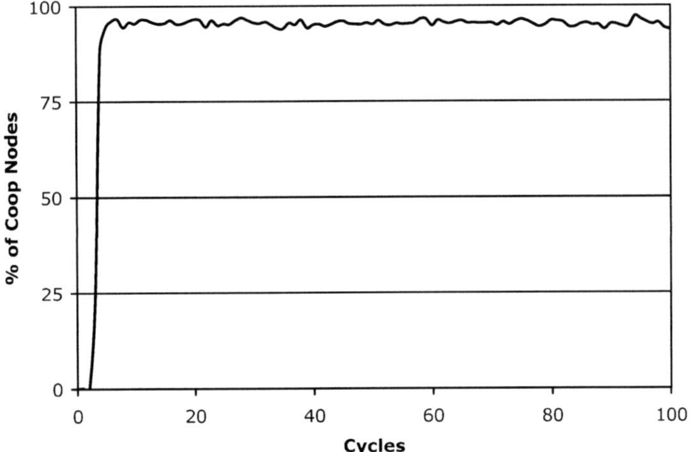

Figure 6. Cooperation with a series of 10000 cycles (single run). Nodes = 4000, MR=MRT=0.001, I=4, payoffs from table 3.

9. Discussion and Conclusion

The results we obtained indicate that high cooperation is produced when nodes follow the SLAC algorithm. Even though the SLAC algorithm implements nodes that behave selfishly in a myopic and greedy way - that is, they copy other nodes in the network that have higher utility - high levels of cooperation are produced in the single round Prisoner's Dilemma (PD) game.

These new results therefore confirm those results previously found[2,4] in similar simulation experiments and this adds confidence that SLAC is robust to different simulation implementation details. Here, for example, we interleaved the reproduction phase with the interaction phase whereas in previous simulations reproduction followed at the end of each cycle of interaction. Also in further experiments (not shown here) we found that similar results were obtained when reproduction was interleaved with interaction in a fully asynchronous way - where each node has a probability of reproduction after interaction.

Interestingly, the results shown in Figure 4 appear to recover some of the reverse scaling properties demonstrated in an early non-network based tag model [5] which appeared to have been lost in our initial network model [2]. However, more runs and analysis are needed to explore this question. However, we can certainly state that in all the experiments so far performed with the SLAC algorithm larger networks do not take longer to converge and often converge more quickly. This is obviously a valuable property for any candidate algorithm for large scale systems.

A further finding of these new results appears to contradict earlier generalizations [3] that tag-type models needed to have higher mutation rates on the "tag" than the strategy - in the case of SLAC this would mean a higher mutation rate on the neighbour list or view which contains the links to neighbour nodes than the behavioural strategy of the node (either to cooperate or defect). But here high levels of cooperation were produced when the mutation rate was the same. This indicates that further work is needed to circumscribe such a generalization since it is currently unclear what difference in implementation has allowed this assumption to be relaxed.

These results also demonstrate that the NEWSCAST protocol can be used to provide the random sampling service required by SLAC but not previously explicitly modelled in simulation. This is important since any actual implementation of SLAC must have access to such a service that is both scalable and robust. NEWSCAST provides such a service[9] with the additional benefit that it has actually been tested in the form of a real implementation [17].

We have argued, and demonstrated previously, that cooperation in the single round PD indicates that cooperation can be produced in other more realistic task domains [4]. We are therefore confident that these results indicate that the PEERSIM implementation could support cooperation in other task domains (such as file sharing or other kinds of resource sharing).

Finally, we note two major issues that could destroy cooperation within SLAC. Firstly, we currently assume that nodes are able to compare utilities correctly, that is, we assume nodes report their utilities honestly when requested to do so by nodes. But what would happen if nodes lied about their utilities or just failed to report anything? This introduces a kind of "second order" free-rider problem at the informational level because if we assume nodes may behave selfishly and / or maliciously then we need to demonstrate individual incentives for supplying correct utility values. Secondly, we also assume nodes will allow themselves to be copied by supplying their behaviour strategy and their current neighbour list or view (containing their node links) to other nodes. Again, this may not be case with malicious and selfish nodes in certain contexts. Both of these issues we aim to address in future work.

Acknowledgements

This work would not have been possible without perceptive discussions with many people, particularly those in the Bologna group including: Mark Jelasity, Alberto Montresor and Simon Patarin. Additionally, we thank the anonymous reviewers of the initial draft of this paper for their comments and suggestions.

References

[1] D. Hales, B. Edmonds Applying a socially-inspired technique (tags) to improve cooperation in P2P Networks *IEEE Transactions in Systems, Man and Cybernetics - Part A: Systems and Humans* pp.385-395 2005

[2] D. Hales. Self-Organizing, Open and Cooperative P2P Societies Ð From Tags to Networks. *Proceedings of the 2nd Workshop on Engineering Self-Organizing Applications (ESOA 2004), LNCS 3464*, pp.123-137. Springer, 2005.

[3] D. Hales. Change Your Tags Fast! – a necessary condition for cooperation? *Proceedings of the Workshop on Multi-Agents and Multi-Agent-Based Simulation (MABS 2004), LNAI 3415*. Springer, 2005.

[4] D. Hales. From selfish nodes to cooperative networks – emergent link based incentives in peer-to-peer networks. In *Proc. of the 4th IEEE International Conference on*

Peer-to-Peer Computing (P2P2004). IEEE Computer Soc. Press, 2004. Available at: http://www.davidhales.com

[5] D. Hales. Cooperation without Space or Memory: Tags, Groups and the Prisoner's Dilemma. In *Moss and Davidsson (eds.) Multi-Agent-Based Simulation*. LNAI 1979:157-166. Springer. Berlin. 2000

[6] G. Hardin. The tragedy of the commons. *Science*, 162, 1243-1248. 1968

[7] J. Holland. The Effect of Lables (Tags) on Social Interactions. *Santa Fe Institute Working Paper 93-10-064*. Santa Fe, NM 1993

[8] M. Jelasity and A. Montresor, Epidemic-Style Proactive Aggregation in Large Overlay Networks, in *Proceedings of the 24th International Conference on Distributed Computing Systems (ICDCS'04)*, March 2004, pp. 102–109, IEEE Computer Society, Available at: http://www.cs.unibo.it/bison/publications/icdcs04.pdf

[9] M. Jelasity and W. Kowalczyk and M. van Steen Newscast Computing *Technical Report IR-CS-006*, Vrije Universiteit Amsterdam, Department of Computer Science, November 2003, Available at: http://www.cs.vu.nl/globe/techreps.html#IR-CS-006.03

[10] J. F. Nash. Equilibrium Points in N-Person Games, *Proc. Natl. Acad. Sci.* USA 36, 48-49, (1950).

[11] R. Riolo, M. D. Cohen, R. Axelrod. Cooperation without Reciprocity. *Nature* 414, 441-443, (2001).

[12] R. Tivers. The evolution of reciprocal altruism. *Q. Rev. Biol.* 46, 35-57. 1971

[13] Peersim Peer-to-Peer Simulator, Available at: http://peersim.sf.net

[14] The BISON Project, http://www.cs.unibo.it/bison

[15] Kazaa Web Site, http://www.kazaa.com

[16] PlanetLab Planetary-Scale Testbed, http://www.planet-lab.org

[17] T. Binci. EpidEm: EPIDemic EMulator, Graduate Thesis in Computer Science, University of Bologna, Department of Computer Science. Available at: http://bincit.web.cs.unibo.it/index.htm

Author Index